Conference Proceedings of the Society for Experimental Mechanics Series

Series Editor

Kristin B. Zimmerman, Ph.D.
Society for Experimental Mechanics, Inc.
Bethel, CT, USA

More information about this series at http://www.springer.com/series/8922

Michael Sutton • Phillip L. Reu
Editors

International Digital Imaging Correlation Society

Proceedings of the First Annual Conference, 2016

Springer

Editors
Michael Sutton
University of South Carolina
Department of Mechanical Engineering
University of South Carolina
Columbia, SC, USA

Phillip L. Reu
Sandia National Laboratory
Albuquerque, NM, USA

ISSN 2191-5644 ISSN 2191-5652 (electronic)
Conference Proceedings of the Society for Experimental Mechanics Series
ISBN 978-3-319-51438-3 ISBN 978-3-319-51439-0 (eBook)
DOI 10.1007/978-3-319-51439-0

Library of Congress Control Number: 2017932545

Printed on acid-free paper

This Springer imprint is published by Springer Nature
The registered company is Springer International Publishing AG
The registered company address is: Gewerbestrasse 11, 6330 Cham, Switzerland

Preface

Digital Image Correlation represents a single volume of technical papers presented at the Annual International DIC Society Conference and SEM Fall Conference organized by the Society for Experimental Mechanics and Sandia National Laboratories and held in Philadelphia, PA, November 7–10, 2016.

The volume presents early findings from experimental, standards development and various other investigations concerning digital image correlation—an important area within Experimental Mechanics.

The area of Digital Image Correlation has been an integral track within the SEM Annual Conference spearheaded by Professor Michael Sutton from the University of South Carolina. In 2016, the SEM and Sandia joined their collaborative strengths to launch a standing fall meeting focusing specifically on developments in the area of Digital Image Correlation.

The contributed papers within this volume span numerous technical aspects of DIC including standards development for the industry.

The conference organizers thank the authors, presenters, and session chairs for their participation, support, and contribution to this very exciting area of experimental mechanics. We are grateful to the SEM and to Sandia National Laboratories for co-sponsoring and/or co-organizing the sessions in this conference. They would also like to acknowledge the SEM support staff for their devoted efforts in accommodating the large number of paper submissions this year, making the 2016 Annual International DIC Society Conference and the published proceedings so successful.

Columbia, SC, USA — Michael Sutton
Albuquerque, NM, USA — Phillip L. Reu

Contents

Chapter 1
Evaluation of Camera Motion in Stereo-DIC

R. Balcaen, P.L. Reu, P. Lava, and D. Debruyne

Abstract The full-field nature of stereo-digital image correlation (stereo-DIC) makes it a widespread technique with vast possibilities. The uncertainty quantification of the technique is however not yet fully understood due to the non-linear optical-numerical measurement chain, limiting analytical research of the technique. Camera motion is a perfect example of this; it is hard to measure and the influence of it is not yet investigated. A simulator (presented in [Balcaen et al. Finite element based image generator for stereo-DIC uncertainty quantification applied on stereo-DIC calibration, 2016] and used in [Balcaen et al. Stereo-DIC uncertainty quantification based on simulated images, 2016]) is used to study how camera motion during a test influences the resulting deformation- and strain-field. Since the exact imposed deformation-field is known in a simulator the resulting errors can be clearly identified. Camera motion between the calibration-stage and the actual measurement is not investigated since this comes down to a poorly calibrated rig. We refer to the literature for more information concerning the influence of poor calibration data on measurement results. Numerous tests indicate that small camera-rotations can introduce a linear displacement-offset and thus also a constant strain-offset, mainly depending on the focal length of the lenses and the angle of rotation. Camera translation on the other side seems to be less pernicious.

Keywords Stereo-DIC • Uncertainty quantification • Optical techniques • Camera motion • Error assessment

1.1 Introduction

The use of stereo-DIC for measuring shape, motion, deformation and material identification (e.g. by using the Virtual Fields Method (VFM) [3]) is increasing tremendously the last few years since the price of scientific cameras is dropping and the computational power is continually increasing. The technique is however not yet fully understood due to its complex optical-numerical nature, constraining the uncertainty quantification research. In pursuance of better knowledge of stereo-DIC, a simulator was developed [1, 2]), offering the possibility to generate images as if they were made during a real experiment (including image noise, lens-distortions, depth of field (DOF), lighting-influences, etc.) and simultaneously knowing the imposed deformation-field. Since everything is exactly known the simulator can be used to determine the influence of an error source uniquely, without having the influence of other, unknown, error sources and without having to do time- and money-consuming experiments. The influence of camera motion on the meteorology of stereo-DIC is a perfect case for the simulator; indeed, camera motion is hard to measure and even though the influence can be quite substantial (as mentioned in [4], but not investigated there) no research into the influence of camera motion on the results could be found by the authors. This paper only covers camera rotation since the influence of camera-rotation is dominant over camera-translation.

R. Balcaen (✉) • D. Debruyne
KU Leuven, Department of Materials Technology, Campus Ghent, Gebroeders Desmetstraat 1, 9000 Ghent, Belgium
e-mail: ruben.balcaen@kuleuven.be; dimitri.debruyne@kuleuven.be

P.L. Reu
Sandia National Laboratories, PO Box 5800, Albuquerque, NM 87185, USA
e-mail: plreu@sandia.org

P. Lava
MatchID, Wijmenstraat 21T, 9000 Ghent, Belgium
e-mail: pascal.lava@matchidmbc.be

M. Sutton, P.L. Reu (eds.), *International Digital Imaging Correlation Society*, Conference Proceedings of the Society for Experimental Mechanics Series, DOI 10.1007/978-3-319-51439-0_1

Table 1.1 Properties of bulging experiment

Parameter	Value
Radius plate [mm]	50
Thickness plate [mm]	3
Young's modulus [GPa]	210
Poisson ratio	0.30
Pressure [Bar]	30
Stereo angle DIC system [°]	30
Focal length lenses-set up 1 [mm]	30
Focal length lenses-set up 2 [mm]	200
Lens distortion factor K_1 [pixels^{-2}]	0.66
Lens distortion factor K_2 [pixels^{-2}]	0.25

Table 1.2 Adopted DIC settings

Parameter	Value
Camera resolution [pixels]	1624 × 1234
Mm per pixel	±0.08
Subset size [pixels]	13
Step [pixels]	6
Interpolation	B-spline
Transformation	Quadratic
Strain window [pixels]	97

1.2 Influence of Camera Rotation

In this section only camera rotation is considered, i.e. both cameras in the stereo-DIC set up can rotate around their respective axis, without translating. Since the simulator is used to generate images, ground truth data (i.e. the imposed deformation-field, the exact imposed angles, etc.) is known and the influence of the rotation can be investigated. The test-case is a fictive bulging experiment, with parameters given in Table 1.1. The simulated images were analyzed with the DIC-settings given in Table 1.2 with MatchID-Stereo [5]. In this fictive experiment a horizontal set up is simulated (with a 30° stereo angle) in which the cameras rotate away from each other along the horizontal axis (i.e. around the respective y-axis of the cameras) with an angle of 0.01° and 0.02°. This virtual experiment is performed with two sets of lenses; once with 30 mm lenses and once with 200 mm lenses.

As can be seen from Fig. 1.1 even these small imposed rotation-angles can have a substantial influence on the obtained displacement- and strain-field. Please note that other parameters than the ones shown below are also influenced by the camera-rotation.

1.3 Conclusion

Camera rotation can have a large influence on the measured displacements and strains, even with small rotation angles and it is therefore advised to give great attention to sturdy mounting of the cameras, as well as avoiding vibrations caused by environmental factors. Displacements are given a linear nature when the cameras in a horizontal set up rotate along the epipolar line, and the strains consequentially obtain a linear offset. More research has to be performed with actual measured camera-motions in order to truly quantify this problem.

Acknowledgements Sandia is a multiprogram laboratory operated by Sandia Corporation, a Lockheed Martin Company, for the United States Department of Energy's National Nuclear Security Administration under contract No. DE-AC04-94AL85000.

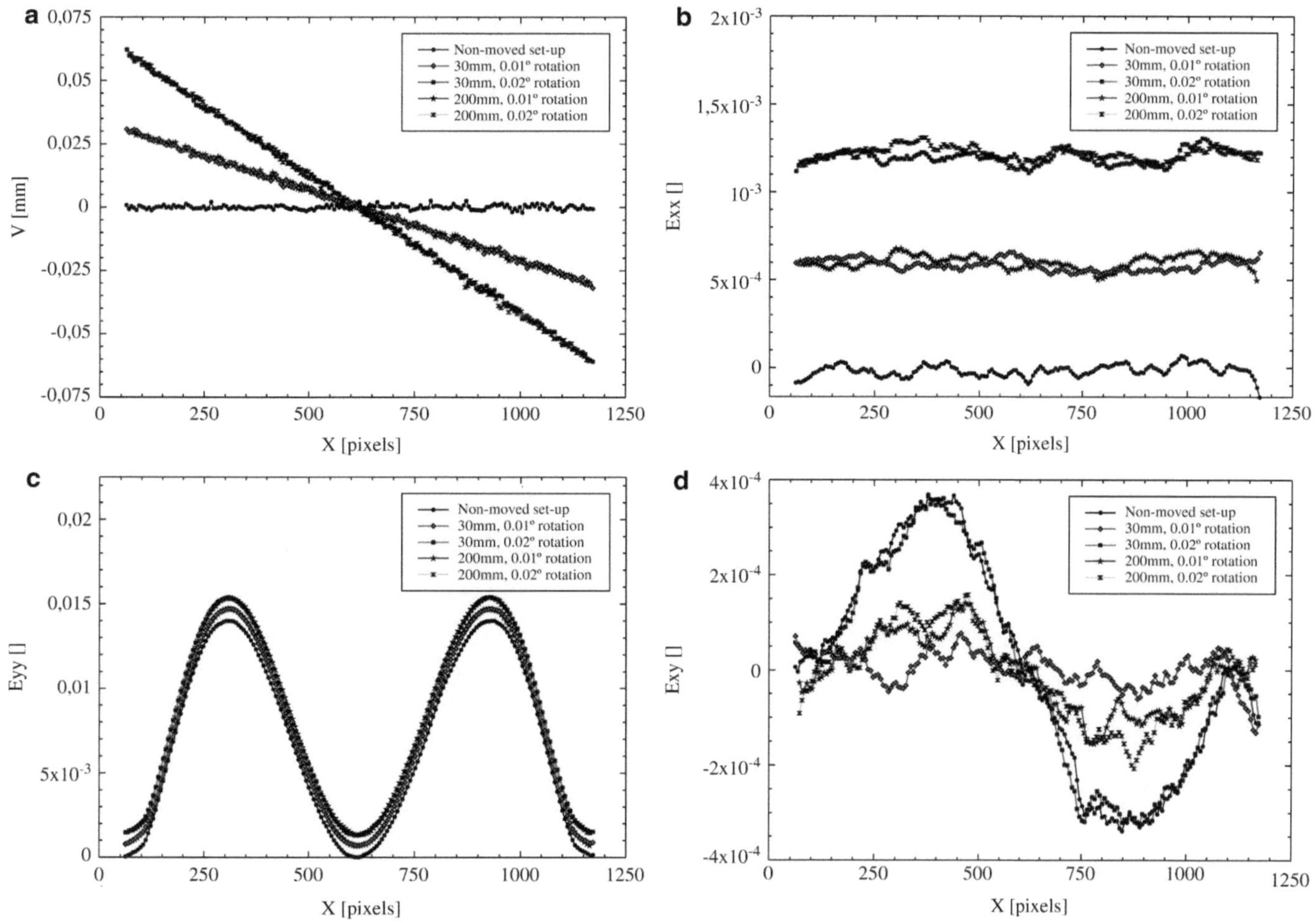

Fig. 1.1 Bias in displacement and strain due to camera rotation. (**a**) Influence on V. (**b**) Influence on E_{xx}. (**c**) Influence on E_{yy}. (**d**) Influence on E_{xy}

References

1. Balcaen, R., Wittevrongel, L., Reu, P.L., Lava, P., Debruyne, D.: Finite element based image generator for stereo-DIC uncertainty quantification applied on stereo-DIC calibration. Submitted to Experimental Mechanics (2016)
2. Balcaen, R., Reu, P.L., Lava, P., Debruyne, D.: Stereo-DIC uncertainty quantification based on simulated images. Submitted to Experimental Mechanics (2016)
3. Pierron, F., Grédiac, M.: The Virtual Fields Method. Springer-Verlag, New York (2011)
4. Ke, X.D., Schreier, H.W., Sutton, M.A., Wang, Y.Q.: Error assessment in stereo-based deformation measurements-Part II: experimental validation of uncertainty and bias estimates. Exp. Mech. 423–441 (2011)
5. MatchID, [Online]. Available: http://www.matchidmbc.com/

Chapter 2
Identifying Dynamic Constitutive Parameters of Bending Plates Using the Virtual Fields Method

Alain Berry, Olivier Robin, and Patrick O'Donoughue

Abstract The identification and mapping of dynamic structural or material parameters is of interest in vibration and acoustic studies. Especially, there is a need for fast, non-destructive and possibly in-situ methods for extracting such parameters. This work investigates the mapping of local bending stiffness and structural loss factor of thin panels using full-field vibration response measurements and the Virtual Fields Method (VFM). VFM is based on an integral form of the dynamic equilibrium. Piecewise virtual displacement functions defined on a small region scanning the whole panel are used to extract the constitutive parameters of the structure. For bending plates, the method requires measuring both the transverse displacement field and bending curvature fields. This is achieved using either scanning Doppler laser vibrometry or full-field optical deflectometry and appropriate spatial finite difference calculations. The presentation will detail the VFM formulation and the experimental, full-field measurement techniques used for this problem. Examples of results will be shown for isotropic (metallic) and orthotropic (composite) panels excited at specific frequencies. In the latter case, the structural parameters agree with those obtained from ASTM standard tests.

Keywords Deflectometry • Constitutive parameters • Virtual Fields Method • Vibration • Acoustics

2.1 Introduction

The virtual work principle applied to identification of material constitutive parameters has resulted in the so-called Virtual Fields Method [1]. This article investigates the Virtual Fields Method combined with full-field Laser Doppler Vibrometry or optical deflectometry to identify bending stiffness parameters of thin panels.

2.2 The Virtual Fields Method for the Extraction of Constitutive Parameters

The Virtual Fields Method (VFM) is used for the identification and mapping of bending stiffnesses of orthotropic panels, on which controlled transverse forces are applied. It is supposed here (without restricting the generality of the approach), that the applied forces are time-harmonic, at angular frequency ω. The equilibrium equation for pure bending of an orthotropic panel whose orthotropy directions 1 and 2 correspond to the x and y axes of the plate, is (in the absence of external forces)

$$\mu \frac{\partial^2 \tilde{w}}{\partial t^2} + \tilde{D}_{11} \frac{\partial^4 \tilde{w}}{\partial x^4} + \tilde{D}_{22} \frac{\partial^4 \tilde{w}}{\partial y^4} + 2\tilde{D}_{12} \frac{\partial^4 \tilde{w}}{\partial x^2 \partial y^2} = 0, \tag{2.1}$$

A. Berry (✉) • O. Robin • P. O'Donoughue
GAUS, Department of Mechanical Engineering, Université de Sherbrooke, Sherbrooke, QC, Canada J1K 2R1
e-mail: alain.berry@usherbrooke.ca

M. Sutton, P.L. Reu (eds.), *International Digital Imaging Correlation Society*, Conference Proceedings of the Society for Experimental Mechanics Series, DOI 10.1007/978-3-319-51439-0_2

where $\tilde{w}$ is the complex transverse displacement, μ is the mass per unit area of the panel and the bending stiffnesses are $\tilde{D}_{11} = \frac{\tilde{E}_1 h^3}{12(1-\upsilon_1\upsilon_2)}$, $\tilde{D}_{22} = \frac{\tilde{E}_2 h^3}{12(1-\upsilon_1\upsilon_2)}$, $\tilde{D}_{12} = \upsilon_2\tilde{D}_{11} + 2\tilde{D}_{66} = \upsilon_1\tilde{D}_{22} + 2\tilde{D}_{66}$ and $\tilde{D}_{66} = \frac{\tilde{G}h^3}{12}$. In these equations, h is the plate thickness, $\tilde{E}_1, \tilde{E}_2$ are the Young's moduli in the principal directions, $\tilde{G}$ is the shear modulus and υ_1, υ_2 are the Poisson's coefficients.

The virtual work principle applied on a region S_v of the plate is [2]

$$\frac{h^3}{12}\int_{S_v} \mathbf{k}^{vT}(\boldsymbol{x})\,\tilde{\mathbf{Q}}(\boldsymbol{x},\omega)\,\tilde{\mathbf{k}}(\boldsymbol{x},\omega)\,d\boldsymbol{x} = \omega^2 h\int_{S_v} \rho w^v(\boldsymbol{x})\,\tilde{w}(\boldsymbol{x},\omega)\,d\boldsymbol{x} + \int_{S_v} w^v(\boldsymbol{x})\,\tilde{q}(\boldsymbol{x},\omega)\,d\boldsymbol{x}, \tag{2.2}$$

where $\tilde{\mathbf{Q}}$ is the 3×3 elastic matrix of the plate, $\tilde{\mathbf{k}}(\boldsymbol{x},\omega) = \{\tilde{k}_1\ \tilde{k}_2\ \tilde{k}_6\} = -\left\{\frac{\partial^2}{\partial x^2}\ \frac{\partial^2}{\partial y^2}\ 2\frac{d^2}{dxdy}\right\}\tilde{w}(\boldsymbol{x},\omega)$ is the bending curvature vector, ρ is the mass density and $w^v(\boldsymbol{x})$ is a *virtual* displacement field defined on S_v. The virtual field w^v in Eq. (2.2) is a C^1 function on S_v. In this work, S_v is chosen to be a small rectangular area (hereafter called "virtual window") scanning the plate surface and w^v is chosen to be a Hermite 16 interpolation polynomial, whose value and first derivatives with respect to x and y vanish over the boundary of the virtual window [2]. Finally, $\tilde{q}$ in Eq. (2.2) is the applied transverse force per unit area and $\mathbf{k}^v(\boldsymbol{x}) = \{k_1^v\ k_2^v\ k_6^v\} = -\left\{\frac{\partial^2}{\partial x^2}\ \frac{\partial^2}{\partial y^2}\ 2\frac{d^2}{dxdy}\right\}w^v(\boldsymbol{x})$ is the *virtual* curvature field related to the virtual displacement w^v.

Assuming that the elastic matrix remains constant over each virtual window, and after some algebraic manipulations, Eq. (2.2) takes the form

$$\tilde{D}_{11}\int_{S_v} k_1^v\tilde{k}_1 d\boldsymbol{x} + \tilde{D}_{22}\int_{S_v} k_2^v\tilde{k}_2 d\boldsymbol{x} + \tilde{D}_{12}\int_{S_v}\left(k_1^v\tilde{k}_2 + k_2^v\tilde{k}_1\right)d\boldsymbol{x} = \omega^2 h\int_{S_v}\rho w^v(\boldsymbol{x})\,\tilde{w}(\boldsymbol{x},\omega)\,d\boldsymbol{x} + \int_{S_v} w^v(\boldsymbol{x})\,\tilde{q}(\boldsymbol{x},\omega)\,d\boldsymbol{x} \tag{2.3}$$

This equation is the basis of the proposed material identification method: given a known external loading $\tilde{q}(\boldsymbol{x},\omega)$ and virtual fields $w^v(\boldsymbol{x})$ and $\mathbf{k}^v(\boldsymbol{x})$, the transverse displacement field $\tilde{w}(\boldsymbol{x},\omega)$ and the corresponding bending curvatures $\tilde{k}_i(\boldsymbol{x},\omega)$ are measured over the plate surface and the various integrals over successive virtual windows in Eq. (2.3) are computed. The process must be repeated for a number of external loadings at least equal to the number of unknown bending stiffnesses $\tilde{D}_{ij}$ to build a linear system of equations whose solution provides the $\tilde{D}_{ij}$ on each virtual window.

If the mapping of elastic parameters is restricted to regions exempt of external loadings, then $\tilde{q}(\boldsymbol{x},\omega) = 0$ and Eq. (2.3) takes a simpler form,

$$\tilde{D}_{11}\int_{S_v} k_1^v\tilde{k}_1 d\boldsymbol{x} + \tilde{D}_{22}\int_{S_v} k_2^v\tilde{k}_2 d\boldsymbol{x} + \tilde{D}_{12}\int_{S_v}\left(k_1^v\tilde{k}_2 + k_2^v\tilde{k}_1\right)d\boldsymbol{x} = \omega^2 h\int_{S_v}\rho w^v(\boldsymbol{x})\,\tilde{w}(\boldsymbol{x},\omega)\,d\boldsymbol{x} \tag{2.4}$$

As a consequence, no knowledge of the excitation is required in this case since the free vibration condition applies.

2.3 Experiments

Experiments were conducted on various aluminium and composite panels, some of them with added masses, stiffeners or damping treatments. In this article, we report results for a 51 cm × 49 cm × 2.3 mm carbon-epoxy composite panel with unidirectional fibers along the x axis. The panel was suspended in free boundary conditions and was set in vibration successively by 4 inertial actuators driven by a broadband input in the 100 Hz–5 kHz range: 3 actuators were positioned close to the plate boundaries and 1 actuator was close to the plate center.

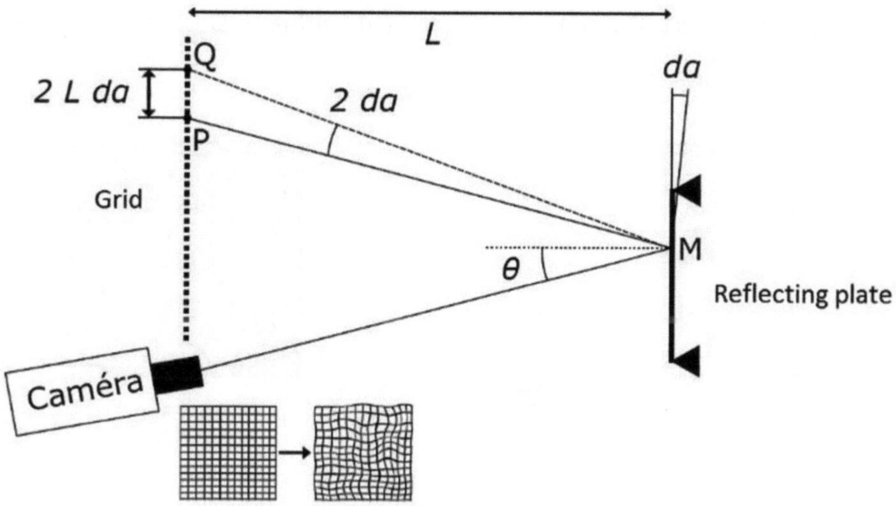

Fig. 2.1 Optical deflectometry principle for a bending plate

The full-field vibration response (transverse displacement and bending curvatures) of the panel was measured either with scanning Laser Doppler Vibrometry (LDV) or optical deflectometry. Scanning LDV can only apply for stationary excitations. In this case, the transverse velocity of the panel is measured at successive positions on a defined grid in the frequency domain. The transverse displacement field is obtained at each frequency by division of the velocity by $j\omega$. The displacement field must be spatially low-pass filtered in order to remove high frequency measurement noise. The bending curvature field is then derived by second order spatial differentiation of the displacement along x and y. The spatial low-pass filtering of the transverse displacement field is crucial to obtain reliable measurements of bending curvatures.

The optical deflectometry set-up is illustrated on Fig. 2.1 [3, 4]. A high-speed camera (model Photron SA-X2 RV, 10,000 fr/s, approximately 20,000 measurement points on the imaged area) observes the specular reflexion by the plate of a grid placed at a distance L. The local bending of the plate induces a local slope da and therefore a displacement $2Lda$ of the corresponding grid point. The method requires a mirror finish of the panel and gives access to local bending slopes $\frac{\partial w}{\partial x}$ and $\frac{\partial w}{\partial y}$ of the panel. Further spatial integration and differentiation operations are necessary to extract the transverse displacement field w and bending curvature field $\mathbf{k}$, respectively. One of the advantages of the deflectometry technique in this case is to avoid second order differentiation operations required in scanning LDV. Therefore, no spatial low-pass smoothing of the measurement displacement fields is necessary in optical deflectometry.

The vibration response of the central region of the carbon-epoxy panel was measured by scanning LDV over a regular rectangular grid of 29 × 27 points. The scanned area excludes the three actuator positions close to plate edges but includes the central actuator position. In the present case, the linear system in terms of $\tilde{D}_{11}$, $\tilde{D}_{12}$ and $\tilde{D}_{22}$ at each point is over-determined since four excitation positions are considered. Figure 2.2 shows the three bending stiffnesses $\tilde{D}_{11}$, $\tilde{D}_{12}$ and $\tilde{D}_{22}$ provided by the VFM over the scanned surface at a frequency of 3 kHz. The values obtained are close to those provided by ASTM standard measurements, $\tilde{D}_{11} = 117$ Pa.m^3, $\tilde{D}_{12} = 10.4$ Pa.m^3 and $\tilde{D}_{22} = 8.8$ Pa.m^3. The singularity observed in the top-right corner of the maps corresponds to the actuator position in the scanned area. This singularity is expected in the VFM since the free vibration equilibrium of the panel does not hold at this position.

Other results (not shown here) obtained for metallic panels with punctual masses or stiffeners and scanning LDV or optical deflectometry full-field measurements confirm the ability of the VFM to identify material and structural parameters of bending plates.

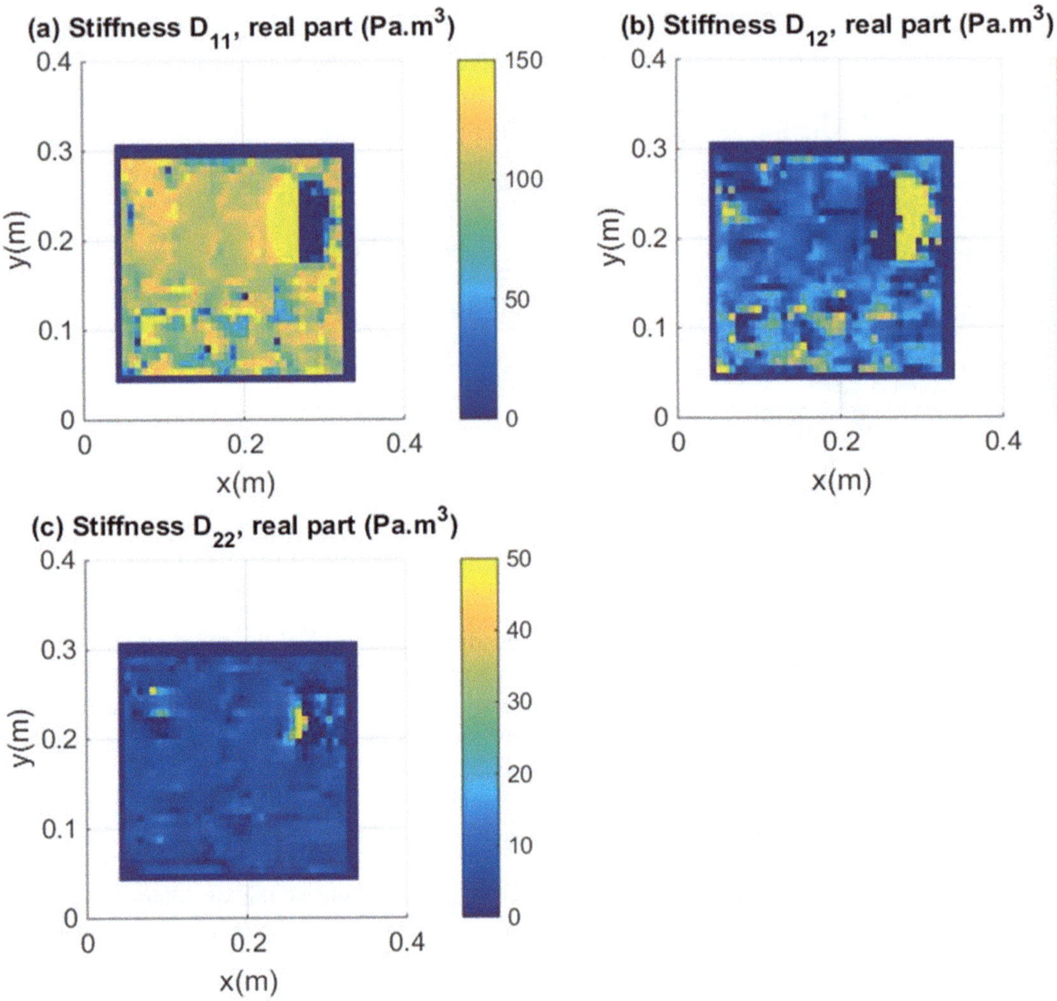

Fig. 2.2 Extracted bending stiffnesses for a carbon-epoxy panel, (a): $\widetilde{D}_{11}$, (b): $\widetilde{D}_{12}$, (c): $\widetilde{D}_{22}$

References

1. Pierron, F., Grédiac, M.: The Virtual Fields Method. Springer, New York (2012)
2. Berry, A., Robin, O., Pierron, F.: Identification of dynamic loading on a bending plate using the Virtual Fields Method. J. Sound Vib. **333**, 643–659 (2014)
3. Surrel, Y.: Deflectometry: a simple and efficient noninterferometric method for slope measurement. In: Xth International Congress on Experimental and Applied Mechanics (2004)
4. Giraudeau, A., Pierron, F.: Measurement of vibrating plate spatial responses using deflectometry and a high speed camera. In: The 9th International Conference on Vibration Measurements by Laser and Noncontact Techniques and Short Course (2010)

Chapter 3
Full-Field Displacement Characterisation of TCW Joint Using Digital Image Correlation and Comparison to Numerical Models

Zhi Bin Tan and Liyong Tong

Abstract In this study, a 2-dimensional digital image correlation (DIC) method is used with smaller speckles to produce displacement and strain field near weld polymer ends of a Thermoset Composite Welding (TCW) joint. The shear and peel strain distribution across the overlap area obtained from the DIC method is compared to the numerical results from finite element modelling. Compared with the traditional single lap joint test, the DIC method provides extra insight into the strain distribution immediately prior to crack propagation, as well as its crack propagation path.

Keywords Small speckles • Strain concentration • Crack path • Strain profiles • Bonded joint

3.1 Introduction

Thermoset Composite Welding (TCW) technology is a technology in which a thermoset composite prepreg layup is surfaced with an integrated layer of thermoplastic material. Cured laminates with this this integrated thermoplastic layer can then be welded together to form a welded assembly. In order to better understand the performance of a TCW joint, it is important to characterise the strain field across the bondline during loading. Digital image correlation (DIC) has been chosen for its ability to provide full field surface displacements and strains field of structures subjected to various types of loading, such as mechanical and thermal loadings. The detailed descriptions of DIC system are widely available in the literature and will not be repeated here for brevity [1–3]. There are several challenges that come with the use of DIC system. Lecompte et al. in one of the earlier research identified the effect of subset size and sizes of speckle pattern have on the accuracy of the correlation result [4]. Their findings show that differences in speckles pattern and the chosen subset size in analysis can clearly affect the displacement accuracy.

This presents an additional difficulty in analysing the strain behaviour of the weld polymer within the bondline of a bonded joint as the adhesive, or in this case the weld polymer only has a thickness in the range of 100–300 μm. In this case, the speckle pattern has to be smaller than the bondline thickness to avoid misrepresenting the carbon-epoxy laminate in the correlation analysis. Consequently, the subset size has to be smaller than the bondline thickness to avoid interpolation of strain field through the bondline, which could significantly alter the true strain behaviour of the weld polymer as the carbon-epoxy laminate is much stiffer than the thermoplastic polymer. While a 3-dimensional digital image correlation (3D DIC) system can provide both in-plane and out-of-plane deformation of a specimen, its magnification is more limited due to the need to maintain cameras distance to ensure concurrent view of the specimen. This restricts the resolution of the captured image and the small speckle pattern along the bondline cannot be captured with enough clarity for the correlation process. In light of this situation, 2-dimensional digital image correlation (2D DIC) is used in this study to allow the area of interest around the bonded area to be studied up close at greater detail. As only one camera is needed in a 2D DIC system, the camera and its lens can be positioned as closely as possible to the specimen to allow high magnification images to be taken.

Z.B. Tan • L. Tong (✉)
School of Aerospace, Mechanical and Mechatronic Engineering, The University of Sydney, Sydney, NSW 2006, Australia
e-mail: liyong.tong@sydney.edu.au

M. Sutton, P.L. Reu (eds.), *International Digital Imaging Correlation Society*, Conference Proceedings of the Society for Experimental Mechanics Series, DOI 10.1007/978-3-319-51439-0_3

3.2 Specimen Preparation

The manufacturing process of the single lap joint using the TCW method can be found here [5]. The overlap area of the single lap joint was first coated with white enamel paint before being speckled with black enamel paint to produce a random surface pattern required for image correlation. In order to create speckle with small diameter, an airbrush with a 0.2 mm nozzle tip spraying at a distance of approximately 300 mm from the specimen was used to apply the black paint on the white coated surface. The speckles diameter created using this method are in the range of 29–75 μm. In order to observe the effect of speckle diameter has on the quality of the result, the speckle pattern on some specimens were produced by using a spray can, resulting in speckles pattern with diameters ranging from 260–530 μm.

3.3 Experimental Method

The 2D DIC system used in this study consists of a CCD camera (Q-400 GigE 9M) with a 28 mm fixed focal lens, HILIS illumination system with red wavelength, and a computer loaded with ISTRA 4D software (Dantec Dynamics) for triggering, acquisition and correlation of the digital images. In order to observe the smaller speckles produced by the airbrush, a 15 mm c-mount extension tube was fitted between the CCD camera and the lens to reduce the minimum allowable working distance. This allows a higher magnification image of the bondline region to be captured by the camera. The acquisition system used in this study allows analogue signal from the INSTRON testing machine to be linked directly to the image captured. With this additional information, the precise loading data from INSTRON machine and corresponding strain data can be analysed in greater details. It is worthwhile to note that the aperture of the lens shall be as small as possible to allow greater depth of field in order to prevent the field of view going out of focus during loading. Strong lighting is needed to compensate for the smaller amount of light allowed due to the reduce aperture size.

Quasi-static tests were conducted on the single lap joint specimens on an INSTRON machine, at a loading rate of 3 mm/min to failure. The images were captured at 5 Hz throughout the loading cycle and were calibrated by using the laminate thickness as a known dimension. A facet size of 25 × 25 pixels was used to carry out the correlation in the ISTRA 4D software. This facet size was chosen to balance both the strain resolution and accuracy of the result, as large subset size improves strain precision but at the expense of the spatial resolution of the field [6]. Figure 3.1 shows the facet size on the surface speckle patterns, the corresponding load output can be seen on the bottom left corner of the image. The setup of the experiment can be seen in Fig. 3.2 below.

3.4 Results and Discussion

3.4.1 Strain Contour

The maximum peel and shear strains were analysed by considering the last available crack-free image prior to the fracture of the single lap joints whenever possible (80% of failure load in this case). Once the crack starts to form within the bondline region of the single lap joint, the surface discontinuities will cause correlation error and no surface strain contour can be generated by the software. Figures 3.3 and 3.4 show the overall peel and shear strain contour of the single lap joint respectively. They were obtained by superimpose the strain field on the undeformed reference image (unloaded state). It can be seen from Fig. 3.3 that the shear strain distribution is clearly visible along the bondline of the lap joint.

As an observation to the effect of speckle size has on the correlation outcome, Fig. 3.5 shows the shear strain contour of a single lap joint where its speckle patterns was prepared by using a spray can, thus having a larger average diameter than those shown in Fig. 3.3. While the correlation outcome from the software still shows a clearly visible shear loaded section along its bondline, the magnitude of the shear strain is much lower than those observed in the specimens with smaller speckle patterns.

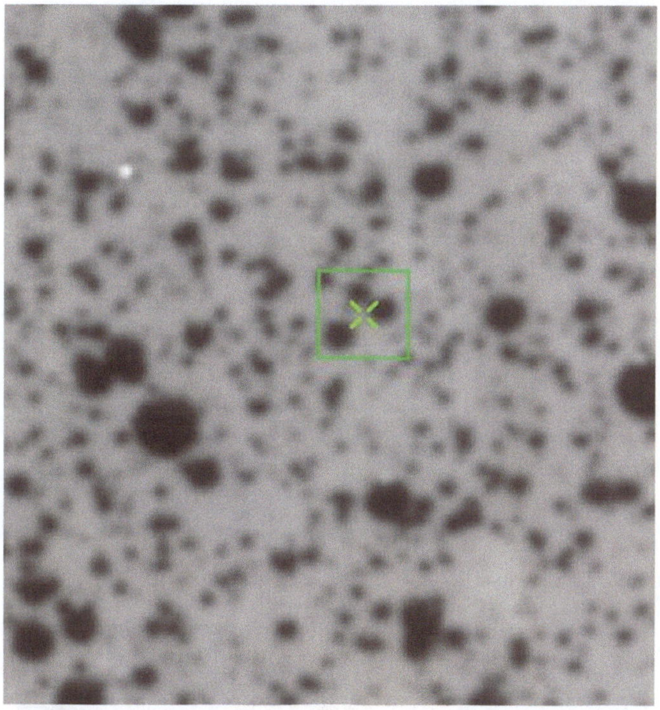

Fig. 3.1 Facet (subset) size used in correlation analysis

Fig. 3.2 Experimental set-up for 2D DIC testing

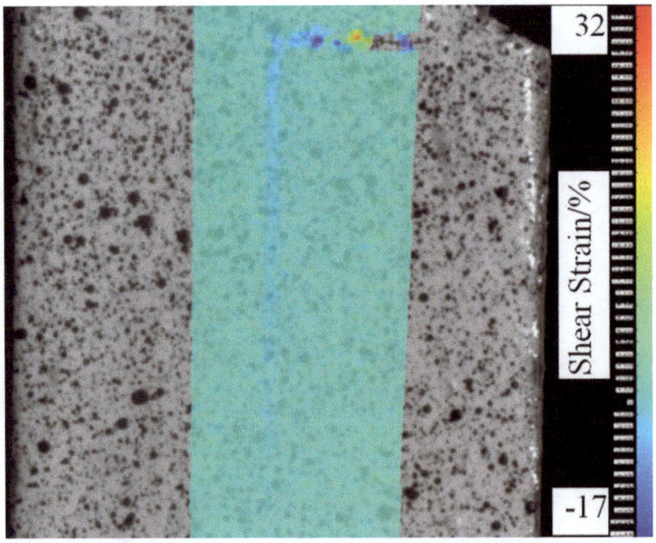

Fig. 3.3 Shear strain contour plot at 80% of failure load

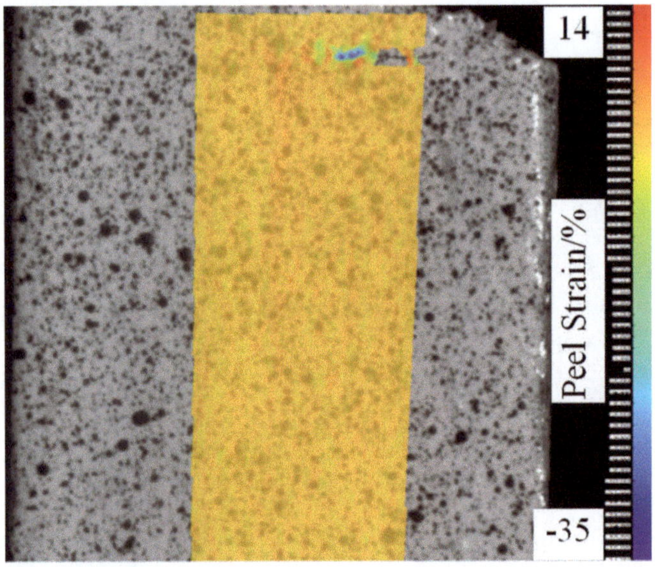

Fig. 3.4 Peel strain contour plot at 80% of failure load

3.4.2 Bondline Surface Strains

The peel and shear strain profiles for the middle path of the bondline are shown in Fig. 3.6, representing the load level prior to crack initiation. The result below shows that there are high strain concentration region at the inner corner of the overlap area. In the subsequent image where cracks have started to form, it can be observed that these high strain concentration regions served as the crack initiation site. By using the high magnification 2D DIC system, the crack propagation process was captured and the crack path is shown in Fig. 3.7 below.

A finite element model has been built for the single lap joint to predict its material properties based on the experimental failure load. The peel and shear strain results within the bondline from the finite element model had been incorporated to the results shown in Fig. 3.6. While there are some discrepancies in terms of the location of the peak shear strain between the finite element analysis result and the values obtained from 2D DIC, the general shape of the strain profile is very similar and both point to the significant built up of strain at the inner corner of the overlap region where it most likely served as the failure initiation site.

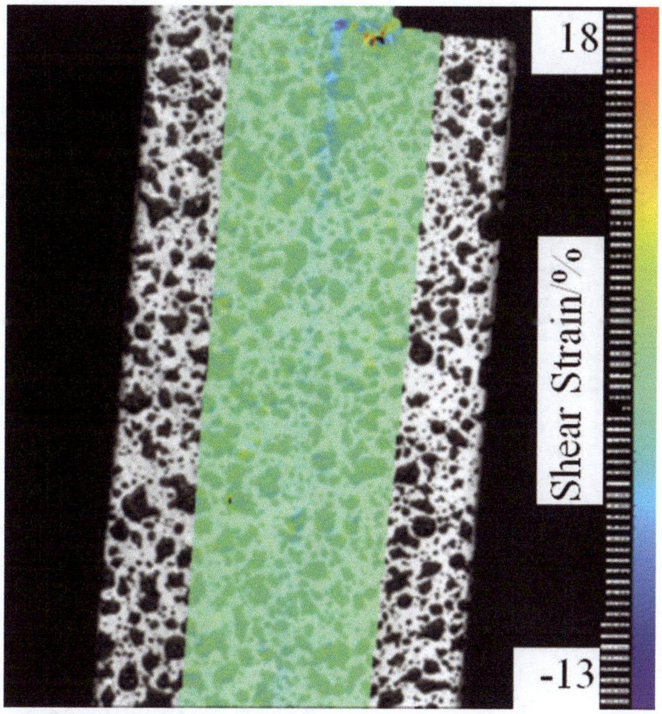

Fig. 3.5 Shear strain contour plot for specimen with larger speckle patterns at a similar load level

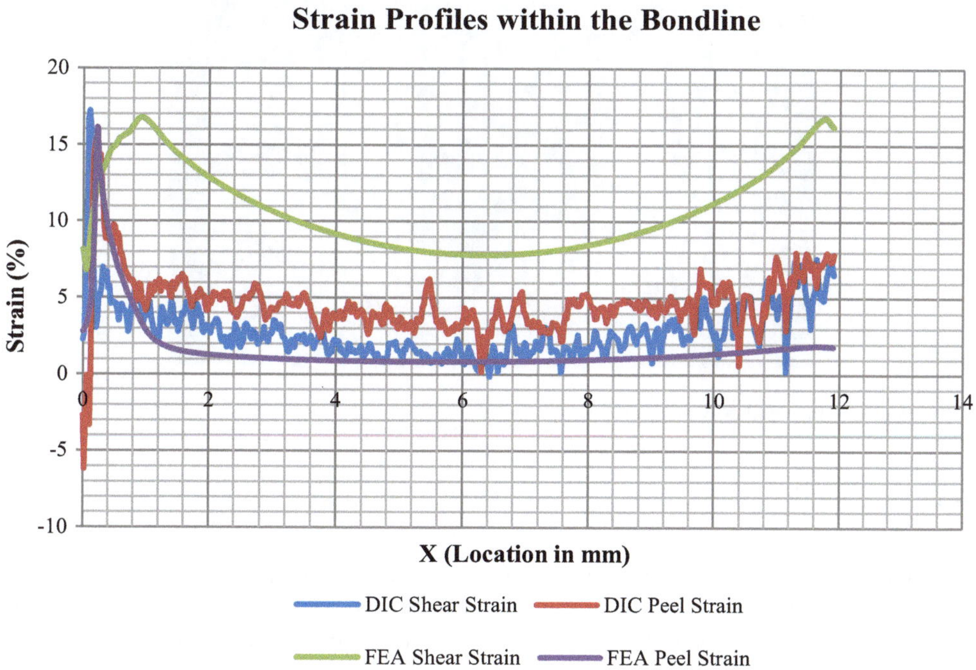

Fig. 3.6 Peel and shear strain profiles within the weld polymer

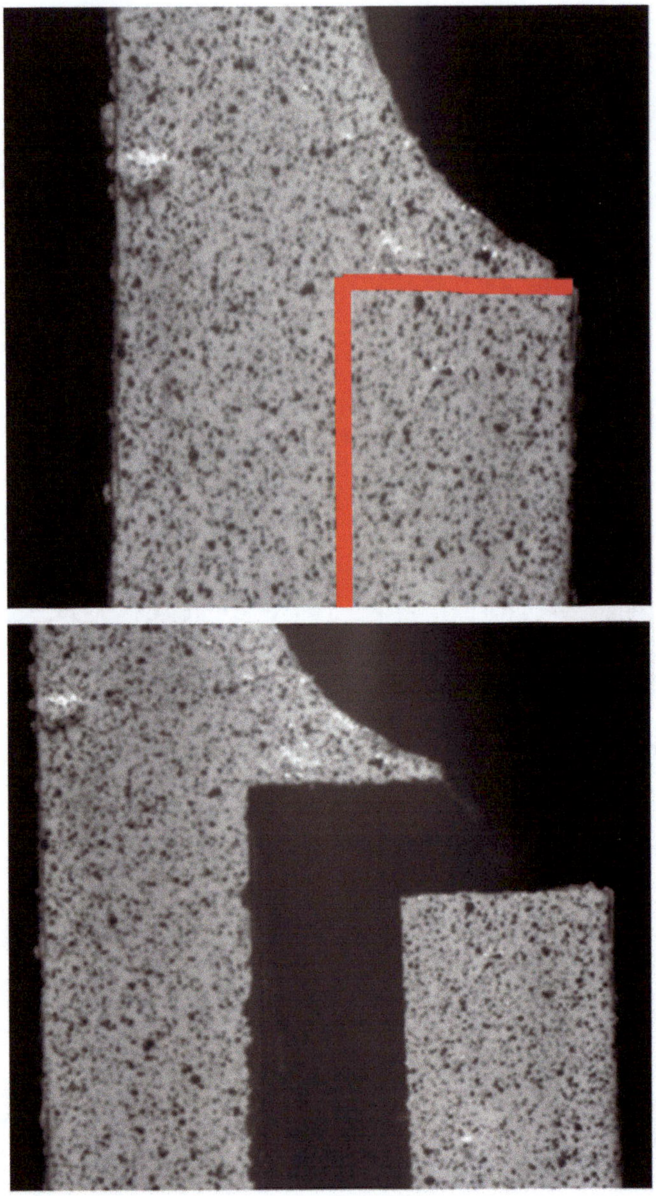

Fig. 3.7 Crack propagation path in the weld polymer bondline

3.5 Conclusions

A 2D DIC setup was chosen to allow high magnification images of a TCW single lap joint to be taken and analysed. In order to maximise the potential of the 2D DIC system, the speckle patterns used has to be small enough to avoid interpolation of results through the laminate layer of the joint. It is clear from the result that the choices of speckle pattern size and the subset size have significant impact on the accuracy of the result. The peel and shear strain profile for the weld polymer shows a spike at the inner corner of the overlap area, which indicates the site of crack initiation due to the high strain concentration. A good correlation between the 2D DIC and the finite element analysis results can be viewed as a validation to the finite element model. In conclusion, this study shows that 2D DIC system, combined with a careful control of speckle pattern size can be a viable method to analyse the built up of localised strain in a joint, and be used to provide a better understanding of the strain distribution within the bondline of a joint.

Acknowledgement Authors would like to acknowledge the support of the Australian Research Council via Discovery-Project Grants Scheme (grant number: DP140104408).

References

1. Comer, A., et al.: Characterising the behaviour of composite single lap bonded joints using digital image correlation. Int. J. Adhes. Adhes. **40**, 215–223 (2013)
2. Kumar, R.V., Bhat, M., Murthy, C.: Experimental analysis of composite single-lap joints using digital image correlation and comparison with theoretical models. J. Reinf. Plast. Compos. **32**(23), 1858–1876 (2013)
3. Hild, F., Roux, S.: Digital image correlation: from displacement measurement to identification of elastic properties – a review. Strain. **42**(2), 69–80 (2006)
4. Lecompte, D., et al.: Quality assessment of speckle patterns for digital image correlation. Opt. Lasers Eng. **44**(11), 1132–1145 (2006)
5. Tan, Z.B., et al.: The effect of post-welding cooling rate on the strength of TCW joints. In: 20th International Conference on Composite Materials, Copenhagen, Denmark (2015)
6. Crammond, G., Boyd, S., Dulieu-Barton, J.: Speckle pattern quality assessment for digital image correlation. Opt. Lasers Eng. **51**(12), 1368–1378 (2013)

Chapter 4
3D Digital Image Correlation Applied to Birdstrike Tests

L. Barrière, O. Cherrier, J.-C. Passieux, M. Bouquet, and J.-F. Ferrero

Abstract The development of new bird strike shielding materials for commercial aircrafts requires test campaigns. During these tests, measurement of the high speed deformation is needed to characterize and compare the mechanical response of the shielding samples and to correlate numerical simulations with experiments. In this work, 3D digital image correlation method is used with high speed (HSP) cameras to compute the displacement and strain fields on a large area (approximately 400 mm wide) of the back side of impacted samples. Compromise on spatial resolution, frame rate of HSP camera and measurement error is discussed.

Keywords 3D DIC • High speed camera • Bird strike • Wide ROI • High velocity impact

4.1 Introduction

Among the quantities of interest measurable during a bird strike test, one is the deformation of the target that could help correlate simulations with experiments and compare different targets one to each other. Deformations are often measured by strain gages [1] but particular attention has been paid recently to 3D digital image correlation (DIC) for HVI or blast [2–4] when region of interest (ROI) remains properly visible during test. The main challenge is to find a compromise between measurement accuracy and size of the ROI because of the lower resolution of high speed cameras.

In the framework of project FUI SAMBA led by Stelia Aerospace, which aims at developing new shielding concepts for protecting the cockpit of aircrafts against bird strike, the back side of shielding samples remains visible during test. The ROI of such one square meter target may span several hundred millimeters side length. In this paper, the ROI reaches 400 mm by 400 mm corresponding to the average length between stringers of the structure which was considered to be protected.

4.2 Preliminary Study

Before bird strike tests, 3D DIC capabilities were evaluated. As shown in Fig. 4.1, two high speed cameras (Photron APX) are used for stereo-vision. At a rate of 30,000 frames per second, resolution was 256 by 256 pixels. A coarse high contrast random pattern (bullet points of diameter approximately equal to 10 mm) has been painted on a mock target plate such that the transition between black and white area corresponds to approximately four pixels [5].

After calibration, correlation was successful in the ROI.[1] Figure 4.2 shows the mock target in reference and rotated positions and its shape reconstruction by stereo-correlation. The maximum out-of-plane displacement that was measured is about 40 mm (Fig. 4.2b). As target plate remains planar and not deformed, computed strain should remain very small [6]. A normal strain of 6×10^{-4} is computed which seems satisfactory at this stage for the application (Fig. 4.2c).

[1]Stereo calibration and 3D DIC are performed with VIC3D software (Correlated Solutions, Inc.).

L. Barrière (✉) • M. Bouquet
Institute of Technology Saint Exupery, 118 Route de Narbonne CS44248, 31432, Toulouse Cedex 4, France
e-mail: ludovic.barriere@irt-saintexupery.com

O. Cherrier • J.-C. Passieux • J.-F. Ferrero
Institut Clément Ader, UFTMP/UMR CNRS 5312, 3 rue Caroline Aigle, 31400, Toulouse, France

M. Sutton, P.L. Reu (eds.), *International Digital Imaging Correlation Society*, Conference Proceedings of the Society for Experimental Mechanics Series, DOI 10.1007/978-3-319-51439-0_4

Fig. 4.1 Stereo-correlation device comprising high speed camera and large scale target with coarse speckle pattern

4.3 Birdstrike Test and Application of 3D-DIC

The bird strike tests are performed at the High Velocity Impact Test Platform in Toulouse, France, which has been co-developed by IRT Saint Exupery and Institut Clément Ader (UMR CNRS 5312) for the last 2 years. A 120 mm diameter gaz launcher is used to shoot 1.5 kg substitute bird (homemade gelatin cylinder) at 170–180 m/s. An aluminum plate (dimensions: 800 mm × 800 mm × 6 mm) is simply supported by two parallel beams of a stiff chassis (see Fig. 4.3a). The camera setup is such that distance to target is approximately 2.5 m and the angle between the axes of cameras is 30°. Cameras are placed outside the shooter enclosure whose backside wall is made of polycarbonate plates (see Fig. 4.3b).

The frame rate of cameras is set to 30,000 fps. The corresponding maximum resolution of 256 by 384 pixels is chosen so that a marker placed on the upper beam of the chassis is in the field of view (Marker tracking on the chassis enables the assessment of the boundary conditions).

After calibration of the stereo system, 3D image correlation of the ROI is performed by VIC3D. As shown in Fig. 4.4, correlation algorithm may not converge in some areas of some images due to light reflection on the deformed target. Nevertheless, the deformation of the plate can be evaluated in most of the ROI for all the images during the impact event. The maximum computed normal displacement reaches then 70 mm.

4.4 Conclusions

The preliminary study has open interesting prospects for the use of 3D DIC for bird strike tests on shielding concepts. It was shown that large region of interest could be considered while keeping a quite good accuracy of measurement of displacement and strain fields. The second step of this work has implemented the stereo-correlation device in bird strike test conditions of the impact test platform which is co-developed by IRT Saint-Exupery and Institut Clément Ader. Further analyses are planned to assess and reduce the measurement error and to correlate with numerical simulations of bird strike.

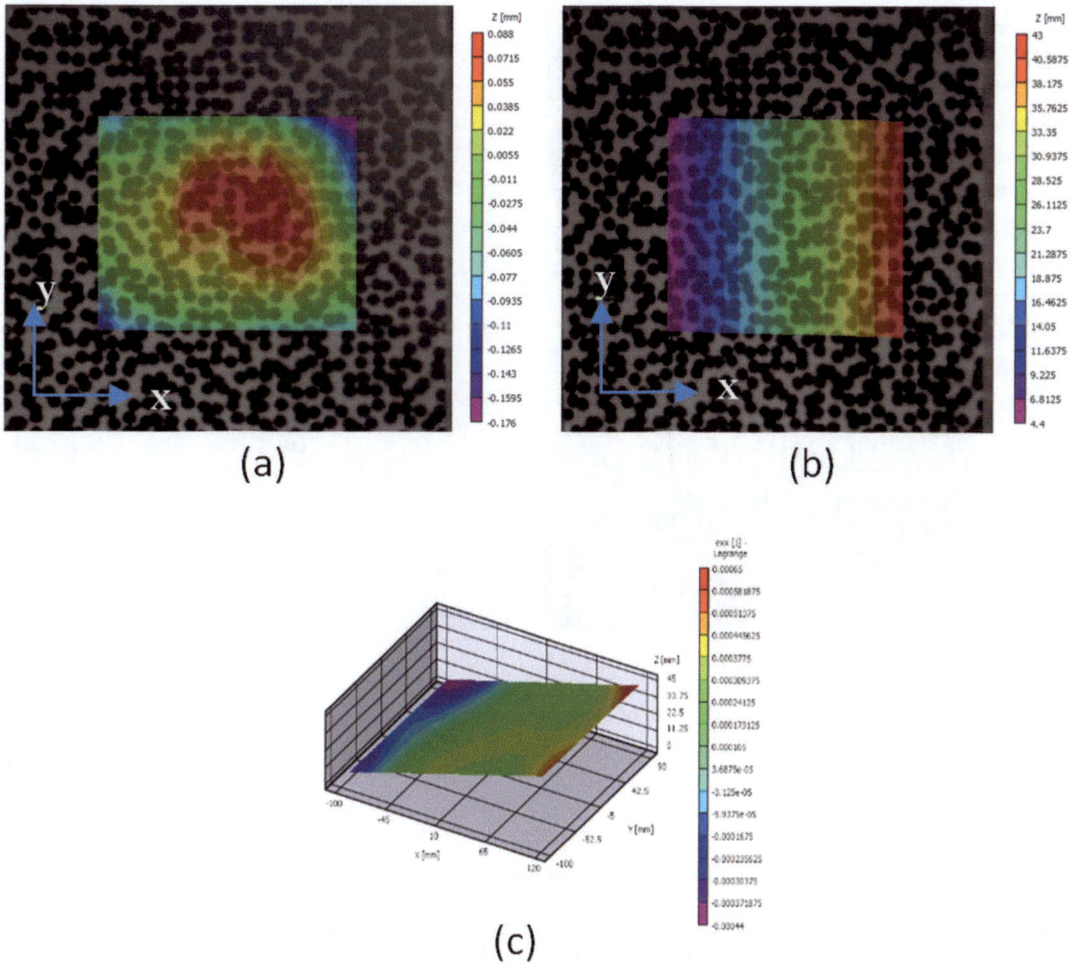

Fig. 4.2 3D surface reconstruction of the target in reference position (**a**) vertically rotated position (**b**) and computation of strain field in rotated position ($\epsilon_{xx} < 6 \times 10^{-4}$) (**c**)

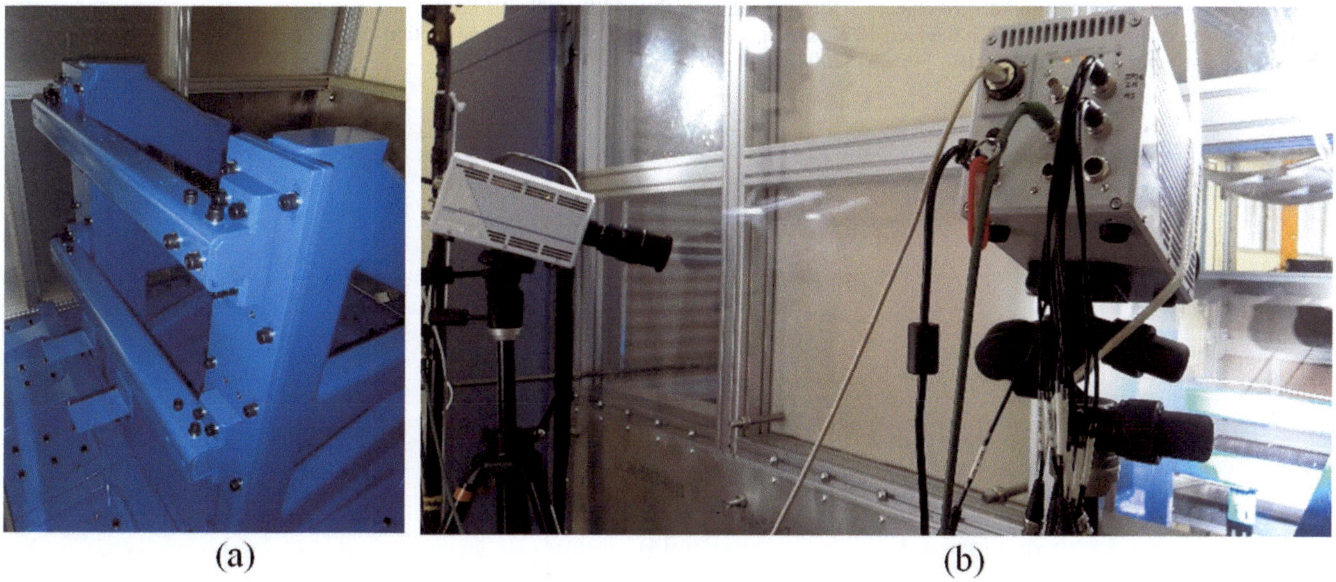

Fig. 4.3 Stiff chassis supporting the target (**a**). High speed camera setup for 3D DIC (**b**)

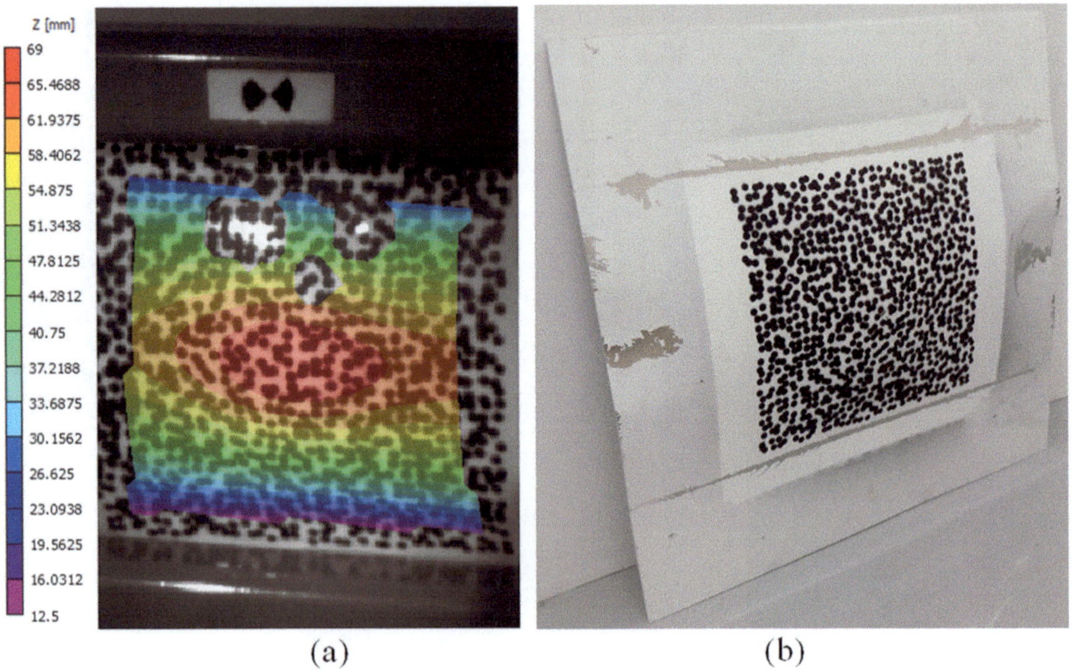

Fig. 4.4 (**a**) 3D DIC of the plate during bird strike, (**b**) state of the plate after impact

References

1. Guida, M., Marulo, F., Meo, M., Russo, S.: Certification by birdstrike analysis on C27J fullscale ribless composite leading edge. Int. J. Impact Eng. **54**, 105–113 (2013)
2. Gupta, S., Parameswaran, V., Sutton, M.A., Shukla, A.: Study of dynamic underwater implosion mechanics using digital image correlation. Proc. R. Soc. Math. Phys. Eng. Sci. **470**(2172), 20140576–20140576 (2014)
3. Tiwari, V., Sutton, M.A., McNeill, S.R., Xu, S., Deng, X., Fourney, W.L., Bretall, D.: Application of 3D image correlation for full-field transient plate deformation measurements during blast loading. Int. J. Impact Eng. **36**(6), 862–874 (2009)
4. Portemont, G., Langrand, B., Mortier, J.M.: An innovative experimental device for ice impac onto composite panel. In: 16th International Conference on Experimental Mechanics (ICEM 16), Cambridge, UK, Jul 2014
5. Ke, X., Sutton, M.A., Lessner, S.M., Yost, M.: Robust stereo vision and calibration methodology for accurate three-dimensional digital image correlation measurements on submerged objects. J. Strain Anal. Eng. Des. **43**(8), 689–704 (2008)
6. Sutton, M.A., Yan, J.H., Tiwari, V., Schreier, H.W., Orteu, J.J.: The effect of out-of-plane motion on 2D and 3D digital image correlation measurements. Opt. Lasers Eng. **46**(10), 746–757 (2008)

Chapter 5
The Ill-Posed Problem in DIC

Rich Lehoucq and Dan Turner

Abstract The contribution of our paper is to present a mixed finite element method for estimation of the velocity in the optical flow constraint, i.e., an advection equation. The resulting inverse problem is well-known to be undetermined because the velocity vector cannot be recovered from the scalar field advected unless further restrictions on the flow, or motion are imposed. If we suppose, for example, that the velocity is solenoidal, a well-defined least squares problem with a minimizing velocity results. Equivalently, we have imposed the constraint that the underlying motion is isochoric. Unfortunately, the resulting least squares system is ill-posed and so regularization via a mixed formulation of the Poisson equation is proposed. Standard results for the Poisson equation demonstrate that the regularized least squares problem is well-posed and has a stable finite element approximation. A numerical example demonstrating the procedure supports the analyses. The example also introduces a closed form solution for the unregularized, constrained least squares problem so that the approximation can be quantified.

Keywords Optical flow • Digital image correlation • Transport equation • Regularization • Mixed finite element method

5.1 Introduction

Our paper describes a constrained, regularized least squares approach for estimating the velocity vector field $\mathbf{v}$ given the scalar image intensity $\phi = \phi(x, y, t)$ for a model given by the advection equation

$$\begin{cases} \phi_t + \mathbf{v} \cdot \nabla\phi = 0 \quad \text{over} \quad \Omega \times (0, T)\,, \\ \phi(x, y, 0) = \phi_0(x, y) \qquad x, y \in \Omega\,, \end{cases} \tag{5.1}$$

including suitable boundary conditions. Such a model represents the so-called optical flow constraint for idealized image motion given the assumption that the image intensity of an object is time independent and that spatial, temporal sampling is sufficiently resolved; see e.g., [7]

A least squares method results when we consider the formal minimization problem: Given intensity data $\hat{\phi}$, find

$$\mathbf{b}^* = \arg\min_{\mathbf{b}} \frac{1}{2} \int_0^T \int_\Omega \left(\hat{\phi}_t + \mathbf{b} \cdot \nabla\hat{\phi}\right)^2 dx\, dt\,, \tag{5.2a}$$

over a class of suitable functions. Proceeding formally and assuming appropriate boundary conditions, the corresponding normal equations are given by the singular linear system

$$\begin{cases} (\nabla\hat{\phi} \otimes \nabla\hat{\phi})\mathbf{b}^* = -\hat{\phi}_t \nabla\hat{\phi} \quad \text{over} \quad \Omega \times (0, T)\,, \\ \hat{\phi}(x, y, 0) = \phi_0(x, y) \qquad x, y \in \Omega\,. \end{cases} \tag{5.2b}$$

R. Lehoucq (✉) • D. Turner
Sandia National Labs, Albuquerque, NM 87185, USA
e-mail: rblehou@sandia.gov

M. Sutton, P.L. Reu (eds.), *International Digital Imaging Correlation Society*, Conference Proceedings of the Society for Experimental Mechanics Series, DOI 10.1007/978-3-319-51439-0_5

Because the coefficient matrix is singular, the least squares problem (5.2a) is undetermined since adding

$$\bar{\mathbf{b}} := \psi J \nabla\hat{\phi}\,, \quad \psi : \Omega \to \mathbb{R}\,, \quad J := \begin{bmatrix} 0 & -1 \\ 1 & 0 \end{bmatrix}, \tag{5.2c}$$

to $\mathbf{b}^*$ is also a minimizer (and also a solution to the normal equations) because $\bar{\mathbf{b}} \cdot \nabla\hat{\phi} = 0$. This indeterminacy is intrinsic and explains that the velocity field cannot, in general, be reconstructed given the intensity and embodies the challenge in attempting to reconstruct a vector parameter from scalar intensity data. An important consequence is that the minimization problem (5.2a) is not well-defined so that the ensuing normal Eq. (5.2b) have an infinite number of solutions. We also remark that because $\phi_t = -\mathbf{v} \cdot \nabla\phi_0$ for the advection equation system (5.1), only the velocity in the direction of $\nabla\phi$ can ultimately be recovered—this is known as the aperture problem.

A linear constraint, however, can be imposed on the velocity so that the resulting normal equations determine a unique velocity from the space of functions defined by $\bar{\mathbf{b}}$ and $\nabla\hat{\phi}$. For instance, we can augment the Eq. (5.2b) with the constraint $\nabla \cdot \mathbf{b}^* = 0$. Equivalently, we have imposed the constraint that the underlying optical flow, or motion is isochoric. Moreover, if the true velocity is indeed solenoidal, then the velocity can be completely reconstructed. We remark, though, that our choice of a solenoidal constraint is illustrative; other constraints are possible. Ultimately, the choice of constraint depends upon the specific problem at hand and whether the choice (along with regularization) leads to a well-posed estimation problem.

The primary contribution of this work is to present a mixed finite element method for the constrained, regularized estimation of the velocity in the optical flow constraint. We show that this method leads to a well-posed problem, both for the infinite and finite dimensional formulations.

The first step is to constrain the minimization problem (5.2a) by considering: Given intensity data $\hat{\phi}$, solve

$$\arg\min_{\mathbf{b}\in\mathcal{B}} \frac{1}{2} \int_0^T \int_\Omega \left(\hat{\phi}_t + \mathbf{b} \cdot \nabla\hat{\phi}\right)^2 dx\, dt\,. \tag{5.3}$$

Our choice of constraint space is given by

$$\mathcal{B} := \{\mathbf{b} \in H_{\mathrm{div},\Gamma}(\Omega) \mid \nabla \cdot \mathbf{b} = 0 \;\text{ over }\; \Omega\,, \mathbf{b} \cdot \mathbf{n} = 0 \text{ over } \Gamma \subset \partial\Omega\}\,, \tag{5.4}$$

where $H_{\mathrm{div},\Gamma}(\Omega) \subset H_{\mathrm{div}}(\Omega)$ is the space of vector functions that are square integrable with zero normal component along Γ and whose divergence is square integrable. However the resulting optimality system is ill-posed. See [2] for an informative review on ill-posed problems in computer vision.

The second step is to regularize the least squares functional with $\nu^2/2 \int_\Omega \mathbf{b} \cdot \mathbf{b}\, dx$ resulting in a well-posed least squares problem. Equivalently, we show that our choice of regularization leads to a saddle point system containing a mixed formulation of the Poisson equation. Standard results for the Poisson equation demonstrate that the regularized system is well-posed and has a stable mixed finite element approximation. The example also introduces a closed form solution for the unregularized, constrained least squares problem so that the approximation can be quantified. In particular, our regularized functional is an instance of an augmented Lagrange method due to Fortin and Glowinski [3]. Again, we emphasize that our choice of constraint space is motivated by physical considerations and mathematical convenience, i.e., the constraint space $\mathcal{B}$ may be replaced by another, suitable, space.

Our approach has application to digital image correlation (DIC) [6] and as the initialization step for the regularized, nonlinear least squares approach introduced in the paper [4] by Ito and Kunisch for estimating the convection coefficient that we considered in the report [5]. The DIC application is that of sequence analysis; the interested reader is referred to the discussion and overview by Aubert and Kornprobst in their textbook [1, pp. 249–256].

References

1. Aubert, G., Kornprobst, P.: Mathematical Problems in Image Processing. Applied Mathematical Sciences, vol. 147, 2nd edn. Springer, New York (2006)
2. Bertero, M., Pogcio, T.A., Torre, V.: Ill-posed problems in early vision. In: Proceedings of the IEEE, pp. 869–889 (1988)
3. Fortin, M., Glowinski, R.: Augmented Lagrangian Methods: Applications to the Numerical Solution of Boundary-Value Problems. Studies in Mathematics and its Applications. Elsevier Science, New York (2000)

4. Ito, K., Kunisch, K.: Estimation of the convection coefficient in elliptic equations. Inverse Prob. **13**(4), 995 (1997)
5. Lehoucq, R.B., Turner, D.Z., Garavito-Garźon, C.A.: PDE constrained optimization for digital image correlation. Technical Report SAND2015-8515, Sandia National Laboratories (2015)
6. Schreier, H., Orteu, J.-J., Sutton, M.A.: Image Correlation for Shape, Motion and Deformation Measurements. Springer, New York (2009)
7. Weickert, J., Schnörr, C.: A theoretical framework for convex regularizers in pde-based computation of image motion. Int. J. Comput. Vis. **45**(3), 245–264 (2001)

Chapter 6
GPU Accelerated High Accuracy Digital Volume Correlation

Tianyi Wang, Lingqi Zhang, Zhenyu Jiang, and Kemao Qian

Abstract A sub-voxel digital volume correlation (DVC) method combining the inverse compositional Gauss-Newton (IC-GN) algorithm with the fast Fourier transform-based cross correlation (FFT-CC) algorithm is proposed to eliminate path-dependence in current iterative DVC methods caused by the initial guess transfer scheme. The proposed path-independent DVC method is implemented on NVIDIA compute unified device architecture (CUDA) for graphic processing unit (GPU) devices. Powered by parallel computing technology, the proposed DVC method achieves a significant improvement in computation speed on a common desktop computer equipped with a low-end graphics card, i.e. up to 23.3 times faster than the sequential implementation and 3.7 times faster than the multithreaded implementation of the same DVC method running on a 6-core CPU. This speedup has no compromise with resolution and accuracy. It benefits from the coarse-grained parallelism that the points of interest (POIs) are processed simultaneously and also from the fine-grained parallelism that the calculation at each POI is performed with multiple threads in GPU.

Keywords Digital volume correlation • Parallel computing • Inverse compositional Gauss-Newton algorithm • Fast Fourier transform-based cross correlation • Graphic processing unit

6.1 Introduction

Digital volume correlation (DVC) has been adopted as an effective technique to study the evolution of internal structure within deformed solid materials. The DVC technique tracks small motion of points of interest (POIs) by minimizing the difference between the POI-centric subvolumes in the reference (un-deformed) volume image and the target (deformed) volume image acquired using 3D imaging techniques. This technique can be considered as a straightforward extension of the well-established digital image correlation (DIC) and shares its simplicity in principles and effectiveness in applications. Compared with DIC, the computational burden of DVC can be several orders of magnitude higher. Thus, high computation speed of DVC has become one of main issues in this area.

The computation speed of DVC could be substantially increased by processing POIs simultaneously on parallel computing device. We proposed a path-independent DIC method [1], which estimated the initial guess for the inverse compositional Gauss-Newton (IC-GN) algorithm at each POI independently using the fast Fourier transform-based cross correlation (FFT-CC) algorithm. Elimination of path-dependency allows the application of parallel computing technology. The parallel computing-powered DIC (paDIC) method, implemented on NVIDIA compute unified device architecture (CUDA) for graphics processing unit (GPU) devices, can reach a speedup of nearly two orders of magnitude without sacrificing high accuracy [2]. In this work, we extend the paDIC to the parallel DVC (paDVC) on GPU devices and assess its performance using computer simulated 3D speckle images.

T. Wang • K. Qian
School of Computer Engineering, Nanyang Technological University, Singapore 639798, Singapore

L. Zhang • Z. Jiang (✉)
Department of Engineering Mechanics, South China University of Technology, Guangzhou 510640, China
e-mail: zhenyujiang@scut.edu.cn

M. Sutton, P.L. Reu (eds.), *International Digital Imaging Correlation Society*, Conference Proceedings of the Society for Experimental Mechanics Series, DOI 10.1007/978-3-319-51439-0_6

6.2 Method and Results

The implementation of the proposed paDVC includes coarse-grained and fine-grained parallelism for all the three major parts, namely FFT-CC algorithm, IC-GN algorithm and the precomputation procedure which can further accelerate the IC-GN algorithm. More details of the algorithms and implementation can be found in Ref. [3].

To quantitatively evaluate the performance of the paDVC method, an 8-bit grayscale 512 × 512 × 512-voxel reference volume image is generated, and then ten target volume images are generated by translating the reference volume image in the Fourier domain, with pre-set sub-voxel displacement along the *z*-direction ranging from 0 to 1 voxel, as shown in Fig. 6.1.

Figure 6.2 shows the mean bias errors and standard deviations of sub-voxel displacement calculated by the paDVC using subvolumes of three sizes, respectively. The mean bias error of the calculated *w*-component falls in a narrow band from -4.3×10^{-3} to 2.9×10^{-3} voxels. The standard deviations of the measured w-component are below 1.42×10^{-3} voxels.

To assess the computational efficiency of the proposed paDVC, the same DVC algorithm with sequential implementation (seDVC) and multithreaded implementation (muDVC) based on OpenMP running merely on CPU are used as benchmarks. Table 6.1 compares the computation time and computation speed among the paDVC, the muDVC and the seDVC with different subvolume sizes. The paDVC achieves a significant speedup, approximately 3.0×∼3.7× over the muDVC and 18.3×∼23.3× over the seDVC.

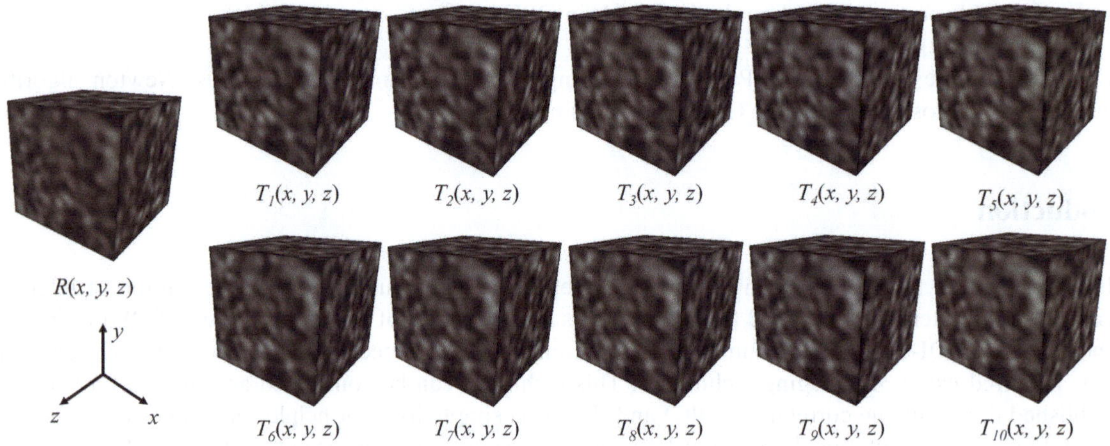

Fig. 6.1 Computer simulated reference volume image *R*(*x*, *y*, *z*) and the 10 translated target volume images from $T_1(x, y, z)$ to $T_{10}(x, y, z)$

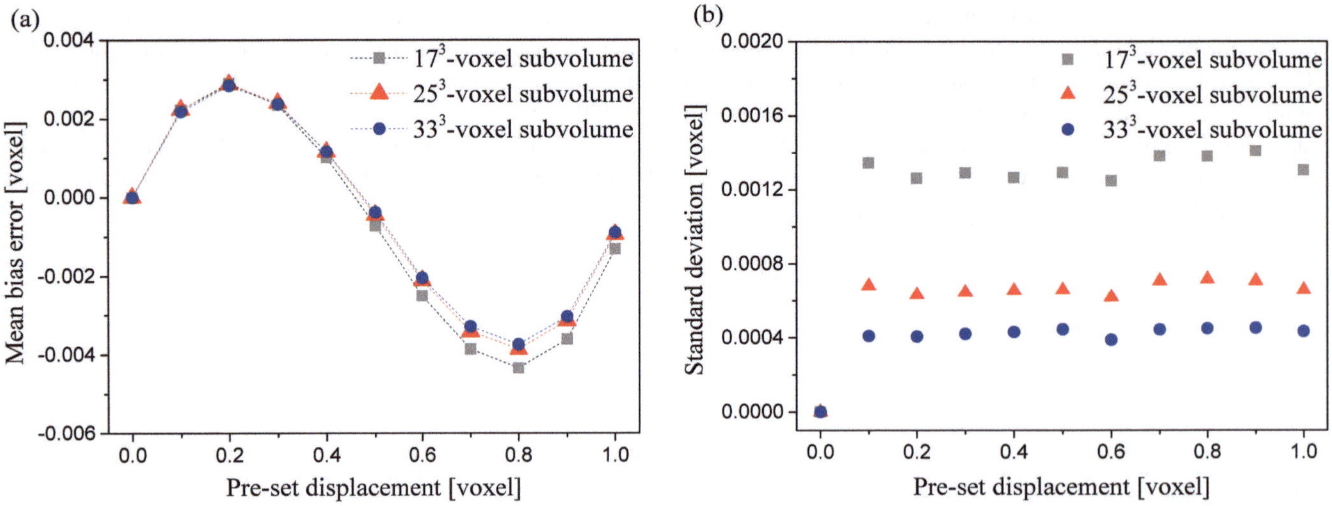

Fig. 6.2 (**a**) Mean bias errors and (**b**) standard deviation of the measured *w*-components

Table 6.1 Average computation time and speed of the paDVC, the muDVC and the seDVC

	Average consumed time [s]			Average iteration number			Average computation speed [POI/s]		
Subvolume [voxel]	paDVC	muDVC	seDVC	paDVC	muDVC	seDVC	paDVC	muDVC	seDVC
17 × 17 × 17	416.5	1261.5	7617.5	4.12	4.12	4.12	1750.4	577.9	95.7
25 × 25 × 25	861.9	2887.9	17862.6	2.91	2.91	2.91	845.8	252.4	40.8
33 × 33 × 33	1641.4	6061.0	38186.6	2.64	2.64	2.64	444.1	120.3	19.1

6.3 Conclusions

A parallel digital volume correlation (paDVC) method is proposed and implemented on GPU devices. The proposed paDVC employs a variety of techniques and strategies to combine parallel computing and batch processing at coarse-grained and fine-grained level, and demonstrates a significantly enhanced computational efficiency compared with the CPU-based implementation of the same DVC method, without sacrificing high accuracy. The paDVC provides the end-users the possibility to carry out intensive DVC computation efficiently on their desktop computers equipped with low or medium level graphics cards for measurement or research work.

Acknowledgements The work is partially supported by a grant, MOE2011-T2-2-037 (ARC 4/12), Ministry of Education, Singapore, the Multi-platform Game Innovation Centre (MAGIC) funded by the Singapore National Research Foundation under its IDM Futures Funding Initiative and administered by the Interactive & Digital Media Programme Office, Media Development Authority, and National Natural Science Foundation of China (NSFC No. 11202081).

References

1. Jiang, Z., Kemao, Q., Miao, H., Yang, J., Tang, L.: Path-independent digital image correlation with high accuracy, speed and robustness. Opt. Lasers Eng. **65**, 93–102 (2015)
2. Zhang, L., Wang, T., Jiang, Z., Kemao, Q., Liu, Y., Liu, Z., et al.: High accuracy digital image correlation powered by GPU-based parallel computing. Opt. Lasers Eng. **69**, 7–12 (2015)
3. Wang, T., Jiang, Z., Kemao, Q., Lin, F., Soon, S.H.: GPU accelerated digital volume correlation. Exp. Mech. **56**(2), 297–309 (2016)

Chapter 7
Optimization of Speckle Pattern for Digital Image Correlation

Zhenning Chen, Xiangyang Xu, Jialin Wu, and Xiaoyuan He

Abstract Since artificial speckle patterns may lead to different computation results, digital speckle pattern generated by computer is adopted for digital image correlation (DIC) to keep the results consistent. A procedure to fabricate controllable and repeatable speckle patterns is proposed, including optimization and transfer of the digital speckle patterns to specimen surfaces. Firstly, through Fast Fourier Transform (FFT) shift experiment, digital speckle pattern in multiple-radius type and single-radius type are discussed considering computation speed, systematic error, standard deviation error and root mean square error. So a speckle pattern with the optimized speckle size is obtained. Then, several indirect transfer methods are compared and utilized to fabricate the optimized digital speckle patterns on specimen surfaces. Hence, speckle pattern for DIC can be controllable and repeatable and this procedure will contribute to DIC standardization.

Keywords Speckle pattern optimization • Standardization • Multiple-radius speckle pattern • Water transfer printing • Digital speckle fabrication

7.1 Introduction

Since first developed in 1990s, digital image correlation (DIC) has acquired high attention due to its capability of simple optical setup and low environmental vulnerability [1]. As a non-contact and full-field method, DIC estimates the variables through a series of images recorded during test. These images should include enough features, such as speckle patterns in white background and dark dots. The performance of speckle patterns is vital to accurate and precious DIC results. Traditional techniques to fabricate the artificial speckle patterns include spray paint, airbrush gun [2], brush, natural textures and so on, and these will result in different speckle patterns and therefore different errors for DIC. Some digital speckle patterns generated from stencil, toner transfer [3], water transfer printing [4], speckle kit and other techniques can keep repeat speckle features and guarantee the speckle errors.

Optimization of speckle pattern is the basement for fabrication since the computer generated speckle patterns can be transferred to specimen surfaces. The past work only takes the single-radius speckle into account, which is not suitable to real case. This paper firstly presents the multiple-radius speckle patterns and its optimization, and then gives some transfer methods for digital speckle patterns fabrication.

7.2 Optimization Experiments

Digital speckle patterns in multiple-radius are generated according to Fig. 7.1. Three dimensions are chosen to generate the 50% coverage multiple-radius speckles, and its corresponding single-radius speckle pattern has 16.7% coverage. Both single-radius and multiple-radius speckle patterns are binary images and have the same distribution. To simulate the speckle pattern better, we use Gaussian filter to imitate the camera imaging, where the same filter window is used. A comparison of the multiple-radius speckle patterns with 50% coverage single-radius speckle patterns is performed by Fast Fourier Transform (FFT) shift experiment considering the computation speed, systematic error, standard deviation error and root mean square error. Here, only computation speed and root mean square error are simply given in Fig. 7.2 since systematic error, standard deviation error nearly has the same tendency. From Fig. 7.2a, the iteration time is the least and its corresponding root mean

Z. Chen • X. Xu • J. Wu • X. He (✉)
School of Civil Engineering, Southeast University, Nanjing 210096, China
e-mail: mmhxy@seu.edu.cn

M. Sutton, P.L. Reu (eds.), *International Digital Imaging Correlation Society*, Conference Proceedings of the Society for Experimental Mechanics Series, DOI 10.1007/978-3-319-51439-0_7

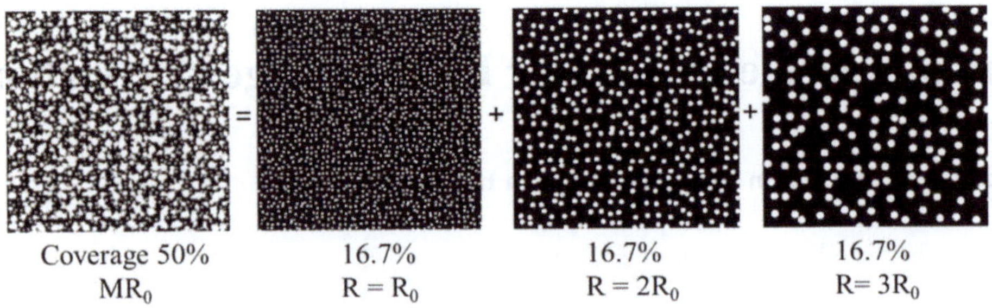

Fig. 7.1 Generation of multiple-radius speckle patterns

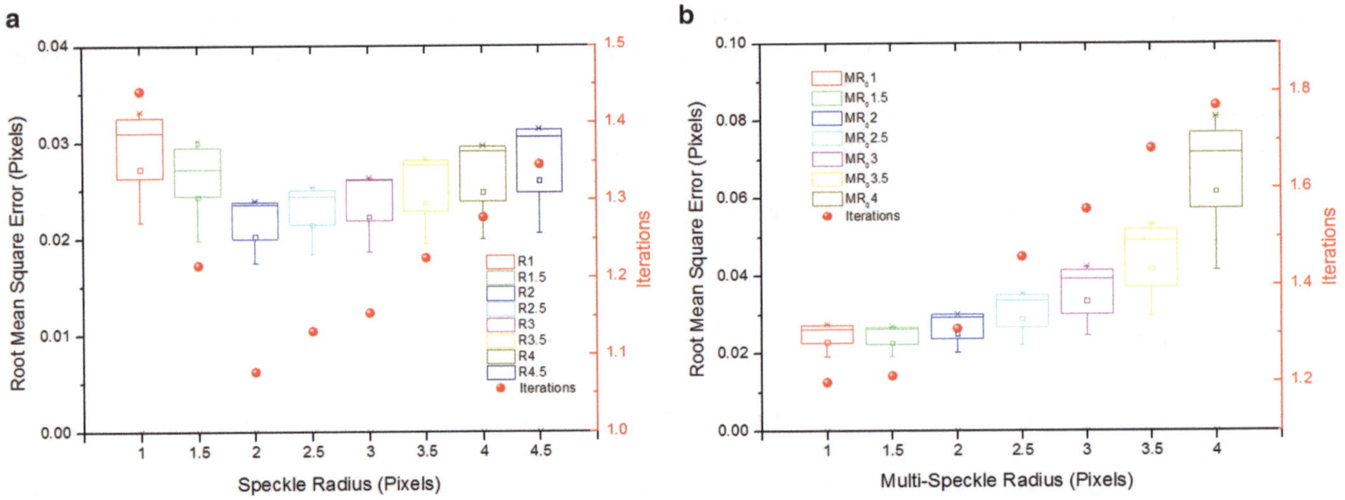

Fig. 7.2 (**a**) Root mean square error and iterations varied with the single-radius speckle radius; (**b**) Root mean square error and iterations varied with the multiple-radius speckle radius

square error is the best when the speckle radius equals to 2 pixels. In Fig. 7.2b, when the base speckle radius R_0 equals to 1–1.5 pixels, the computation speed and root mean square error achieve the best. Compare these results, we can obtain that the accuracy of single-radius (2 pixels) speckle pattern is prior to any other multiple-radius speckle patterns.

7.3 Fabrication Techniques

The vital technique in speckle fabrication is to obtain the consistent speckle in small or large view of fields and obtain speckle pattern radius equal to 2 pixels. Here we simply present several indirect transfer speckle patterns, including screen printing, photosensitive seal, laser engrave and water transfer printing technique (Fig. 7.3). Physical sizes of the speckles and range of application depend on the fabrication procedures.

7.4 Conclusion

In this paper, we present the speckle pattern optimization and fabrication method. Multiple-radius speckle patterns are considered to simulate real speckle patterns better. FFT shift experiments are performed to estimate the error and computation speed. The results show that single-radius speckle pattern is prior to multiple-radius type when the speckle radius equal to 2 pixels. Combination of the digital speckle pattern fabrication techniques, the computer generated speckle patterns can be transferred to specimen surfaces controllable and repeatable, which may help the procedures of DIC standardization.

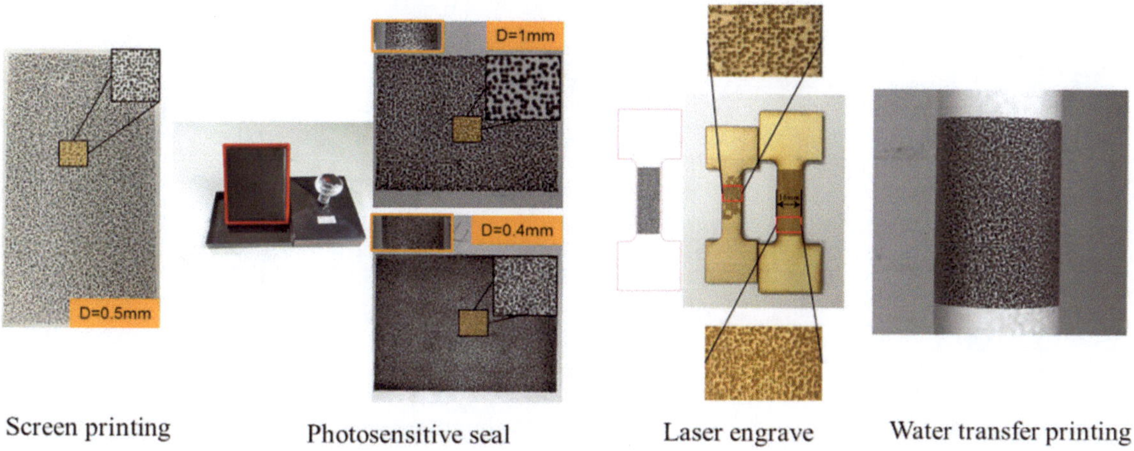

Fig. 7.3 Techniques for speckle fabrication

References

1. Sutton, M.A., Orteu, J.J., Schreier, H.: Image Correlation for Shape, Motion and Deformation Measurements: Basic Concepts, Theory and Applications. Springer Science & Business Media (2009)
2. Lionello, G., Cristofolini, L.: A practical approach to optimizing the preparation of speckle patterns for digital-image correlation. Meas. Sci. Technol. **25**(10), 107001 (2014)
3. Mazzoleni, P., Zappa, E., Matta, F., et al.: Thermo-mechanical toner transfer for high-quality digital image correlation speckle patterns. Opt. Lasers Eng. **75**, 72–80 (2015)
4. Chen, Z., Quan, C., Zhu, F., et al.: A method to transfer speckle patterns for digital image correlation. Meas. Sci. Technol. **26**(9), 095201 (2015)

Chapter 8
A Speckle Patterning Study for Laboratory-Scale DIC Experiments

Sharlotte Kramer, Phillip Reu, and Sarah Bonk

Abstract A "good" speckle pattern enables DIC to make its full-field measurements, but oftentimes this artistic part of the DIC setup takes a considerable amount of time to develop and evaluate for a given optical configuration. A catalog of well-quantified speckle patterns for various fields of view would greatly decrease the time it would take to start making DIC measurements. The purpose of this speckle patterning study is to evaluate various speckling techniques we had readily available in our laboratories for fields of view from around 100 mm down to 5 mm that are common for laboratory-scale experiments. The list of speckling techniques is not exhaustive: spray painting, UV-printing of computer-designed speckle patterns, airbrushing, and particle dispersion. First, we quantified the resolution of our optical configurations for each of the fields of view to determine the smallest speckle we could resolve. Second, we imaged several speckle patterns at each field of view. Third, we quantified the average and standard deviation of the speckle size, speckle contrast, and density to characterize the quality of the speckle pattern. Finally, we performed computer-aided sub-pixel translation of the speckle patterns and ran correlations to examine how well DIC tracked the pattern translations. We discuss our metrics for a "good" speckle pattern and outline how others may perform similar studies for their desired optical configurations.

Keywords Digital Image Correlation (DIC) • Speckle pattern • speckling techniques • Laboratory-scale DIC experiments • Comparison of speckles

8.1 Introduction

A "good" speckle pattern enables DIC to make its full-field measurements, but oftentimes this artistic part of the DIC setup takes a considerable amount of time to develop and evaluate for a given optical configuration. A catalog of well-quantified speckle patterns for various fields of view would greatly decrease the time it would take to start making DIC measurements. The purpose of this speckle patterning study is to evaluate various speckling techniques we had readily available in our laboratories for fields of view from around 100 mm down to 5 mm that are common for laboratory-scale experiments. The list of speckling techniques is not exhaustive: spray painting, UV-printing of computer-designed speckle patterns, airbrushing, and particle dispersion. First, we quantified the resolution of our optical configurations for each of the fields of view to determine the smallest speckle we could resolve. Second, we imaged several speckle patterns at each field of view. Third, we quantified the average and standard deviation of the speckle size, speckle contrast, and density to characterize the quality of the speckle pattern. Finally, we performed computer-aided sub-pixel translation of the speckle patterns and ran correlations to examine how well DIC tracked the pattern translations. We discuss our metrics for a "good" speckle pattern and outline how others may perform similar studies for their desired optical configurations.

8.2 Methodology

To achieve the 5–100-mm field of view (FOV) size, we utilized two different optical setups. For the larger FOVs, we used a 12-MP Grasshopper3 CCD camera with a 35-mm fixed focal length lens. For the small FOVs below 25-mm, we used the 12-MP Grasshopper3 CCD camera with a long-distance working lens, the Navitar 6000 Zoom lens with 0.67x adapter tube

S. Kramer (✉) • P. Reu • S. Bonk
Sandia National Laboratories, 1515 Eubank Blvd SE, Albuquerque, NM 87123, USA
e-mail: slkrame@sandia.gov

M. Sutton, P.L. Reu (eds.), *International Digital Imaging Correlation Society*, Conference Proceedings of the Society for Experimental Mechanics Series, DOI 10.1007/978-3-319-51439-0_8

Fig. 8.1 Magnified views for four speckle patterns (896 × 484 pixels each): (*upper left*) airbrush; (*upper right*) carbon powder deposition; (*lower left*) toner powder deposition; and (*lower right*) UV-printed pattern for a large FOV

and 1.5x lens attachment. Note that the FOV length here denotes the vertical height of the image. We imaged a 1951 United States Air Force (USAF) Resolution Target for each FOV to determine the smallest resolvable feature, using the metric of a minimum of 20% contrast between dark and bright adjacent lines. We used a metric that a speckle had to either be a minimum of 5 pixels or at least the width of the smallest resolvable object, whichever was larger, to ensure no aliasing of the speckles. We aimed for speckles 5–10 pixels in diameter, if that could be resolved with the setup.

We then created several speckle patterns using spray paint or the UV printer for FOVs greater than 25-mm and airbrushing and particle deposition for FOVs smaller than 25-mm. Spray-painted and the UV-printed, computer-designed speckle patterns proved to be too large for the small FOVs. The airbrushed and particle-deposited speckle patterns were too small for the larger FOVs. We then imaged these patterns and evaluated the speckles using a custom-written LabVIEW program that estimated the speckle size, the standard deviation of speckle size, the range of speckle size, the number of speckles in a region of interest, and the percent coverage. This program assumed a dark speckle on a bright background. This same LabVIEW program was used to perform sub-pixel image shifting, which then allowed us to correlate the images using Correlated Solutions VIC-2D software to detect the sub-pixel shifts of the different speckle patterns. Figure 8.1 shows magnified views for four different speckle patterns. A researcher would have difficulty selecting an appropriate speckle pattern without quantitative measures, hence the impetus for this study.

8.3 Results and Conclusions

For FOVs greater the 25-mm, we determined that printing a computer-design speckle pattern gave the most consistent patterns with good contrast, size, coverage, and distribution. Though not considered in this study, we did see in other studies that printed speckles did not necessarily have as large a range of deformation before pattern delamination as freshly-painted spray-paint patterns. For the small FOVs, the airbrush and particle deposition techniques produced good speckles sizes and coverage, but black toner powder had the best contrast compared to speckles from airbrushing and carbon-powder deposition. The particle deposition techniques were the most difficult to create repeatable patterns; these techniques required a deft touch to appropriately agitate the particles through filter paper using pressured air. Overall, this speckle study created the methodology for systematic and quantitative evaluation of speckle patterns for various FOVs, allowing us to gradually build our catalog of patterning techniques that will help to expedite our DIC setup.

Acknowledgments Sandia National Laboratories is a multi-program laboratory managed and operated by Sandia Corporation, a wholly owned subsidiary of Lockheed Martin Corporation, for the U.S. Department of Energy's National Nuclear Security Administration under contract DE-AC04-94AL85000.

Chapter 9
Combining Full-Field Measurements and Inverse Techniques for Smart Material Testing

E.M.C. Jones, J.D. Carroll, K.N. Karlson, S.L.B. Kramer, R.B. Lehoucq, P.L. Reu, and D.Z. Turner

Abstract Traditionally, material properties (such as Young's modulus, yield stress, etc.) are determined from a series of simple tensile and shear tests performed under various temperatures, strain rates, etc. This process is time-consuming and expensive. Additionally, material properties determined from simplified stress states may not adequately describe material behaviour under more complex loading conditions. In the past decade, full-field deformation measurements such as Digital Image Correlation (DIC) and inverse techniques such as the Virtual Fields Method (VFM) have reached a maturity level that takes them from the realm of development to the realm of application. This paper will present a concerted effort beginning at Sandia National Laboratory to explore a new experimental protocol for high-throughput, high-quality material identification by combining DIC, full-field temperature measurements, and VFM. Our current thrust is focused on identifying visco-plastic material parameters of 304L stainless steel. In particular, we will discuss the question of uniqueness of the material model and how parameter covariance affects material identification.

Sandia is a multiprogram laboratory operated by Sandia Corporation, a Lockheed Martin Company, for the United States Department of Energy's National Nuclear Security Administration under contract No. DE-AC04-94AL85000.

Keywords Material identification • Digital Image Correlation (DIC) • Virtual Fields Method (VFM) • Visco-plasticity • 304L stainless steel

9.1 Introduction

Simulations of material deformation under a given loading provide invaluable information to design engineers investigating components under complex loading conditions. Accurate simulation results rely on well characterized material properties such as elastic constants (e.g. Young's modulus and Poisson's ratio) and plastic parameters (e.g. yield stress and hardening behaviour). As the material becomes more complex (e.g. anisotropy) and/or the loading conditions become more complex (e.g. multi-axial stress state, varying strain-rate or temperature, etc.), the number of material parameters that are required for a high-fidelity material model increases dramatically.

Traditionally, material properties are determined from standard tensile tests of dog-bone specimens where the states of stress and strain are constant in a gage section, which allows for simple interpretation of the material response. In order to capture variation of the material parameters as a function of loading conditions such as strain-rate and temperature, a series of tensile tests under different loading conditions must be performed. This process is time-consuming and expensive. Additionally, material properties determined from a single, simple stress state may not adequately describe material behaviour under more complex stress states.

To tackle these limitations, we are exploring a new experimental protocol for high-throughput, high-quality material identification. Our current thrust is focused on identifying visco-plastic material properties of 304L stainless steel. First, we replace the standard dog-bone specimen with a more complex specimen geometry, in which a heterogeneous, multi-axial stress state is induced during loading and in which multiple strain-rates are induced for a single given cross-head velocity. Next, we utilize digital image correlation (DIC) to measure surface displacements on both sides of the specimen. In-plane strains on both surfaces are directly computed, and through-thickness strains are interpolated from measured out-of-plane displacements of each surface. Finally, we employ the virtual fields method (VFM) in combination with a pre-selected material model to identify material parameters given the experimentally measured displacements and strains.

E.M.C. Jones (✉) • J.D. Carroll • K.N. Karlson • S.L.B. Kramer • R.B. Lehoucq • P.L. Reu • D.Z. Turner
Sandia National Laboratories, 1515 Eubank, Albuquerque, NM 87123, USA
e-mail: emjones@sandia.gov

M. Sutton, P.L. Reu (eds.), *International Digital Imaging Correlation Society*, Conference Proceedings of the Society for Experimental Mechanics Series, DOI 10.1007/978-3-319-51439-0_9

In the current paper, we present a unique specimen geometry and initial results validating the VFM implementation using simulated experimental deformation data from a Finite Element Model (FEM). We then discuss uniqueness concerns for the parameter set used to describe material behaviour and finally present preliminary experimental data for 304L stainless steel.

9.2 Specimen Geometry

We designed a novel specimen geometry (Fig. 9.1) with the goal of providing better material identification. The main design criterion was to maximize the stress and strain-rate heterogeneity in the specimen. By performing material identification using a specimen undergoing a multi-axial stress state, rather than a simple tensile or simple shear stress, we propose that the identified material parameters will have higher fidelity to actual material behaviour of components subjected to complex loading conditions. Additionally, covering a range of strain-rates for a single cross-head velocity allows for characterization of the rate-dependence of a material with fewer experimental tests, thus reducing the time and expense associated with traditional material identification. Finally, since DIC measures only surface deformation, we limited the specimen to a planar sample, but allowed arbitrarily complex 2D geometry. We then measured surface displacements on both sides of the sample and interpolated the displacements through the thickness of the specimen in order to generate quasi-3D displacements.

9.3 Non-unique Parameter Sets

In order to validate our implementation of VFM, we simulated experimental deformation data through FEM. The idea was to create a closed loop, in which a particular material model was chosen and reference values for model parameters were identified via traditional material identification techniques. Displacements of the specimen were solved for using FEM, and these "simulated experimental" displacements were then fed into the VFM algorithm. With no experimental noise and no model-form error, we expected the VFM to recover the reference parameters input into the FE model exactly, closing the loop.

$$\sigma_f(p, \dot{p}, \xi) = \sigma_o \left\{ 1 + \operatorname{asinh}\left[\left(\frac{\dot{p}}{b_\sigma} \right)^{1/m_\sigma} \right] \right\} + \frac{H}{R_d} \left[1 - \exp(-R_d p) \right] \left\{ 1 + \operatorname{asinh}\left[\left(\frac{\dot{p}}{b_\kappa} \right)^{1/m_\kappa} \right] \right\} \tag{9.1}$$

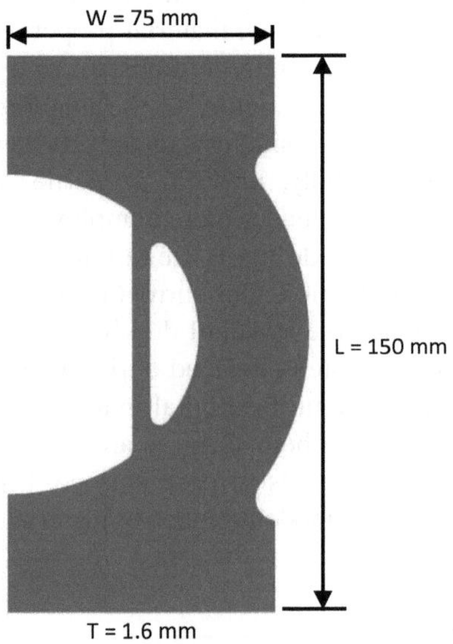

Fig. 9.1 Specimen geometry inducing heterogeneous stress and strain-rate state during uniaxial loading

Table 9.1 Material identification results

Parameter	Reference value	Identified value	Percent error
σ_o	307.9 MPa	312.6 MPa	−1.5
H	2.692 GPa	2.629 GPa	2.2
R_d	2.601	2.613	−0.5
m_σ	3.169	2.601	17.9
m_κ	3.169	3.719	−17.4
b_σ	16.25 s^{-1}	14.50 s^{-1}	10.8
b_κ	16.25 s^{-1}	14.48 s^{-1}	10.9

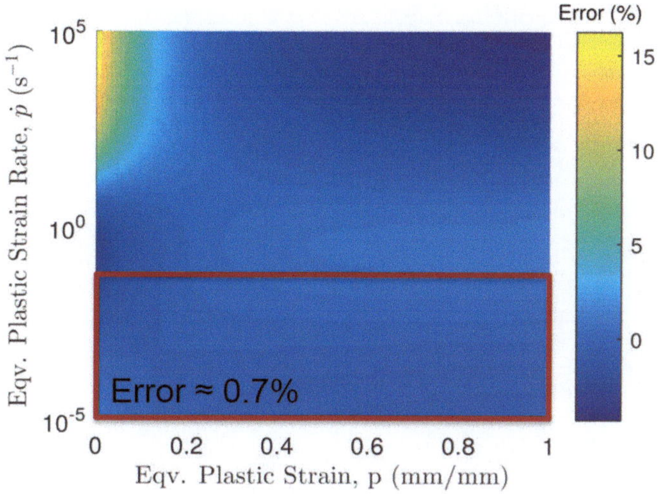

Fig. 9.2 Error in flow stress calculated with the reference parameters compared to the identified parameters

Equation (9.1) presents the flow stress for a modified version of the Bammann-Chiesa-Johnson Microstructural Evolution Model (Brown and Bammann, 2012, *Int. J. Plasticity*) developed at Sandia National Labs, where p is the equivalent plastic strain, $\dot{p}$ is the equivalent plastic strain rate, and $\xi = \{\sigma_o, H, R_d, b_\sigma, m_\sigma, b_\kappa, m_\kappa\}$ are the model parameters. Table 9.1 presents the reference values for the model parameters that were input into the FE model and the identified values found from VFM. Surprisingly, the identified parameters deviated from the reference values by up to 18%. However, the flow stress calculated using the two different parameter sets differed by less than 1% over a range of plastic strain from 0 to 1 and a range of plastic strain rate from 10^{-5} to 10^{-1} s^{-1} (Fig. 9.2). These results indicate that, while individual parameters may vary significantly, two parameter sets as a whole may produce essentially the same flow stress. Thus, the parameter set for a particular material model may not be unique. The significance of this result is currently being explored.

9.4 Conclusions

Accurate characterization of material properties is critical for reliable and trustworthy computational studies of materials and components. As materials and/or loading conditions become more complex and the number of material parameters required to describe material behaviour increases, traditional dog-bone tests are no longer able to satisfactorily characterize the full material behaviour. Here, we build upon previous works in an emerging field of combining full-field measurements of sample deformation with inverse techniques in order to more accurately identify complex material properties. Our particular work focuses on exploring the ramifications of non-unique parameter sets in both material identification and in predictive computational modelling.

Acknowledgements Sandia is a multiprogram laboratory operated by Sandia Corporation, a Lockheed Martin Company, for the United States Department of Energy's National Nuclear Security Administration under contract No. DE-AC04-94AL85000.

Chapter 10
Distortion of Full-Field Surface Displacements from Heat Waves

Elizabeth M.C. Jones and Phillip L. Reu

Abstract One nearly ubiquitous, but often overlooked, source of measurement error in Digital Image Correlation (DIC) arises from imaging through heat waves. "Heat waves" is a colloquial term describing a heterogeneous refractive index field caused by temperature (and thus density) gradients in air. Many sources of heat waves exist in a typical DIC experiment, including hot lights, a heated sample, sunlight, or even a hot camera. This paper presents a detailed description of the error introduced to DIC measurements as a result of heat sources being present in the system. We present characteristic spatial and temporal frequencies of heat waves, and explore the relationships between the location of the heat source, the focal length of the lens, and the stand-off distance between the camera and the imaged object. Finally, we conclude with suggested methods of mitigating the effects of heat waves first by careful design of the experiment and second through data processing.

Sandia is a multiprogram laboratory operated by Sandia Corporation, a Lockheed Martin Company, for the United States Department of Energy's National Nuclear Security Administration under contract No. DE-AC04-94AL85000.

Keywords Digital Image Correlation (DIC) • Heat waves • Uncertainty quantification

10.1 Introduction

Digital image correlation (DIC) is an optical, non-contact technique for measuring full-field displacements on either an object's surface (as in 2D or stereo DIC) or through an object's volume (as in 3D DIC). Since the initial introduction of DIC in the late 1980s, the field of DIC has been significantly developed, and today, bundled commercial packages make the deployment of DIC relatively easy and straight-forward. As a result, DIC is used in a myriad of applications, from classic tensile specimens to biological tissues, battery electrodes, and explosive devices, to name a few. Each non-traditional experimental system presents unique challenges to utilizing DIC. The present work focuses on utilizing DIC in high-temperature applications, and on the displacement errors that are introduced when imaging through heat waves.

Figure 10.1 shows a simplified test setup in which a region of air with a heterogeneous refractive index field is adjacent to the plane of the sample being imaged. The heterogeneous refractive index field could be caused, for instance, by air turbulence emanating from a heated sample. A light ray leaving the object plane is deflected as it passes through the heterogeneous refractive index field. The ray arrives to the recording plane (e.g. the camera) at a location Q^* that is displaced from the undistorted location Q_o at which the light ray would have arrived if it had passed through a homogeneous refractive index field. The displacement between the undistorted location Q_o and the distorted location Q^* can be estimated using the Fermat principle if the spatial variation of the refractive index field is known [1]. Additionally, the temperature dependence of the refractive index of air is described empirically by the Ciddor equations [2], and can be approximated with a simple linear fit over the temperature range 20–150 °C. Thus, the errors introduced into DIC displacement measurements can be estimated if the temperature field through which the images are captured is known. For illustrative purposes, let the temperature vary sinusoidally across the field of view in the x-direction between 25 and 150 °C. For $W = 0.1$ m and $L = 0.5$ m, the resulting DIC displacement error will be ca. 300 μm. Errors of this magnitude could completely wash out the measurement of interest. Thus, understanding the errors introduced from imaging through heat waves and how to mitigate these errors is of critical importance to anyone utilizing DIC in thermally heterogeneous environments.

E.M.C. Jones (✉) • P.L. Reu
Sandia National Laboratories, 1515 Eubank, Albuquerque, NM 87123, USA
e-mail: emjones@sandia.gov

M. Sutton, P.L. Reu (eds.), *International Digital Imaging Correlation Society*, Conference Proceedings of the Society for Experimental Mechanics Series, DOI 10.1007/978-3-319-51439-0_10

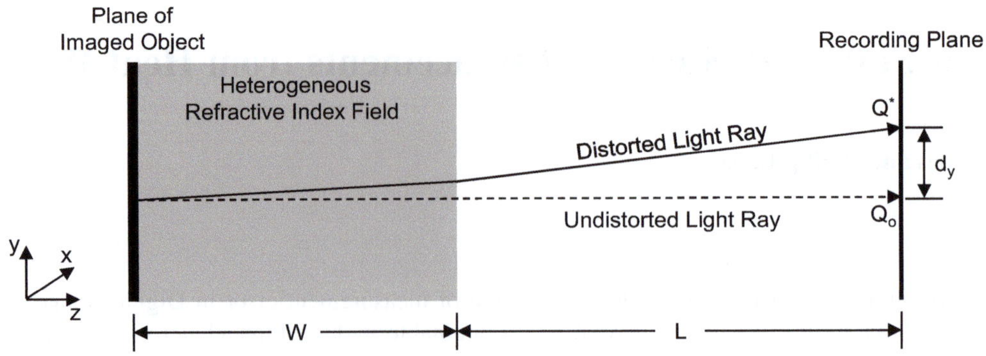

Fig. 10.1 Schematic of simplified experimental configuration in which an object is imaged through a heterogeneous refractive index field, inducing distortions in the images

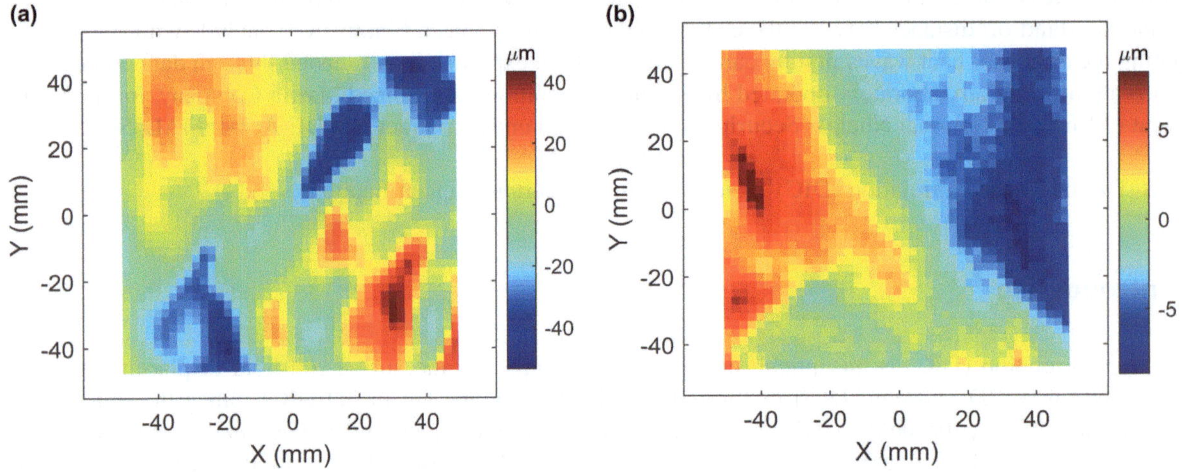

Fig. 10.2 Contour plot of horizontal displacement error. (**a**) Images were captured through a heterogeneous temperature field generated by the natural convection pattern from a hot plate with surface temperature of 150 °C. (**b**) A fan was directed at the hot plate in order to break up the heat waves, thus reducing the displacement error

10.2 Experimental Method

A simplified experimental set-up was used to interrogate the distortions arising from imaging through heat waves. A pair of stereo, high-speed cameras was placed approximately 1 m from a planar, static sample, and a hot plate with a surface temperature of 150 °C was placed in between the sample and the cameras. Natural convection from the hot plate generated a heterogeneous temperature field through which images were captured at 250 Hz. Images were correlated and strains were calculated using Vic3D software, with a subset size of 37 pixels (3.1 mm), step size of 20 pixels (1.7 mm), and strain window of 15 points, leading to a virtual strain gage size of 281 pixels (23.9 mm).

10.3 Results and Discussion

Correlation of images captured through a homogeneous temperature field (i.e. without the presence of the hot plate) show that the baseline horizontal displacement noise for the system is ca. 6.3 μm, leading to a baseline strain noise of ca. 52 micro-strain. Figure 10.2a shows a contour of the displacements measured when the sample was imaged through a heterogeneous temperature field. False displacements due to imaging through heat waves are clearly observed. The absolute value plus three standard deviations of the false displacements is ca. 52.4 μm, leading to false strains of ca. 2700 micro-strain.

There are two general tactics are available to mitigate the effects of heat waves: (1) modify the experimental set up in order to reduce the presence of heat waves a priori and (2) filter the data after images have been collected. With regards to the first option, Fig. 10.2b shows a contour of the displacements measured when a fan was blowing over the hot plate. By introducing forced convection, the displacement error was reduced to ca. 10.0 μm and the strain error was reduced to ca. 370 micro-strain. While this strategy was effective in this simplified experimental configuration, placing a fan on a sample is not always practical in a real experimental setup. Therefore, a second strategy was employed in which the groups of 50 images, captured over a window of 0.2 s, were averaged together before they were correlated. Temporal averaging in the image space reduced the displacement error to ca. 49.2 μm and the strain error to ca. 1710 micro-strain. While temporal averaging of images distorted by heat waves does not reduce the error as much as reducing the heat waves experimentally, it does provide an option for reducing noise if modification to the experimental setup is not possible and/or if distorted images have already been captured.

10.4 Conclusions

Digital image correlation (DIC) is being used in increasingly complex experimental conditions that bring new challenges to obtaining precise and accurate measurements. Heat waves emanating from hot samples can cause significant distortions in images, resulting in large displacement and strain errors. Reducing heat waves experimentally before images are captured is the best route to minimizing distortions, but this is not always practical. Temporal filtering of images after they have been captured can also help to reduce noise resulting from imaging through a heterogeneous temperature field.

Acknowledgements Sandia is a multiprogram laboratory operated by Sandia Corporation, a Lockheed Martin Company, for the United States Department of Energy's National Nuclear Security Administration under contract No. DE-AC04-94AL85000.

References

1. Merzkirch, W.: Flow Visualization, 2nd edn., pp. 123–126. Academic Press, Inc (1987)
2. Ciddor, P.E.: Appl. Opt. **35**(9), 1566–1573 (1996)

Chapter 11
Development of a 2D DIC Experimental Tool for Piezoelectric Strains Measurements

Valentin Segouin, Mathieu Domenjoud, Yves Bernard, and Laurent Daniel

Abstract Piezoelectric materials are used in a wide range of actuation and sensing applications. The characterisation of their non-linear coupled electro-mechanical behaviour is a real challenge due to the heterogeneity of materials, the complexity of boundary conditions, and safety issues related to severe loading conditions (electric field, stress, temperature). In this context, Digital Image Correlation (DIC) is a promising tool for piezoelectric materials characterisation since this non-contact method enables remote full-field measurements of strains.

In the present work, a 2D DIC bench has been developed for high resolution displacement measurements. The proposed experimental set up uses the long distance microscope Questar QM100 and a low speed 9 megapixels camera. Since a micrometric speckle is needed, Badger Sotar 20/20 airbrush combined with nano-pigment paints has been used. In a first step, the setup has been characterised in order to identify the optimal speckle size for our application. In a second step, an optical characterization tool has been developed to minimize errors from experimental setup and image acquisition.

Based on our studies, the DIC bench was calibrated and measurement noise has been assessed experimentally. For this purpose, a piezoelectric sample was deformed and characterised under a quasi-static electrical loading. From these experimental results, the noise level is estimated and discussed.

Keywords 2D-DIC • Speckle • Piezoelectric strain • Ferroelectric • Multiphysics

11.1 Introduction

Due to excellent electromechanical efficiency, piezoelectric materials have been widely integrated in sensors, energy harvesting devices, actuators or ultrasonic motors. The development of such devices requires piezoelectric behaviour characterisation. These characterisations must cover complex loadings so as to be representative for realistic loading conditions (temperature, electric field, stress and coupled loadings). Besides, the complexity of boundary conditions during tests also creates experimental difficulties. Moreover, the level of piezoelectric strains is typically of the order of 10^{-5} which is very small for standard strain measurement techniques. Thus, the accurate determination of strains in piezoelectric materials subjected to complex loadings remains an experimental challenge.

DIC is a promising tool to address this challenge. In the recent past, piezoelectric strains have already been successfully studied using DIC [1, 2]. These works show good consistency between DIC and other strain measurement methods. They were limited to electric field loading and use the natural texture of the material as the basis for the correlation procedure. The work discussed in this paper aims at building an experimental setup for the characterisation of piezoelectric strains under coupled electro-mechanical loadings using DIC. A specific attention is paid to the size of the speckle so as to optimise the resolution of the measurement.

This paper describes the conditions required to obtain the displacement resolution needed for the characterisation of standard piezoceramics. In a first part, the experimental set up and corresponding procedure is presented. This part includes details about the calibration process and the speckle pattern preparation. The quality of the image is then investigated. For this purpose, the optical transfer function of the optical device was experimentally characterised, and used in a simulation study to determine an optimal size for the speckle. Then, an experimental procedure was set up to produce an optimised micrometric speckle. Finally, samples were prepared according to this procedure and strain measurement were performed

V. Segouin (✉) • M. Domenjoud • Y. Bernard • L. Daniel
GeePs|Group of Electrical Engineering - Paris, UMR CNRS 8507, CentraleSupélec, Univ. Paris-Sud, Université Paris-Saclay, Sorbonne Universités, UPMC Univ Paris 06, 3 & 11 rue Joliot-Curie, Plateau de Moulon, 91192 Gif-sur-Yvette CEDEX, France
e-mail: valentin.segouin@centralesupelec.fr

M. Sutton, P.L. Reu (eds.), *International Digital Imaging Correlation Society*, Conference Proceedings of the Society for Experimental Mechanics Series, DOI 10.1007/978-3-319-51439-0_11

under quasi-static electrical loadings. The results are analysed in terms of measurement uncertainty so as to conclude on the capability of the experimental setup to provide accurate measurement of strains in piezoceramics.

11.2 Experimental Procedure

The piezoelectric sample is a rectangular cuboid Noliac NCE41. It was subjected successively to a 1.2 and 3 kV/cm DC electric field along the polarisation direction. Due to the high voltage conditions, the piezoceramic was isolated with expansive foam and teflon plates, leaving only one visible surface on which the speckle was deposited. To measure in-plane strains, the sample was placed on a table, in front of the long distance microscope (Questar QM100) that can operate from 14.5 to 37 cm from the sample surface. The QM100 is used with a Ximea MD091MU camera (monochrome 14 bits CCD, 3384×2708 px with 3.69 microns pixel size). This optical system leads to a physical distance from 0.57 μm/px (at 14.5 cm) to 1.18 μm/px (at 37 cm) on the digital image. Regarding the correlation process, the software CorreliQ4, based on a FFT calculation method, was used [3].

An image acquisition program was developed using MATLAB software to capture images from the camera. Numerical tools were implemented into this program to assist the calibration step of the optical set up. The first main tool is about lighting: a real time histogram of the previewed images is calculated with the possibility to evaluate the contrast of the picture. This contrast was defined as the width of 95% of the histogram and expressed as a percentage of the full grayscale. As focalisation is very sensitive for this kind of optics, an algorithm based on image gradient calculation was programmed to find the point of maximum image clarity. This algorithm was then extended to quantify and minimise perpendicularity defects between the measured surface and the optical axis. This has been done with a bidirectional analysis of the image gradient and led us to ensure perpendicularity defects lower than 0.2°.

Due to limited contrast and surface imperfections, the natural texture of the piezoceramic was not usable. Furthermore, no surface treatment were employed to avoid possible changes in the constitutive behaviour of the sample. The surface were prepared by deposit of a speckle. The process first consists in covering the sample surface with a white acrylic paint. In a second step, black toner powder is mixed with ethanol and spread out on the surface. Projections are performed using the Badger Sotar 20/20 airbrush.

11.3 Image Quality Evaluation: Theoretical Study

Several researches have already been conducted to point out correlation error sources. For both spatial and frequency domain correlation methods, it has been brought to light that correlation errors were particularly related to the image mean intensity gradient (MIG) [3–6]. In this work, the MIG has been used as the parameter to assess the image quality. A numerical study based on a random generation of images was carried out to define the parameters of a speckle allowing a maximisation of the MIG.

The numerical images representative for the speckles have been defined as a pattern of black droplets on a white background. A few parameters have been selected to characterise and describe uniquely the speckle. The droplets are defined by their shape and dimension. The shape has been chosen to be circular because of its isotropic autocorrelation (error should be identical in any direction on a single droplet). Consequently, the studied dimension is the radius of a disk. As the droplet is repeated inside a pattern, we must take into account the density of the disks and some repetition condition (overlapping permitted or not). In the following, overlapping was permitted. A complete study showed the consistence between the results with our without overlapping and even with periodic speckles. The last parameter is the type of distribution (how the disks are spread on the pattern). The presented work shows simulations results for a uniform random distribution. Figure 11.1 presents the studied speckles for different configurations. The statistical influence of a variation of the disk radius for a given pattern has also been analysed. This is why an uncertainty of 10% on the radius is indicated in the figures.

Another parameter was analysed to take into account image degradation due to optical system resolution. The analysis consisted in filtering the images with a low-pass filter for different cut-off frequencies (see Fig. 11.1 image d). To identify which type of filter where the most convenient, the optical transfer function (OTF) of the optical system was measured for different working distances and lenses using a calibrated test pattern as the input image. The OTF was then compared to numerical filters (Gaussian, first and second order filters). Comparisons showed that our optic can be represented with a second order transfer function of the form:

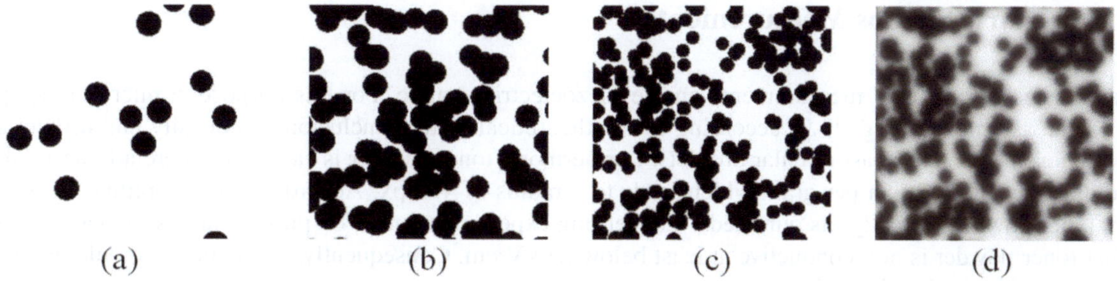

Fig. 11.1 Simulated speckles, uniform random distribution with density 0.1 and 10 px mean radius ± 10% (**a**), density 0.5 and 10 px mean radius ± 10% (**b**), density 0.5 and 6 px mean radius ± 10% (**c**), image c filtered with the low pass filter H (Fc = 0.05 lp/px) (**d**)

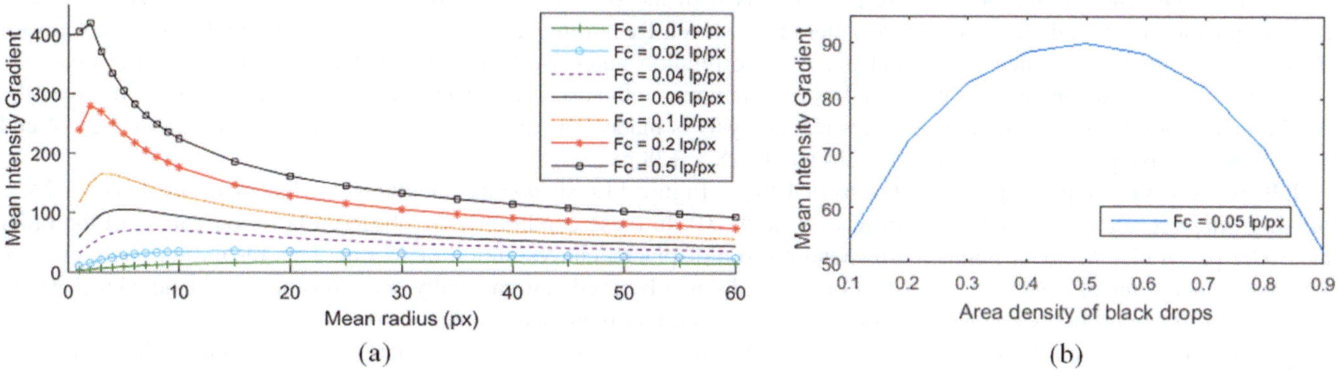

Fig. 11.2 MIG versus mean radius of disks for a density of 0.5 and for different cut-off frequencies of the low pass filter (**a**), MIG versus density of droplets for mean radius of 8 px (**b**)

$$H\left(f_x, f_y\right) = \frac{1}{1 + \left(\frac{\sqrt{f_x^2 + f_y^2}}{F_c}\right)^2} \tag{11.1}$$

where Fc is the numerical cut-off frequency (in lp/px) and fx and fy are spatial frequencies along direction x and y.

From these definitions, various speckle patterns were simulated with different radius and density. As our optical resolution is likely to change depending on the material and samples tested, simulated speckles were filtered with a wide range of cut-off frequencies. All of these images were analysed using the MIG. Results are presented in Fig. 11.2. Figure 11.2a represents MIG versus radius of the disks in the speckle. This relationship was plotted for different cut-off frequencies. When the cut-off frequency increases (lower filtering), the optimal speckle size decreases (radius of 1–2 px for Fc = 0.5 lp/px). A similar work has already been conducted using Mean Subset Fluctuation [7] instead of MIG, but without connecting the curve peaks (optimal size for a given filtering) to the optical resolution.

Since the OTF of the optics was measured for different configurations (lenses and working distances), a set of corresponding cut-off frequencies could be estimated. According to these measurements, the cut-off frequency remains between 0.04 and 0.06 lp/px. As a consequence, the optimal speckle should have a mean radius between 5 and 7 px, depending on the optical configuration.

The second graph, Fig. 11.2b, reveals another feature to optimise for the speckle: the maximum MIG value is obtained for a density of 0.5. The curve is expected to be symmetric since the MIG of a monochrome image and its negative are equal. The slight dissymmetry observed in Fig. 11.2b is a consequence of the image creation procedure: images were created independently for each density so that images at density "d" are not the negatives of the ones at density "1 – d".

11.4 Piezoelectric Strains Measurements

The last step is to perform DIC measurements on the piezoelectric sample. For this purpose, a microscopic speckle was deposited on the sample (see Fig. 11.3) according to the theoretical study conclusions on density and droplet size. Toner powder was chosen to obtain quasi-circular droplets. Furthermore, toner powder is easily commercially accessible and has an average diameter of ~10 μm per grain which leads to a radius of 4–9 px. According to the optimal sizes obtained by simulation, toner powder perfectly fits our needs. Concerning experimental tests on piezoceramics, previous measurements revealed that toner powder is non-conductive at least below 25 kV/cm. Consequently, this material should not affect results for measurement of piezoelectric strains.

The piezoelectric sample with speckle presented in Fig. 11.3 was deformed by applying an electric field of 1200 and 3000 V/cm. Measurements were carried out at 0.6 μm/px and DIC computations were made using Fig. 11.3a as reference image. The correlation software provided us with the statistical evaluations of two displacement error sources: the interpolation error and the error due to camera noise level. Regarding the first one, the software applies known subpixel displacements to the reference image and makes a correlation analysis. The standard deviation of the results gives the statistical error. The second error is computed by adding noise to the original image. Then a correlation is made between the original and noised images and another statistical analysis is made. The analysis gave a standard deviation of 7.2e−4 and 5e−3 px respectively on displacements for subsets of 128 × 128 px.

DIC was conducted using subsets and steps of 128 px. Figure 11.4 shows the calculated displacement maps with a 2 × 2 binning (maps result in 13 × 10 points instead of 26 × 20). In these conditions, the observed noise on data is of the order of ±0.02 px on displacement points, corresponding to a displacement resolution of 12 nm. This resolution could even be improved since some of the observed fluctuations or peaks are observed systematically whatever the electric field level. These apparent errors are probably connected with sample defects rather than measurement error.

The longitudinal displacement is expected to evolve linearly in the y-direction and to be constant in the x-direction since the loading is axial (poling along y-direction) and supposed to be uniform. Longitudinal strains were computed from displacements using Q4 method from Correli. Similarly to displacements, strains were 2 × 2 binned to obtain a 13 × 10 maps. Figures 11.5 and 11.6 show results for 1200 V/cm and 3000 V/cm loadings. Here again, some peaks are systematically observed at some locations, probably corresponding to material defects. They participate to a large standard deviation (visible on the density distribution of data on Figs. 11.5b and 11.6b. However, even with an error bar of 10^{-3} on strains, the average value gives significant information of global strain level in the material (due to the number of subsets, the random error is divided here by $\sqrt{130}$). In addition, mean strains for both electric fields seems consistent with the expected behaviour of the material.

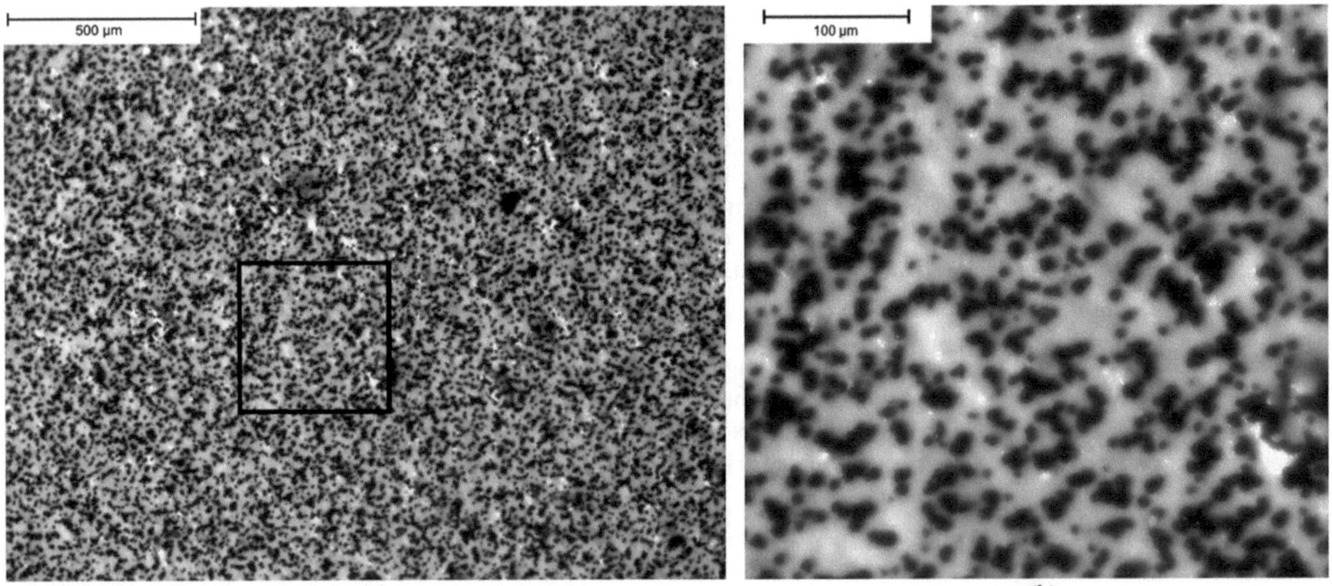

(a) (b)

Fig. 11.3 Photo of a speckle made on a piezoelectric material with white paint and toner powder, droplets density of 0.6 and mean radius ~5 px (**a**); Zoom inside the image a (*black square*) (**b**)

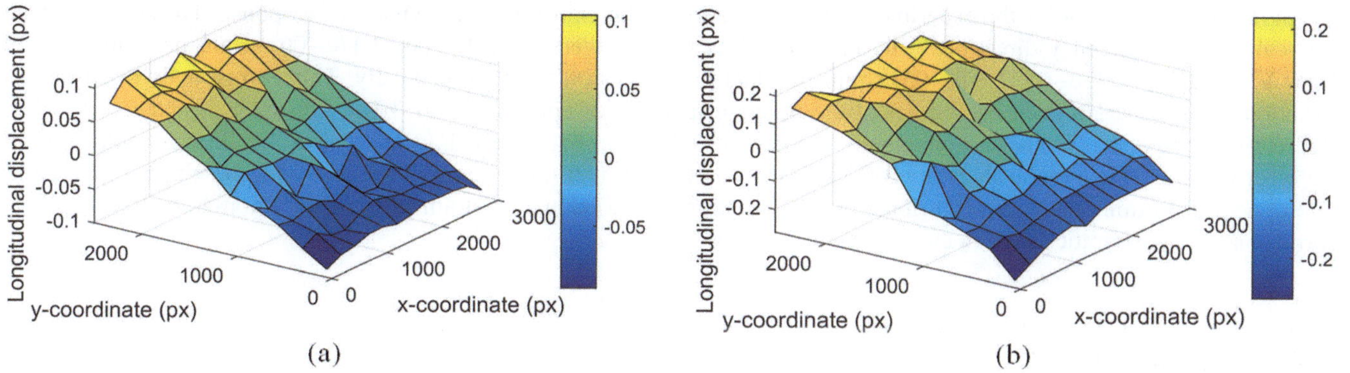

Fig. 11.4 Longitudinal displacements maps of a Noliac NCE41 piezoelectric sample (10 mm between electrodes and 10 × 20 mm^2 electrode area) at—1200 V/cm (**a**) and 3000 V/cm (**b**)—in the polarisation direction

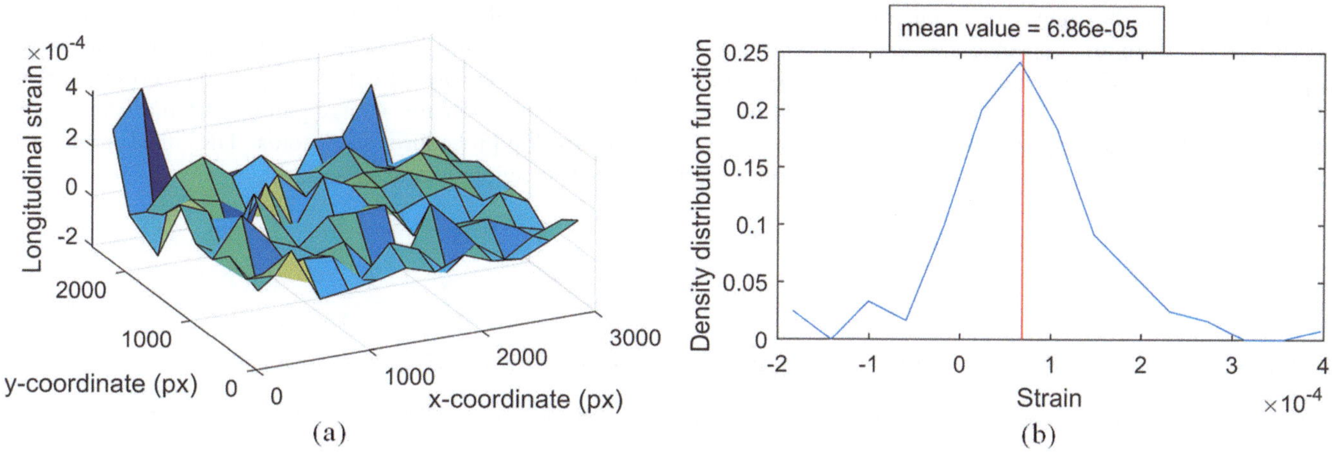

Fig. 11.5 Longitudinal strain map (**a**) and corresponding density distribution function (**b**)—Noliac NCE41 piezoelectric sample at 1200 V/cm along the polarisation direction

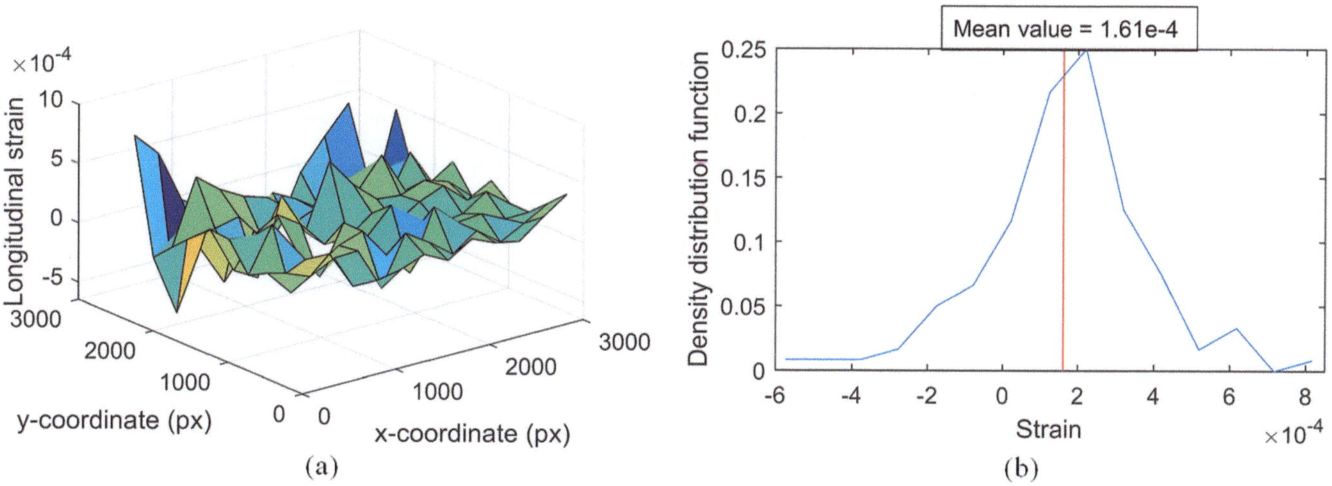

Fig. 11.6 Longitudinal strain map (**a**) and corresponding density distribution function (**b**)—Noliac NCE41 piezoelectric sample at 3000 V/cm along the polarisation direction

An important indicator for the experimental test evaluation is the uniformity of the material response. The strain gradient was calculated in x and y directions. The obtained slopes (x; y) were (−4.11e−06; 2.47e−06) strain per data point at 1200 V/cm and (−1.33e−05; 1.84e−05) strain per data point at 3000 V/cm respectively. This corresponds to an overall variation (x; y) of strain from one corner of the image to another of approx. (5.3×10^{-5}; 2.5×10^{-5}) and (1.7×10^{-4}; 1.8×10^{-4}) respectively. These numbers are significant, which probably reveals a non-uniformity in the application of the electric field. This hypothesis is reinforced by the observation that the variation increases with the electric field level. The optical distortion could also contribute to this variation in the calculated strain [8]. Indeed a non-zero strain gradient corresponds to a quadratic shape in the displacement map.

11.5 Conclusion

A microscale 2D-DIC bench was developed for the measurement of strains in piezoelectric materials. The optical setup was calibrated using a focalisation and angle defect minimisation procedure. Based on an analysis of images using Mean Intensity Gradient, the requirements on droplet size and density for an optimal speckle were defined. A deterioration parameter was introduced in the study to account for the optical transfer function of the setup. This analysis showed that for each configuration, an optimal speckle size can be found, and that the optimal density is always 0.5. Consequently, a speckle deposit process was developed. 2D-DIC measurements were then carried out on piezoelectric samples. The observed noise was of the order of 5×10^{-4} on strains. This level of noise is high but actually contains a large part of systematic errors that can be attributed to the experimental conditions. The next step is to develop procedures so as to use DIC measurements to control the experimental conditions before performing experiments for behaviour characterisation.

References

1. Malakooti, M.H., Sodano, H.A.: Noncontact and simultaneous measurement of the d33 and d31 piezoelectric strain coefficients. Appl. Phys. Lett. **102**, 061901 (2013)
2. Malakooti, M.H., Sodano, H.A.: Direct measurement of piezoelectric shear coefficient. J. Appl. Phys. **113**, 214106 (2013) [Online]. Available: http://dx.doi.org/10.1063/1.4809636
3. Besnard, G., Hild, F., Roux, S.: "Finite-element" displacement fields analysis from digital images: application to portevin - le chatelier bands. Exp. Mech. **46**(6), 789–803 (2006)
4. Pan, B., Xie, H., Wang, Z., Qian, K., Wang, Z.: Study on subset size selection in digital image correlation for speckle patterns. Opt. Express. **16**(10), 7037–7048 (2008)
5. Reu, P.: All about speckle: contrast. Exp. Tech. **39**(1), 1–2 (2015)
6. G. Crammond, S. Boyd, and J. Dulieu-Barton, Speckle pattern quality assessment for digital image correlation. Opt. Lasers Eng. **51**(2013), 1368–1378 (2013) [Online]. Available: http://dx.doi.org/10.1016/j.optlaseng.2013.03.014
7. Hua, T., Xie, H., Wang, S., Hu, Z., Chen, P., Zhang, Q.: Evaluation of the quality of a speckle pattern in the digital image correlation method by mean subset fluctuation. Opt. Laser Technol. **43**, 9–13 (2010)
8. J.-E. Dufour, F. Hild, and S. Roux, Integrated digital image correlation for the evaluation and correction of optical distortions, Opt. Lasers Eng., vol. 56, pp. 121–133, 2014.

Chapter 12
Applications of DIC in the Mechanics of Collective Cell Migration

Aashrith Saraswathibhatla and Jacob Notbohm

Abstract During wound repair, embryonic development, and cancer invasion, cells migrate in cooperative groups and clusters. Collective migration has been studied for decades, for example with experiments to determine how quickly a sheet of cells migrates to close a wound. Until recently, those experiments have been limited to investigating molecular mechanisms due to a lack of experimental tools to quantify the kinematics and forces. Now that digital image correlation has gained wide acceptance in engineering, physics, and biology, new possibilities exist to reveal quantitative understanding of the connections between signaling, motions, and forces. Here we review applications of digital image correlation in measuring cell velocities, cell-to-substrate tractions, and cell-to-cell stresses. The data show that forces and motions follow no typical constitutive relationship such as that of an elastic solid or a viscous fluid; in many cases even the orientations of force and motion are misaligned. Ongoing research is seeking to connect force and motion, often by modeling the active processes associated with cell signaling. While it remains unclear how to reduce the complicated and numerous signaling pathways into a physical picture of collective cell migration, the experimental tools now available offer a useful place to start.

Keywords Digital image correlation • Traction force microscopy • Monolayer stress microscopy • Cell mechanics • Cell migration

12.1 Introduction

In pattern formation, would healing, and invasive cancer, cells move most often not as individual entities but as collective sheets, ducts, strands, or clusters. Groups of collectively moving cells also form within a cellular monolayer, resulting in fluctuations in velocity and local cellular density [1]. Together with these fluctuations in kinematics, fluctuations in the tensile stress between cells result in a rugged stress landscape [2]. The fluctuations have been linked to a change in phase from a fluid-like unjammed state to a solid-like jammed one [1–3]. In the unjammed state, cells can move freely past their neighbors whereas in the jammed state, each cell within the monolayer remains stuck in place. To quantify this collective motion, we use digital image correlation on phase contrast images of the cell monolayer. The force required to move is generated internally by each cell and applied to both the substrate beneath the cell and the cell's immediate neighbors. To begin to relate the active cell forces to the motions, we describe experimental techniques for quantifying both force and motion here.

12.2 Experimental Methods

Here we give a summary of our experimental procedures, which include digital image correlation (DIC), traction force microscopy, and monolayer stress microscopy. Full details are in our recent publication [4]. Cells are typically micropatterned into circular monolayers of diameter 1 mm by using standard micropatterning techniques to adhere adhesive collagen to the substrate. The substrate is a compliant gel, typically polyacrylamide, embedded with fluorescent particles. The particles serve as a high contrast pattern to quantify the substrate's displacements using DIC. We centrifuge the gel upside down during polymerization, which is sufficient to move the particles to a thin layer at the top of the gel. This thin layer allows for the microscope to focus exactly on the particles and prevents errors due to imaging particles that are out of the focal plane.

A. Saraswathibhatla • J. Notbohm (✉)
Department of Engineering Physics, University of Wisconsin-Madison, 1500 Engineering Drive, Madison, WI 53706, USA
e-mail: jknotbohm@wisc.edu

M. Sutton, P.L. Reu (eds.), *International Digital Imaging Correlation Society*, Conference Proceedings of the Society for Experimental Mechanics Series, DOI 10.1007/978-3-319-51439-0_12

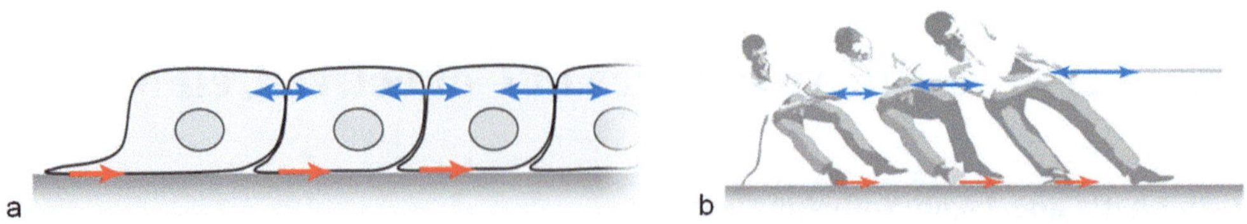

Fig. 12.1 Stresses within the monolayer are computed with monolayer stress microscopy. (**a**) Each cell within the monolayer applies traction to the substrate beneath it (*red arrows*) and stress to its neighbors (*blue arrows*). (**b**) The intercellular stresses are computed by summing the tractions, similar to how tension builds up in the rope of a tug-of-war game. Image adapted from Ref. [10]

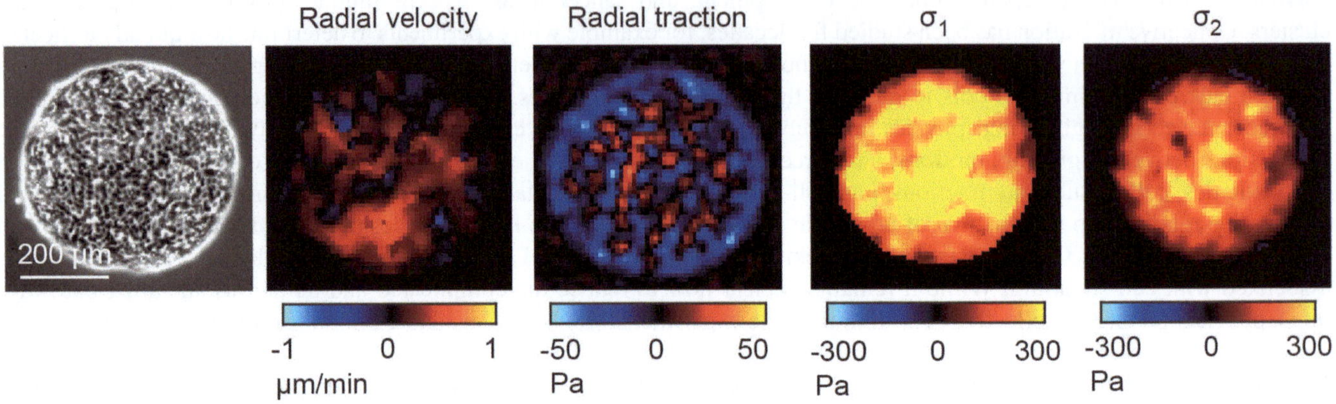

Fig. 12.2 Typical results showing cell image, radial component of velocity, radial component of traction, first principal stress (σ_1) and second principal stress (σ_2). For velocity and traction, positive is defined as outward from the center of the cell island. Notice that both principal stresses are positive, indicating there are no compressive stresses within the monolayer

The displacements in the substrate result from the tractions applied by the cells. To compute tractions from displacements we used the Boussinesq-Cerruti solution for the displacement field induced by a point force on a semi-infinite half space [5] as a Green's function. Solving for tractions required an inverse solution, which we have implemented in the Fourier domain [6] with a correction for a substrate of finite thickness [7, 8]. We then apply monolayer stress microscopy [2, 9] to compute the two dimensional (2D) stress tensor σ_{ij} within the monolayer. The basic idea of monolayer stress microscopy is illustrated in Fig. 12.1. Stresses within the monolayer are computed from a summation of the tractions, much like stress builds up in the rope of a tug-of-war game. In two dimensions, stresses are computed using the stress equilibrium equations, $\partial\sigma_{ij}/\partial x_j + b_i = 0, i, j = 1, 2$, with body force b_i. For a 2D monolayer, the traction t_i, applied by the substrate to the monolayer, acts as the body force, which is given by $b_i = t_i/h$ where h is the height of the monolayer [2, 9]. The 2D stress tensor has three unique components whereas there are only two equilibrium equations. To solve for all three unique components of the stress tensor, a third equation is required. The third equation we use is compatibility, which assumes the displacement or velocity field of the monolayer is continuous. Note that the earliest version of monolayer stress microscopy assumed the monolayer to be a linear elastic solid, but we have recently shown that the results are identical if the monolayer is a viscous fluid [4]. Thus the only assumptions are that the monolayer can be treated as a continuum and that there is a linear relationship between gradients in cell velocity and stress. It has been shown that dependence on these assumptions in minimal [9], which makes sense given that two of the three equations are simply force equilibrium which make no assumptions on the constitutive properties of the monolayer.

12.3 Results

Typical results are shown in Fig. 12.2. Cell velocities and tractions are both vector quantities; the components in the radial direction are plotted here. The stress is a tensor representing the in-plane normal and shear stresses within the monolayer. The principal stresses are shown in the figure. The two principal stresses are nearly always positive, indicating that there is only tension, no pressure, within the monolayer. This finding is consistent with the notion that cells are contractile and therefore generate tensile stresses.

References

1. Angelini, T.E., Hannezo, E., Trepat, X., Marquez, M., Fredberg, J.J., Weitz, D.A.: Proc. Natl. Acad. Sci. U. S. A. **108**, 4714–4719 (2011)
2. Tambe, D.T., Hardin, C.C., Angelini, T.E., Rajendran, K., Park, C.Y., Serra-Picamal, X., Zhou, E.H., Zaman, M.H., Butler, J.P., Weitz, D.A., Fredberg, J.J., Trepat, X.: Nat. Mater. **10**, 469–475 (2011)
3. Sadati, M., Taheri Qazvini, N., Krishnan, R., Park, C.Y., Fredberg, J.J.: Differentiation. **86**, 121–125 (2013)
4. Notbohm, J., Banerjee, S., Utuje, K.J., Gweon, B., Jang, H., Park, Y., Shin, J., Butler, J.P., Fredberg, J.J., Marchetti, M.C.: Biophys. J. **110**, 2729–2738 (2016)
5. Landau, L.D., Lifshitz, E.M.: Theory of Elasticity, 3rd edn. Pergamon Press (1986)
6. Butler, J.P., Tolic-Norrelykke, I.M., Fabry, B., Fredberg, J.J.: Am. J. Physiol.-Cell Phys. **282**, C595–C605 (2002)
7. del Alamo, J.C., Meili, R., Alonso-Latorre, B., Rodriguez-Rodriguez, J., Aliseda, A., Firtel, R.A., Lasheras, J.C.: Proc. Natl. Acad. Sci. U. S. A. **104**, 13343–13348 (2007)
8. Trepat, X., Wasserman, M.R., Angelini, T.E., Millet, E., Weitz, D.A., Butler, J.P., Fredberg, J.J.: Nat. Phys. **5**, 426–430 (2009)
9. Tambe, D.T., Croutelle, U., Trepat, X., Park, C.Y., Kim, J.H., Millet, E., Butler, J.P., Fredberg, J.J.: PLoS One. **8**, e55172 (2013)
10. Trepat, X., Fredberg, J.J.: Trends Cell Biol. **21**, 638–646 (2011)

Chapter 13
Original Methodology Using DIC to Characterize Friction Materials Compression Behavior

Itziar Serrano-Munoz, Vincent Magnier, Ruddy Mann, and Philippe Dufrénoy

Abstract Friction materials are expected to fulfill several requirements such as adequate friction coefficient, low noise, long life and low wear rate. In order to achieve these properties, friction materials are generally heterogeneous mixtures of fillers incorporated in a more or less continuous matrix. Therefore, a good understanding of the influence of row constituents (i.e., formulation, shape and size) and fabrication process on the mechanical behavior of as-produced heterogeneous materials is necessary to better customize the final products. Nowadays, the customizing process is purely empiric involving costly efforts to characterize the resulting performances. Moreover, the determination of the effect of each component is usually hampered because the behavior of row constituents can evolve during the fabrication process. An in situ compression test using X-ray laboratory tomography is performed in order to study the micro-mechanical behavior of a heterogeneous material. A DIC technique using the reconstructed images provides the displacement fields that are subsequently introduced into a FE simulation which will reproduce the material behavior. This novel methodology allows to identify the micro-constituents mechanical properties of the as-produced material as well as to optimize the numeric material constitutive law.

Keywords Heterogeneous friction material • Processing-formulation-property relationships • Compression testing • DIC and DVC techniques • Inverse identification of microstructural properties

13.1 Introduction

Friction materials, such as brake-pads, have severe performance expectations (e.g., friction level and stability, wear resistance, low noise) as they are critical to transportation safety [1–3]. In order to meet such requirements, most friction materials are complex composites of many components (usually up to 30) with a wide range of sizes and morphologies. The search for performance improvements as well and the changing environmental regulations require the development of new brake-pad formulations.

So far, these formulations are empirically designed. Therefore, the interest of modelling the influence of the formulation on the as-produced material performances is clear. However, realistic models of disk/pads systems require good knowledge of the phenomena occurring at the microstructure level. Characterising heterogeneous materials is particularly challenging because of the numerous components and scales (macro, meso and micro) that are involved. This is why the development of new characterisation tools is necessary in order to improve the experimental/numerical relationships.

Due to the complexity of the task at hand (multi-scale and multi-physic phenomena), the material characterisation has been dissociated depending on the properties (e.g., tribological, mechanical, thermal) that are to be studied. Moreover, the material of study is produced using a simplified formulation where only the most interesting components are retained (nine in total). The aim of this study is to present a novel methodology (named inverse identification) that, using DIC data as input, allows determining the influence of components on the material compressive behaviour.

I. Serrano-Munoz (✉) • V. Magnier • R. Mann • P. Dufrénoy
Polytech-Lille, Laboratoire de Mécanique de Lille, Bd. Paul Langevin, Villeneuve d'Asq Cedex 59655, France
e-mail: itziar.serrano.munoz@gmail.com

M. Sutton, P.L. Reu (eds.), *International Digital Imaging Correlation Society*, Conference Proceedings of the Society for Experimental Mechanics Series, DOI 10.1007/978-3-319-51439-0_13

13.2 Material and Methods

13.2.1 Material

As afore mentioned, in this study the number of components (also known as raw materials) is reduced to nine. The formulation is as follows (weight %): 70% of metallic materials, 10% ceramic 1&2 and 20% graphite 1&2. Graphite 2 particles are the largest ones, with sizes ranging between 550 μm and 1800 μm and the remaining eight components have sizes <500 μm. Prior to the mould compacting (up to 1000 tones), the components are weighed and mixed. The final step consists in sintering the material in a furnace during 8 h and using a protective atmosphere with temperatures up to 1100 °C. This manufacturing process involves producing a composite containing certain amount of porosity, characteristic that is sought as it improves the tribological properties.

13.2.2 In Situ Compression Tests Coupled with DIC

A cube of 20 mm^3 is carved out from the as-produced material. This size is judged to be representative of the material macroscopic compressive behavior. The top and bottom faces (where the compressive load is applied) of the cube are rectified to minimize misalignment defects. The compression test is carried out using an INSTRON-5500 machine equipped with a force sensor and where the load is applied following the compacting direction (which also is the braking direction). In order to perform Digital Image Correlation (DIC), one of the cube four free faces is painted in black to produce a homogeneous, non-reflective background. White paint is sprayed over the dry black paint to create a speckle pattern. Optical fibers are used to deliver a good quality lighting and image acquisition (every 250 ms) is done with a XIMEA camera (four million pixels) allowing a resolution of ~300 μm. At this resolution, only graphite 2 particles are clearly distinguishable and the other eight components are considered to form a composite, hereafter named as matrix. Finally, three strain gauges are glued to the other three free faces to validate: (1) the DIC results (obtained using YaDICs [9] software) and (2) that the cube compressive behavior is similar throughout the entire volume.

13.2.3 Ex Situ Compression Tests Coupled with DVC

Laboratory tomography is used to study the 3D compressive behavior of the material. The setup using the in situ testing machine only allows a voxel size of 21 μm. In order to achieve higher resolution (voxel size = 4 μm for a cylinder of Ø = 5 mm, also carved out from an as-produced brake-pad, different from the one used to produce the 20 mm^3 cube) and be able to study the microstructure evolution, an ex situ compression test is performed. In this test, the cylinder is scanned prior and after the compressive load which is applied using the same INSTRON-5500 machine. It is necessary to apply 30 MPa (the highest loads reached during the braking are 2 MPa) to the cylinder to induce enough residual plasticity. This is because the pattern provided by material microstructure is coarser (by a factor of four) and so, the Digital Volume Correlation (DVC) results (also obtained using YaDICs [9] software) are less accurate than those obtained with the 2D speckle pattern. Another advantage of using this voxel size is that it allows a realistic 3D meshing of the microstructure (using AVIZO [4] and GMSH [5] software). These complex meshes (up to 8×10^6 elements) are subsequently used in FEA [6] where the material properties previously obtained (in situ compression tests) can be introduced.

13.3 Results

13.3.1 In Situ Compression and DIC Results

Both the strain gauges and the DIC give a macroscopic compressive modulus of $E_{exp\text{-}brake\text{-}pad} = 5500$ MPa. The DIC results show that the deformation is not homogeneous along the cube face and that some deformation bands are formed (shown in red in Fig. 13.1). Based on the microstructure image, it is obserbed that the deformation bands are mainly correlated to the presence of graphite 2 particles.

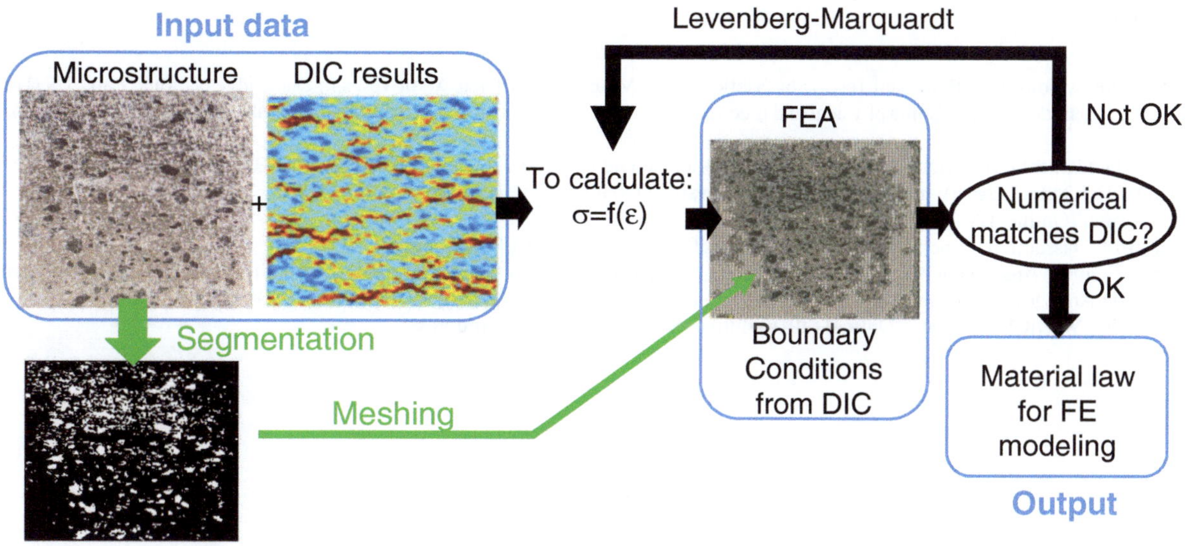

Fig. 13.1 Schematic illustration of the inverse identification method

13.3.2 Inverse Identification of Components Properties

This 2D technique requires, as input, an image of the microstructure (taken prior to the speckle pattern painting) and the DIC results (see Fig. 13.1). The microstructure image is first segmented (the matrix is shown in black and graphite 2 particles in white) to subsequently create a 2D fine mesh using OOF2 [7] free software. The boundary conditions for the FEA (applied on the top and bottom of the mesh) are obtained using the DIC displacement fields. This method seeks to iteratively minimize the cost function ($f = |\varepsilon_{DIC} - \varepsilon_{FEA}|^2$), optimizing the gap between DIC strain fields (ε_{DIC}) and the calculated ones (ε_{FEA}). The optimizing algorithm used in this loop is the one proposed by [8]. In this FEA (performed using ABAQUS [6]), the unknown values to solve are two compressive modulus ($E_{graphite\ 2}$ and E_{matrix}). The calculated values are $E_{graphite\ 2} = 2500$ MPa and $E_{matrix} = 11{,}000$ MPa.

13.3.3 Tomographic and DVC Results

Which indicate that the formation of deformation bands requires clusters of at least three particles of graphite 2; the deformation produced by isolated graphite 2 particles is less important and does not reach the red values of Fig. 13.1 DIC results. The calculated $E_{graphite\ 2}$ and E_{matrix} values are introduced in the 3D FEA where only the matrix and graphite 2 particles are meshed (following the 2D mesh example shown in Fig. 13.1) even though a more detailed mesh (meshing also graphite 1 and ceramics particles) is possible. The 3D FEA results gives a $E_{exp\text{-}brake\text{-}pad} = 5300$ MPa which is in good agreement with the $E_{exp\text{-}brake\text{-}pad} = 5500$ MPa.

13.4 Conclusions and Perspectives

A novel methodology has been developed allowing the numerical characterisation of as-produced microstructural properties for a highly heterogeneous brake pad material. This technique is crucial to improve the experimental/numerical relationships and so, for the development of realistic models of disk/pads systems. The implementation of this technique to 3D cases (currently in progress) requires tomographic in situ compressive testing. Nevertheless, other experimental techniques such as in situ SEM compression tests are contemplated to be able to characterize the smallest components.

References

1. Blau, P.J.: Compositions, Functions, and Testing of Friction Brake Materials and Their Additives. s.l.: Oak Ridge National Lab., TN (US) (2001)
2. Degwa, I.M., Ibhedode, A.O.A.: Some physical and mechanical properties of asbestos-free experimental brake pad. J. Raw Mater. **3**((2)), 94–103 (2006)
3. Singer, I.L.: Fundamentals of Friction: Macroscopic and Microscopic Processes. s.l.: Springer, Netherlands (1992)
4. AVIZO 3D, FE. [Online] (2016). http://www.fei.com/software/avizo3d/
5. GSMH software. [Online] (2016). http://gmsh.info/
6. ABAQUS 6.14, Dassault. [Online] (2016). http://abaqus.software.polimi.it/v6.14/index.html
7. OOF: Finite Element Analysis of Microstructures. NIST US Department of Commerce. [Online] (2016). http://www.ctcms.nist.gov/oof/
8. Marquardt, D.: An algorithm for least-squares estimation of nonlinear parameters. SIAM J. Appl. Math. **11**, 431–441 (1963)
9. Lille University, YADICS Software. [Online] (2016). http://yadics.univ-lille1.fr/wordpress/

Chapter 14
Forming Limit Diagram Determination Using Digital Image Correlation: A Review

Junrui Li, Xin Xie, Guobiao Yang, Changqing Du, and Lianxiang Yang

Abstract The forming limit diagram (FLD) is one of the most important criteria in sheet metal forming because the formability of materials can be assessed by the FLD. Recently, different measurement methods based on Digital Image Correlation (DIC) have been developed for FLD. This paper presents a review of DIC for FLD determination, the principle of each method is descripted, and the pros and cons of these methods are discussed.

Keywords Multi--Camera DIC • Tensile Testing • Thickness Strain • Volume Change • Cluster Approach

14.1 Introduction

The idea and the basic methodology of assessing the formability of materials by FLD were first proposed by Keeler in 1963 using punch test [1]. Now, many different standard punch tests are existing, such as Keeler, Nakazima, Hasek, Marciniak tests [2–5]. The Nakazima test uses a hemispherical punch and circular die to fulfill the test, while the different specimen shape introduces different strain paths. Both the setup and specimen shape are quite simple, which lead this method quite popular in the industry. Another important method is the Marciniak test. In this test, a flat punch is used in the Marciniak test with an additional washer to introduce the pure in-plane strain. A more linear strain path is obtained due to this advantage. The major and minor strains need to be measured no matter which method is used to simulate the strain paths. The conventional method to measure the strains are described by Keeler using circle method, in which circles are etched onto the sample surface to obtain the major and minor strains at different locations after specimen is deformed.

Recently, the conventional surface circle etching method has been replaced by Digital image correlation (DIC) in strain measurement of the FLD determination. DIC is an optical measurement method which has been widely used in many industrial applications [6, 7]. DIC has the advantage of being a full-field, non-contact measurement method to provide accurate strain distribution as well as the deformation. Many FLD determination methods using DIC are proposed in recent years. These methods can typically be divided into two categories: time-dependent method and position-dependent method [8]. The time-dependent methods are quite straightforward and have more physical meaning. Since the strain in the forming limit curve is defined as the strain when the necking starts, the time-dependent methods try to determine the onset of localized necking through the whole strain/deformation history and the strain at the onset point is defined as the strain in FLC. Whereas, the position-dependent methods are more like an empirical method which uses the strain distribution when the fracture occurs and estimate the strain in FLC based on this strain distribution. Both methods have their advantages and disadvantages respectively. Time-dependent methods have explicit physical meaning. However, it often contains hassles in "onset" of necking determination, which leads to a big variation of the determined strain on FLC. Position-dependent methods provide definitive result in strain determination. However, many researchers question the validation of the estimation. In this paper, several typical methods in FLD determination using DIC is reviewed, and the pros and cons of these methods are discussed.

J. Li • G. Yang • L. Yang (✉)
Applied Optical Laboratory, Department of Mechanical Engineering, Oakland University, Rochester, MI 48309, USA
e-mail: Yang2@oakland.edu

X. Xie
Department of Mechanical Engineering, Lawrence Tech University, Southfield, MI 48075, USA

C. Du
Advance Stamping Manufacturing Engineering, FCA US LLC, Auburn Hills, MI 48326, USA

M. Sutton, P.L. Reu (eds.), *International Digital Imaging Correlation Society*, Conference Proceedings of the Society for Experimental Mechanics Series, DOI 10.1007/978-3-319-51439-0_14

14.2 Time-Dependent Method

The first category of time-dependent methods uses the information of strain or strain rate to define the onset of local necking. Vacher first proposed this idea in 1999 [9]. The basic idea in Vacher's method is tracking the whole-field strain increment between two successive images ("strain velocity" in his article) through the whole process. The onset of localized necking is defined at the time when the strain increment keeps almost zero in majority area except some narrow bands (neck bands). Once the onset of localized necking is determined, the strain in FLC is naturally acquired. the definition quite follows the physical meaning of necking. However, the onset of localized necking quite depends on the selection of the strain incremental threshold. The different threshold could produce quite a different onset of localized necking point. Rather than using the whole field strain increment in Vacher's method, Huang proposed a similar method in onset necking determination [10]. The center point in the neck zone is selected, and the strain history through this point is exported. The strain rate at this point is calculated based on the strain history, and the onset of localized necking is defined when the strain rate starts to increase dramatically. However, the same problem in this method is that the selection of the strain rate threshold varies the onset of necking a lot.

Another kind of time-dependent methods uses the surface topography to determine the onset of necking. This idea is first proposed by Wang and Oakland University from 2013 to 2014 separately [11, 12]. The basic idea of this method can be traced back to Keeler's method. Since there is lacking reliable method in determining the onset of necking in 1980s, Keeler uses his finger to "feel" the onset of necking. In this kind of methods, the 3D surface topography is measured using DIC, and the z-direction height history along the line perpendicular to the fracture direction is exported. The maximum height difference in this line is calculated through the whole history, and the onset of necking is defined as the point when the height difference radical increased. The strain in FLC is automatically obtained with the result of the onset of necking. The idea of this kind of methods is most similar to the basic idea in FLD determination proposed by Keeler. However, the sensitivity and the accuracy of the out-of-plane measurement using DIC is one of the limitations of this method. There could be a lot of noise in the height difference history, which could lead to an incorrect result in the onset of necking determination. Besides that, the height difference at the onset of necking point varies with the type of material or the thickness of the specimen, the threshold of the height difference could also be a problem in these methods.

14.3 Position-Dependent Method

The most popular position-dependent method is ISO standard method, which is published in 2008 [13]. As a position-dependent method, the strain in FLC is estimated based on the strain distribution when the fracture occurs. The whole-field strain map is obtained by DIC. The major and minor strain distribution in the step before fracture along the line perpendicular to the fracture direction is plotted, and an inverse parabola is constructed to fit the curve in the out of neck area portion. The strain in FLC is defined based on the extremum in these curves. The FLC determined using this method do not need to identify the onset of necking, which makes it avoid the hassles in the onset of necking determination. However, there are questions about its reliability because the strain in FLC is determined by a pure empirical calculation. Besides that, the ISO standard method assumes there is only one neck band exists on the specimen, whereas, multiple neck bands occur prevalently for aluminum alloys especially in the Marciniak test. As a result, the ISO method is difficult to apply for this kind of material tests. To answer the questions to the lacking of physical meaning, Hogström proposed a modified ISO method by determining the strain in FLC at the onset of necking step. To solve the multiple necking problems, Li proposed another modified ISO method by seeking the major necking area and its neighborhood, and the inverse-parabola is constructed in this restricted area [15].

14.4 Summary

Each method for FLD determination has its own advantages and disadvantages. Generally speaking, the time-dependent methods are straight forward and have more physical meaning. These methods often have difficulty in determining the accurate onset of necking point, and provides a zone for the onset of necking. Different people or different threshold could obtain different onset of necking points based on the same experimental data. This reduces the repeatability of the time-dependent methods. Compared to the time-dependent method, position-dependent methods such as the ISO method have good repeatability from each single test result and minimizes the impact of the human decision. However, these methods always face the question about the lacking of physical meaning.

References

1. Keeler, S.P., Backofen, W.A.: Plastic instability and fracture in sheets stretched over rigid punches. Asm. Trans. Q. **56**(1), 25–48 (1963)
2. Keeler, S.P.: Circular grid system—a valuable aid for evaluating sheet metal formability. No. 680092. SAE Technical Paper, 1968
3. Nakazima, K., Kikuma, T., Hasuka, K.: Study on the formability of steel sheets. Yawata Tech Rep No. 264, 8517–8530 (1968)
4. Hasek, V.V.: An evaluation of the applicability of theoretical analysis to the forming limit diagram. ICF4, Waterloo (Canada) 1977 (2013)
5. Marciniak, Z., Kuczyński, K., Pokora, T.: Influence of the plastic properties of a material on the forming limit diagram for sheet metal in tension. Int. J. Mech. Sci. **15**(10), 789–800 (1973)
6. Xie, X., et al.: Tensile test for polymer plastics with extreme large elongation using quad-camera digital image correlation. No. 2016-01-0418. SAE Technical Paper, 2016
7. Xu, W., et al.: Strain analysis of pressure vessels contained pits based on digital image correlation method. In: Seventh International Symposium on Precision Mechanical Measurements. International Society for Optics and Photonics (2016)
8. Banabic, D.: Formability of Metallic Materials: Plastic Anisotropy, Formability Testing, Forming Limits. Springer Science & Business Media, Berlin (2000)
9. Vacher, P., Haddad, A., Arrieux, R.: Determination of the forming limit diagrams using image analysis by the corelation method. CIRP Ann. Manuf. Technol. **48**(1), 227–230 (1999)
10. Huang, G., Sriram, S., Yan, B.: Digital image correlation technique and its application in forming limit curve determination. In: Proceedings of the IDDRG 2008 International Conference (2008)
11. Wang, K., et al.: Measuring forming limit strains with digital image correlation analysis. J. Mater. Process. Technol. **214**(5), 1120–1130 (2014)
12. Li, H., et al.: Benchmark 1–nonlinear strain path forming limit of a reverse draw. AIP Conf. Proc. **1567**, 15 (2013)
13. ISO/DIS 12004-2, 2008. Metallic Materials – Sheet and strip – Determination of Forming Limit Curves – Part 2: Determination of Forming Limit Curves in the Laboratory. International Organization for Standardization 20087
14. Hogström, P., Ringsberg, J.W., Johnson, E.: An experimental and numerical study of the effects of length scale and strain state on the necking and fracture behaviours in sheet metals. Int. J. Impact Eng. **36**(10), 1194–1203 (2009)
15. Li, J., et al.: Experimental study of FLD0 for aluminum alloy using digital image correlation with modified ISO method. Int. J. Mater. Res. **107**(3), 245–253 (2016)

Chapter 15
High-Accuracy and High-Efficiency Compensation Method in Two-Dimensional Digital Image Correlation

Xiaohai Xu and Qingchuan Zhang

Abstract The traditional two-dimensional (2D) digital image correlation (DIC) method often leads to results without enough accuracy due to the out-of-plane motions of the specimens in the deformation measurements of plate specimens. To remove the effects of the out-of-plane motions, a compensation method for the 2D DIC method has been developed. The compensation method sharply improves the measuring accuracy of the 2D DIC method. However, the existing compensation method often deduces compensation parameters with strong fluctuations. To solve these problems, an improved compensation method is proposed. The proposed method employs an extra single camera calibration to determine the distortion parameters so that the corresponding coefficient matrix is not affected by the experimental noise. For simplicity, the extensometer line measurement is taken as an example to compare the measuring accuracy of the two compensation methods. Static experiments have been carried out to analyze effects of different measuring parameters. Tensile experiments indicate that the measuring results of the improved compensation method are not only comparative with the results measured by the 3D DIC method but also accord with the results of the strain gage. Besides, a real-time extensometer measuring system with a measurement accuracy of 5 $\mu\varepsilon$ at a rate of 28 fps, has been developed.

Keywords Digital image correlation • Compensation method • High-accuracy • High-efficiency • Real-time extensometer

15.1 Introduction

Since the results measured by the two-dimensional (2D) digital image correlation (DIC) method [1] is affected by out-of-plane motions which are difficult to be avoided, the three-dimensional (3D) DIC method has been proposed and developed [2]. However, the 3D DIC method may be restricted by its experimental setup which is more complex and expensive than the 2D DIC experimental setup. Recently, Pan has proposed a compensation method which can guarantee both the measurement accuracy and the simplicity of the experimental setup [3]. However, the existing compensation method has two drawbacks: (1) it often gains compensation parameters with strong fluctuations; (2) the coefficient matrix will be singular when the deformation of the specimen is small enough. What' more, the effects of the factors by human choice, such as the position of the extensometer, the subset size of the extensometer, the subset size of the compensation points, the positions of the compensation points, the number of the compensation points, the length of the extensometer and the thickness of the compensation panel, need to be systematically studied.

This work is organized as follows. Section 15.2 introduced the existing compensation method and the improved compensation method. Section 15.3 demonstrates the accuracy the improved compensation method using tensile experiments. Section 15.4 concludes the whole work.

15.2 Principle

Considering the effects of lens distortion [3] and out-of-plane motions [4], the existing compensation method can be expressed as Eq. (15.1).

X. Xu • Q. Zhang (✉)
CAS Key Laboratory of Mechanical Behavior and Design of Materials, University of Science and Technology of China, Hefei 230027, China
e-mail: zhangqc@ustc.edu.cn

M. Sutton, P.L. Reu (eds.), *International Digital Imaging Correlation Society*, Conference Proceedings of the Society for Experimental Mechanics Series, DOI 10.1007/978-3-319-51439-0_15

$$\begin{bmatrix} 1 & x_{c1} & y_{c1} & x_{c1}^2 & x_{c1}y_{c1} & 0 & 0 & 0 & 0 & 0 & x'_{c1}\left(x'^2_{c1}+y'^2_{c1}\right)-x_{c1}\left(x_{c1}^2+y_{c1}^2\right) \\ 1 & x_{c2} & y_{c2} & x_{c2}^2 & x_{c2}y_{c2} & 0 & 0 & 0 & 0 & 0 & x'_{c2}\left(x'^2_{c2}+y'^2_{c2}\right)-x_{c2}\left(x_{c2}^2+y_{c2}^2\right) \\ \vdots & \vdots & \vdots & \vdots & \vdots & \vdots & \vdots & \vdots & \vdots & \vdots & \vdots \\ 1 & x_{cn} & y_{cn} & x_{cn}^2 & x_{cn}y_{cn} & 0 & 0 & 0 & 0 & 0 & x'_{cn}\left(x'^2_{cn}+y'^2_{cn}\right)-x_{cn}\left(x_{cn}^2+y_{cn}^2\right) \\ 0 & 0 & 0 & 0 & 0 & 1 & x_{c1} & y_{c1} & x_{c1}y_{c1} & y_{c1}^2 & y'_{c1}\left(x'^2_{c1}+y'^2_{c1}\right)-y_{c1}\left(x_{c1}^2+y_{c1}^2\right) \\ 0 & 0 & 0 & 0 & 0 & 1 & x_{c2} & y_{c2} & x_{c2}y_{c2} & y_{c2}^2 & y'_{c2}\left(x'^2_{c2}+y'^2_{c2}\right)-y_{c2}\left(x_{c2}^2+y_{c2}^2\right) \\ \vdots & \vdots & \vdots & \vdots & \vdots & \vdots & \vdots & \vdots & \vdots & \vdots & \vdots \\ 0 & 0 & 0 & 0 & 0 & 1 & x_{cn} & y_{cn} & x_{cn}y_{cn} & y_{cn}^2 & y'_{cn}\left(x'^2_{cn}+y'^2_{cn}\right)-y_{cn}\left(x_{cn}^2+y_{cn}^2\right) \end{bmatrix} \begin{bmatrix} a_0 \\ a_1 \\ a_2 \\ a_3 \\ a_4 \\ b_0 \\ b_1 \\ b_2 \\ b_3 \\ b_4 \\ k_1 \end{bmatrix} = \begin{bmatrix} u_{c1} \\ u_{c2} \\ \vdots \\ u_{cn} \\ v_{c1} \\ v_{c2} \\ \vdots \\ v_{cn} \end{bmatrix} \tag{15.1}$$

$$\begin{bmatrix} 1 & x_{c1} & y_{c1} & x_{c1}^2 & x_{c1}y_{c1} & 0 & 0 & 0 & 0 & 0 \\ 1 & x_{c2} & y_{c2} & x_{c2}^2 & x_{c2}y_{c2} & 0 & 0 & 0 & 0 & 0 \\ \vdots & \vdots & \vdots & \vdots & \vdots & \vdots & \vdots & \vdots & \vdots & \vdots \\ 1 & x_{cn} & y_{cn} & x_{cn}^2 & x_{cn}y_{cn} & 0 & 0 & 0 & 0 & 0 \\ 0 & 0 & 0 & 0 & 0 & 1 & x_{c1} & y_{c1} & x_{c1}y_{c1} & y_{c1}^2 \\ 0 & 0 & 0 & 0 & 0 & 1 & x_{c2} & y_{c2} & x_{c2}y_{c2} & y_{c2}^2 \\ \vdots & \vdots & \vdots & \vdots & \vdots & \vdots & \vdots & \vdots & \vdots & \vdots \\ 0 & 0 & 0 & 0 & 0 & 1 & x_{cn} & y_{cn} & x_{cn}y_{cn} & y_{cn}^2 \end{bmatrix} \begin{bmatrix} a_0 \\ a_1 \\ a_2 \\ a_3 \\ a_4 \\ b_0 \\ b_1 \\ b_2 \\ b_3 \\ b_4 \end{bmatrix} = \begin{bmatrix} u_{c1}-k_1\left(x'_{c1}\left(x'^2_{c1}+y'^2_{c1}\right)-x_{c1}\left(x_{c1}^2+y_{c1}^2\right)\right) \\ u_{c2}-k_1\left(x'_{c2}\left(x'^2_{c2}+y'^2_{c2}\right)-x_{c2}\left(x_{c2}^2+y_{c2}^2\right)\right) \\ \vdots \\ u_{cn}-k_1\left(x'_{cn}\left(x'^2_{cn}+y'^2_{cn}\right)-x_{cn}\left(x_{cn}^2+y_{cn}^2\right)\right) \\ v_{c1}-k_1\left(y'_{c1}\left(x'^2_{c1}+y'^2_{c1}\right)-y_{c1}\left(x_{c1}^2+y_{c1}^2\right)\right) \\ v_{c2}-k_1\left(y'_{c2}\left(x'^2_{c2}+y'^2_{c2}\right)-y_{c2}\left(x_{c2}^2+y_{c2}^2\right)\right) \\ \vdots \\ v_{cn}-k_1\left(y'_{cn}\left(x'^2_{cn}+y'^2_{cn}\right)-y_{cn}\left(x_{cn}^2+y_{cn}^2\right)\right) \end{bmatrix} \tag{15.2}$$

To improve the compensation method, the 11th column of the coefficient matrix can be moved to the right side of the equal sign, as shown in Eq. (15.2). Thus, the errors due to the correlation method would no longer affect the coefficient matrices.

15.3 Experiments

The results of tensile experiment illustrate that the C0 and C1 compensation methods can compensate the measurement errors due to out-of-plane motions and lead to accurate results.

15.4 Conclusion

This paper systematically studies the measurement accuracy and efficiency of the two compensation methods. The drawbacks of the C0 compensation method are firstly discussed: (1) the coefficient matrices of the C0 compensation method are affected by the experimental noises and correlation errors; (2) the coefficient matrices are singular when the specimen keeps static or at the beginning of the experiment. To solve the problems, the C1 compensation method with high-accuracy and high-efficiency is proposed. The cond parameter is introduced to analyze the stability of the two compensation methods. Static experiments illustrate that the C1 compensation method is not affected by the subset size of the compensation points, nor the positions of the compensation points, nor the number of the compensation points, nor the thickness of the compensation panel. The measurement accuracy of the C1 compensation method reaches 5 $\mu\varepsilon$. Experimental results show that the lengths of the extensometers heavily affect the measurement accuracy. Tensile experiments demonstrate that the C1 compensation method has the same measurement accuracy as the 3D DIC method and the strain gage. A video extensometer is realized with a high measurement accuracy of 5 $\mu\varepsilon$ at a rate of 28 fps (Fig. 15.1).

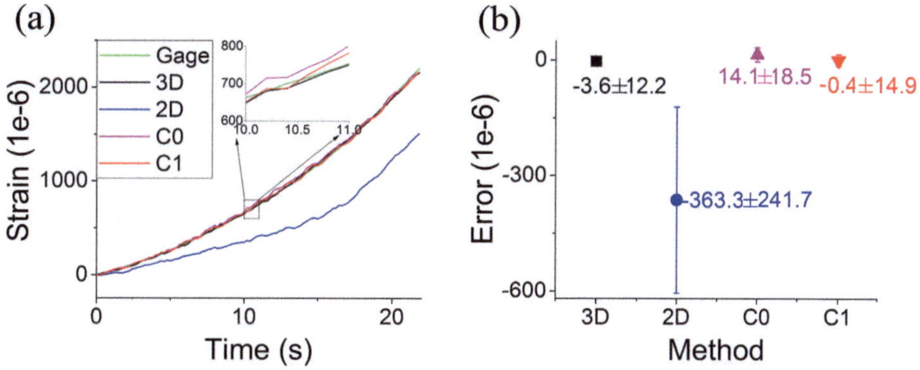

Fig. 15.1 The results and statistical analyses of a typical tensile experiment using the realized video extensometer. (**a**) Strain results measured by the strain gauge and the four DIC methods (3D DIC, 2D DIC, C0 compensation and C1 compensation). (**b**) Strain errors compared with the results measured by the strain gauge for the four DIC methods

References

1. Peters, W.H., Ranson, W.F.: Digital imaging techniques in experimental stress analysis. Opt. Eng. **21**(3), 427–432 (1982)
2. Gao, Y., Cheng, T., Su, Y., Xu, X., Zhang, Y., Zhang, Q.: High-efficiency and high-accuracy digital image correlation for three-dimensional measurement. Opt. Lasers Eng. **65**, 73–80 (2015)
3. Pan, B., Yu, L., Wu, D.: High-accuracy 2D digital image correlation measurements using low-cost imaging lenses: implementation of a generalized compensation method. Meas. Sci. Technol. **25**, 025001 (2014)
4. Sutton, M.A., Yan, J.H., Tiwari, V., Schreier, H.W., Orteu, J.J.: The effect of out-of-plane motion on 2D and 3D digital image correlation measurements. Opt. Laser Eng. **46**(10), 746–757 (2008)

Chapter 16
Theoretical and Numerical Analyses of Systematic Errors in Local Deformations

Xiaohai Xu and Qingchuan Zhang

Abstract There are displacement systematic errors due to undermatched shape functions and strain systematic errors due to undermatched surface fitting functions using the digital image correlation (DIC) method. The measured displacement and strain results are heavily affected by the calculation parameters (such as the subset size, the grid step, and the strain window size) due to undermatched shape functions and surface fitting functions. To evaluate the systematic errors, theoretical and numerical analyses of displacement and strain systematic errors have been carried out when the first- and second-order shape functions, and the quadric surface fitting functions are used. When the forth-order Taylor expansion is employed for the displacement, the results come out: (1) the approximate displacement systematic errors are functions of the second- and forth-order displacement gradients when first- and second-order shape functions are used, respectively; (2) the approximate strain systematic errors are functions of the third-order displacement gradients when first-order shape functions are used, and functions of the third- and fifth-order displacement gradients when second-order shape functions are used.

Keywords Digital image correlation • Shape functions • Surface fitting functions • Displacement systematic errors • Strain systematic errors

16.1 Introduction

The measurement accuracy of the digital image correlation (DIC) method has become a major concern for the researchers and users. There are many reports on the measurement errors in uniform deformations [1–3]. Studies indicate that the measurement errors include random errors and systematic errors. The random errors are caused by the image noise and other experimental noise [1]. The systematic errors are mainly caused by the imperfect sub-pixel interpolation algorithm including the interpolation bias [2] and the noise-induced bias [3]. For heterogeneous deformations, the undermatched shape functions heavily affect the measurement errors [4], especially in local deformations with high gradients [5]. Schreier built a theoretical model for quantifying the systematic errors due to undermatched shape functions, and employed a second-order deformation to analyze the relationship between the systematic errors and the subset size [6]. However, further solutions for local deformations had not be discussed yet.

This work is organized as follows. Section 16.2 deduces the theoretical estimations and approximations of the displacement and strain systematic errors. Section 16.3 validates the theoretical estimations and approximations through simulated experiments. Section 16.4 draws conclusions from this work.

16.2 Derivation

The measured displacements by the first- and second-order shape functions are expressed in Eqs. (16.1) and (16.2), respectively. The parameter M represents half width of the subset.

X. Xu • Q. Zhang (✉)
CAS Key Laboratory of Mechanical Behavior and Design of Materials, University of Science and Technology of China, Hefei 230027, China
e-mail: zhangqc@ustc.edu.cn

M. Sutton, P.L. Reu (eds.), *International Digital Imaging Correlation Society*, Conference Proceedings of the Society for Experimental Mechanics Series, DOI 10.1007/978-3-319-51439-0_16

$$u^{(1)}(x,y) = \frac{\sum\limits_{i,j=-M}^{M} u(x+i, y+j)}{(2M+1)^2} \tag{16.1}$$

$$u^{(2)}(x,y) = -\frac{15\sum\limits_{i,j=-M}^{M}\left(i^2+j^2\right)u(x+i,y+j) + \left(14M^2+14M-3\right)\sum\limits_{i,j=-M}^{M} u(x+i,y+j)}{(2M-1)(2M+1)^2(2M+3)} \tag{16.2}$$

The displacement systematic errors are shown in Eqs. (16.3) and (16.4). Note that the third- and fifth-order Taylor expansions for the displacement field *u(x,y)* are used when the first- and second-order shape functions are employed, respectively.

$$u_{err}^{(1)}(x,y) \approx \frac{M(M+1)}{6}\left(\frac{\partial^2}{\partial x^2}+\frac{\partial^2}{\partial y^2}\right)u(x,y) \tag{16.3}$$

$$u_{err}^{(2)}(x,y) \approx -\frac{1}{280}(M-1)M(M+1)(M+2)\left(\frac{\partial^4}{\partial x^4}+\frac{\partial^4}{\partial y^4}\right)u(x,y) - \frac{1}{36}M^2(M+1)^2\frac{\partial^4}{\partial x^2\partial y^2}u(x,y) \tag{16.4}$$

The strain systematic errors are shown in Eqs. (16.5) and (16.6). The parameter G represents the grid step, and the parameter N represents half width of the strain window.

$$exx_{err}^{(1)}(x,y) \approx \left(\frac{G^2\left(3N^2+3N-1\right)}{30}+\frac{M(M+1)}{6}\right)\frac{\partial^3 u(x,y)}{\partial x^3}+\frac{G^2N(N+1)}{6}\frac{\partial^3 u(x,y)}{\partial x\partial y^2} \tag{16.5}$$

$$\begin{aligned} exx_{err}^{(2)}(x,y) \approx\ & \frac{G^2\left(3N^2+3N-1\right)}{30}\frac{\partial^3 u(x,y)}{\partial x^3} + \left(\frac{G^4N(N+1)\left(3N^2+3N-1\right)}{180}-\frac{M^2(M+1)^2}{36}\right)\frac{\partial^5 u(x,y)}{\partial x^3\partial y^2} \\ & + \left(\frac{G^4\left(3N^4+6N^3-3N+1\right)}{840}-\frac{(M-1)M(M+1)(M+2)}{280}\right)\frac{\partial^5 u(x,y)}{\partial x^5}+\frac{G^4N(N+1)\left(3N^2+3N-1\right)}{360}\frac{\partial^5 u(x,y)}{\partial x\partial y^4} \end{aligned} \tag{16.6}$$

16.3 Validation

The displacement and strain systematic errors when a sine function shaped deformation is employed are shown in Fig. 16.1a, b, respectively. The systematic errors due to second-order shape functions are much smaller than that due to first-order shape functions.

16.4 Conclusion

This work mainly focuses on the systematic errors due to undermatched shape functions and surface fitting functions in local deformations. To evaluate the systematic errors, theoretical estimations have been deduced when undermatched shape functions and quadric surface fitting functions are employed. The results indicate: (1) the approximate displacement systematic errors are proportional to the second- and forth-order displacement gradients when first- and second-order shape functions are used, respectively; (2) the approximate strain systematic errors are functions of the third-order displacement gradients when first-order shape functions are used, and functions of the third- and fifth-order displacement gradients when second-order shape functions are used. The results have been demonstrated through simulated experiments.

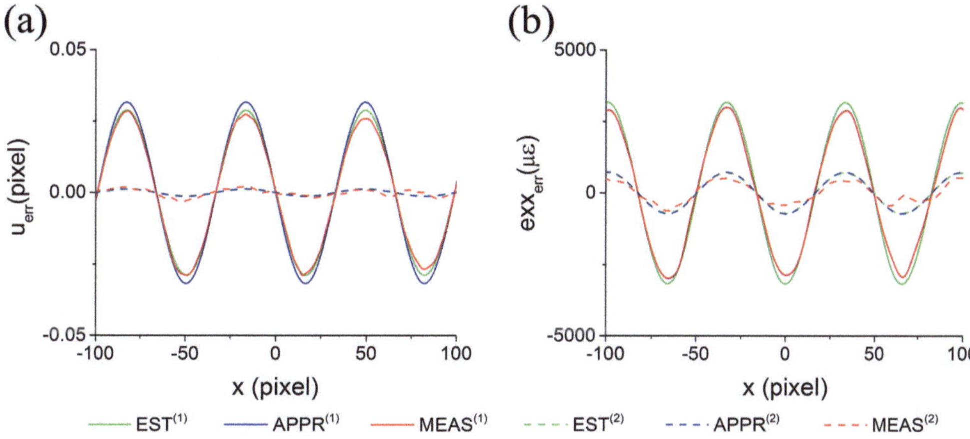

Fig. 16.1 Analysis of displacement and strain systematic errors when different shape functions are used. Estimated, approximated and measured (**a**) displacement errors and (**b**) strain errors

References

1. Gao, Z.R., Xu, X.H., Su, Y., Zhang, Q.C.: Experimental analysis of image noise and interpolation bias in digital image correlation. Opt. Lasers Eng. **81**, 46–53 (2016)
2. Su, Y., Zhang, Q.C., Gao, Z.R., Xu, X.H., Wu, X.P.: Fourier-based interpolation bias prediction in digital image correlation. Opt. Exp. **23**(15), 19242–19260 (2015)
3. Su, Y., Zhang, Q.C., Gao, Z.R., Xu, X.H.: Noise-induced bias for convolution-based interpolation in digital image correlation. Opt. Exp. **24**(2), 1175–1195 (2016)
4. Bornert, M., Brémand, F., Doumalin, P., et al.: Assessment of digital image correlation measurement errors: methodology and results. Exp. Mech. **49**(3), 353–370 (2009)
5. Xu, X.H., Su, Y., Cai, Y.L., Cheng, T., Zhang, Q.C.: Effects of various shape functions and subset size in local deformation measurements using DIC. Exp. Mech. **55**(8), 1575–1590 (2015)
6. Schreier, H.W., Sutton, M.A.: System errors in digital image correlation due to undermatched subset shape functions. Exp. Mech. **42**(3), 303–310 (2002)

References

Chapter 17
Quality Assessment of Speckle Patterns by Estimating RMSE

Yong Su and Qingchuan Zhang

Abstract The primary objective of this proceeding is to properly assess the quality of speckle patterns utilized in digital image correlation (DIC). Since a good pattern should associate with a small measurement error, the overall error can be used to assess the quality of a speckle pattern. The overall error consists of interpolation bias, noise-induced bias, and uncertainty due to image noise. Considering all these factors, this proceeding presents two assessment parameters, the maximum and the quadratic mean of the overall error respectively, to characterize the quality of a speckle pattern. Analytical formulae of proposed parameters were derived; estimating strategy was developed. Numerical simulations show the efficiency of proposed estimating strategy and demonstrate that the proposed parameters are more preferable than existing parameters due to the consideration of both systematic errors and random errors.

Keywords Digital image correlation • Speckle patterns • Interpolation bias • Noise-induced bias • Measurement uncertainty

17.1 Introduction

Digital image correlation (DIC) [1] is a widely used optical metrology for shape, motion, and deformation measurements. This technique generally needs premade artificial speckle patterns onto the specimen. Because the measurement performance is influenced by the patterns considerably, it is important to assess the quality of a pattern properly. Although several assessment parameters have been proposed during the past decade, these parameters neither lack a solid theoretical foundation [2] nor overlook the effects of interpolation [3]. In order to overcome these deficiencies, in this work, an analytical formula of the root mean square error (RMSE) is derived, two quality assessment parameters of speckle patterns based on RMSE are presented, and the superiorities of these parameters are proved.

17.2 Theoretical Analysis of the Quality of Speckle Patterns

A good pattern should lead to a small error, and thus the overall error deserve as a characteristic of the quality of a speckle pattern. The overall error comprises *interpolation bias*, *noise-induced bias*, and *random error* [4].

The *interpolation bias* u_{ib} can be approximated as [5]

$$\begin{gathered}
u_{ib} \approx C_{ib} \sin 2\pi u_0, \\
C_{ib} = \frac{1}{2\pi} \frac{\displaystyle\int_{-1/2}^{1/2} E_{\text{SSD}}(\nu) \left|\widehat{f}(\nu)\right|^2 \mathrm{d}\nu}{\displaystyle\int_{-1/2}^{1/2} \nu^2 \widehat{\varphi}(\nu) \left|\widehat{f}(\nu)\right|^2 \mathrm{d}\nu}, \\
E_{\text{SSD}}(\nu) = (\nu-1)\widehat{\varphi}(\nu-1) - (\nu+1)\widehat{\varphi}(\nu+1) + \widehat{\varphi}(\nu)\widehat{\varphi}(\nu+1) + \widehat{\varphi}(\nu)\widehat{\varphi}(\nu-1),
\end{gathered} \tag{17.1}$$

Y. Su • Q. Zhang (✉)
CAS Key Laboratory of Mechanical Behavior and Design of Materials, University of Science and Technology of China, Hefei 230027, China
e-mail: zhangqc@ustc.edu.cn

M. Sutton, P.L. Reu (eds.), *International Digital Imaging Correlation Society*, Conference Proceedings of the Society for Experimental Mechanics Series, DOI 10.1007/978-3-319-51439-0_17

where u_0 is the actual displacement, C_{ib} is the amplitude of interpolation bias determined by the power spectrum $\left|\widehat{f}(\nu)\right|^2$ and the interpolation transfer function $\widehat{\varphi}(\nu)$. The *noise-induced bias* u_{nb} can be approximated as [6]

$$u_{nb} \approx C_{nb} \sin 2\pi u_0, \quad C_{nb} = \frac{\sigma^2}{\overline{f'^2}} a_1, \quad a_1 = 4\pi \int_{-\infty}^{\infty} \nu \widehat{\varphi}(\nu) \widehat{\varphi}(\nu - 1)\, d\nu, \tag{17.2}$$

where C_{nb} is the amplitude of noise-induced bias, σ is the standard deviation of the noise of deformed image, and $\overline{f'^2}$ is the mean of squared intensity gradients. The *random error* Var(u_e) can be approximated as [4, 7]

$$\mathrm{Var}(u_e) \approx \frac{2\sigma^2}{\sum f'^2}, \tag{17.3}$$

where σ is the noise level in reference and deformed image, $\Sigma f'^2$ is the *sum of square intensity gradients* (SSSIG).

The overall error is generally characterized by the *root mean square error* (RMSE). A combination of Eqs. (17.1), (17.2), and (17.3) yields that

$$\mathrm{RMSE}(\widehat{u}; u_0) = \sqrt{\mathrm{E}\left((\widehat{u} - u_0)^2\right)} \approx \sqrt{\frac{2\sigma^2}{\sum f'^2} + (C_{ib} + C_{nb})^2 \sin^2 2\pi u_0}. \tag{17.4}$$

where $\widehat{u}$ is an estimation of the actual displacement u_0. Hence, a formula of the overall error (RMSE) is derived.

According to Eq. (17.4), RMSE is a function of the actual displacement u_0. Two parameters, the maximum and the quadratic mean of RMSE respectively, can be utilized to characterize RMSE irrespectively of u_0

$$V_{\max}^2 = \max\left(\mathrm{RMSE}^2(\widehat{u}; u_0)\right) \approx \frac{2\sigma^2}{\sum f'^2} + (C_{ib} + C_{nb})^2, \quad V_{\mathrm{avg}}^2 = \int_0^1 \mathrm{RMSE}^2(\widehat{u}; u_0)\, du_0 \approx \frac{2\sigma^2}{\sum f'^2} + \frac{1}{2}(C_{ib} + C_{nb})^2. \tag{17.5}$$

The superiority of these parameters is the direct linkage with the overall error and thus have a solid theoretical foundation and take both systematic and random errors into account automatically.

17.3 Numerical Experiment

Numerical experiments were conducted to make comparison of the proposed parameters with MIG [3], the widely used assessment parameter.

Ten speckle patterns shown in Fig. 17.1 are utilized. Each pattern was translated using Fourier shifting to generate 21 sub-pixel translated images, with a displacement of 0.05 pixels for successive images. Gaussian white noise with $\sigma = 2.55$ was added using MATLAB. 8281 calculation points were chosen and the subset size was 29 × 29. The DIC program employed SSD criterion, zero-order shape function, cubic B-Spline, and forward Gauss–Newton method. Displacements of the 8281 data points were analyzed to evaluate the systematic errors, the random errors, the overall errors, and the proposed parameter $V_{\max}$ and V_{avg}. The theoretical estimation of proposed parameters are evaluated by Eq. (17.5). Figure 17.2a illustrates both the numerical and the theoretical evaluation of $V_{\max}$ and V_{avg}, which show good agreements.

These numerical simulation clearly point out the limitation of the MIG criterion. MIG depicted in Fig. 17.2b indicates the best pattern is *Pattern b*. Although *Pattern b* have large image gradient, its interpolation bias is large, so that the overall error is large, and thus it is not the optimal speckle pattern. The best pattern should be *Pattern a*. An increase in image gradients can decrease the random errors and noise-induced bias, but generally increase the interpolation bias. Hence, the proposed parameters are more preferable.

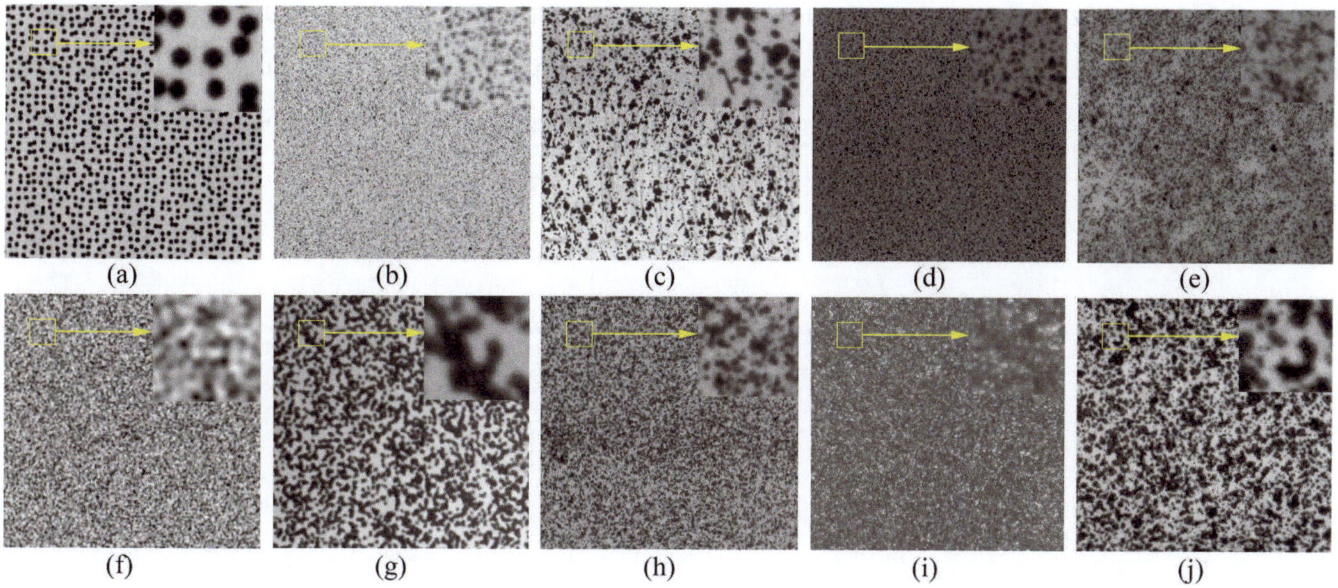

Fig. 17.1 Speckle patterns utilized for numerical simulation

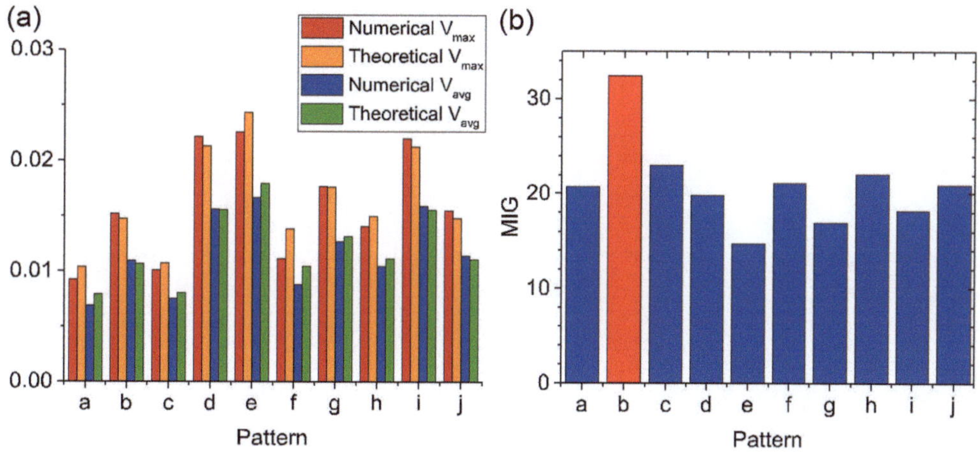

Fig. 17.2 (**a**) Proposed parameters; (**b**) mean intensity gradients (MIG)

17.4 Conclusion

This work presents a quality assessment of speckle patterns for DIC by estimating the RMSE. A formulae of RMSE is derived, two speckle assessment parameters are presented. Compared to existing assessment parameters, parameters presented in this paper are more preferable due to the consideration of both systematic errors and random errors.

References

1. Sutton, M.A., Orteu, J.J., Schreier, H.W.: Image Correlation for Shape, Motion and Deformation Measurements. Springer, New York (2009)
2. Lecompte, D., Smits, A., Bossuyt, S., Sol, H., Vantomme, J., Van Hemelrijck, D.: Quality assessment of speckle patterns for digital image correlation. Opt. Lasers Eng. **44**(11), 1132–1145 (2006)
3. Pan, B., Lu, Z., Xie, H.: Mean intensity gradient: an effective global parameter for quality assessment of the speckle patterns used in digital image correlation. Opt. Lasers Eng. **48**(4), 469–477 (2010)
4. Wang, Y., Sutton, M., Bruck, H., Schreier, H.: Quantitative error assessment in pattern matching: effects of intensity pattern noise, interpolation, strain and image contrast on motion measurements. Strain. **45**(2), 160–178 (2009)

5. Su, Y., Zhang, Q., Gao, Z., Xu, X., Wu, X.: Fourier-based interpolation bias prediction in digital image correlation. Opt. Express. **23**(15), 19242–19260 (2015)
6. Su, Y., Zhang, Q., Gao, Z., Xu, X.: Noise-induced bias for convolution-based interpolation in digital image correlation. Opt. Express. **24**(2), 1175–1195 (2016)
7. Pan, B., Xie, H., Wang, Z., Qian, K., Wang, Z.: Study on subset size selection in digital image correlation for speckle patterns. Opt. Express. **16**(10), 7037–7048 (2008)

Chapter 18
Statistical Error Analysis of the Inverse Compositional Gauss-Newton Algorithm in Digital Image Correlation

Xinxing Shao and Xiaoyuan He

Abstract In this paper, a theoretical model of the inverse compositional Gauss-Newton (IC-GN) algorithm was derived based on the sum of squared differences correlation criterion and linear interpolation. The model indicates that the IC-GN algorithm has better noise robustness than the forward additive Newton-Raphson (FA-NR) algorithm, and shows no noise-induced bias if the gray gradient operator is chosen properly. Both numerical simulations and experiments show good agreements with the theoretical predictions. Based on the proposed theoretical model, a new statistical error formulation of digital image correlation is presented.

Keywords Digital image correlation • Inverse compositional Gauss-Newton algorithm • Statistical error analysis • Noise-induced bias • Noise robustness

18.1 Introduction

The technique of digital image correlation [1] (DIC) has been widely used for non-contact deformation measurement in the field of experimental mechanics. In DIC, the selection of the sub-pixel registration algorithm is vitally important. At present, the most commonly used sub-pixel registration algorithm is the forward additive Newton-Raphson (FA-NR) algorithm [2], namely the famous Lucas-Kanade algorithm [3]. To improve the computational efficiency of the Lucas-Kanade algorithm, Baker and Matthews proposed the inverse compositional Gauss-Newton (IC-GN) algorithm [4] which can avoid repeated calculation of Hessian matrix. Sutton et al. introduced the IC-GN algorithm to DIC community in Ref. [1].

The performance of the IC-GN algorithm and the FA-NR algorithm have been analysed by Baker and Matthews in detail [5]. For noise robustness, their conclusion is that the FA-NR algorithm is slightly more robust to noise added to template and the input image [5]. In this paper, through theoretical analysis of the IC-GN algorithm, we will show that Baker and Matthews's conclusion is not complete. Moreover, we compare the statistical error formulation of the IC-GN algorithm with that of the FA-NR algorithm. The statistical error formulation of the FA-NR algorithm has been analysed by Sutton in detail in his book [1]. After comparisons, we find that the IC-GN algorithm has better noise robustness than the FA-NR algorithm, and shows no noise-induced bias if the gray gradient operator is chosen properly [6].

18.2 Results

Numerical analyses of six sets of synthetic images were carried out in this study. In this first image set, translated speckle images for each speckle pattern were generated by applying the appropriate shift in the Fourier domain according to the shift theorem [7]. The sub-pixel displacements applied in the x direction varied from 0 to 0.95 pixels. The remaining five image sets were generated by adding various levels of white Gaussian noise to all images in the first image set. The standard deviations of the added noise levels were 1, 3, 5, 8 and 10% of the full 8-bit grayscale. Figure 18.1 shows the mean bias error of the measured u displacement for FA-NR and IC-GN using the same bicubic spline interpolation.

Equation (18.1) is the statistical error formulation of the IC-GN algorithm, where $h(x_i)$ is the gray interpolation error, f_x is the first-order derivatives of the grayscale intensities of reference image, σ_1 and σ_2 are the standard deviations of the added Gaussian noise. Equation (18.1) shows that the noise-induced bias of IC-GN is eliminated.

X. Shao • X. He (✉)
Institute of Mechanics, Southeast University, Nanjing 210096, China
e-mail: mmhxy@seu.edu.cn

M. Sutton, P.L. Reu (eds.), *International Digital Imaging Correlation Society*, Conference Proceedings of the Society for Experimental Mechanics Series, DOI 10.1007/978-3-319-51439-0_18

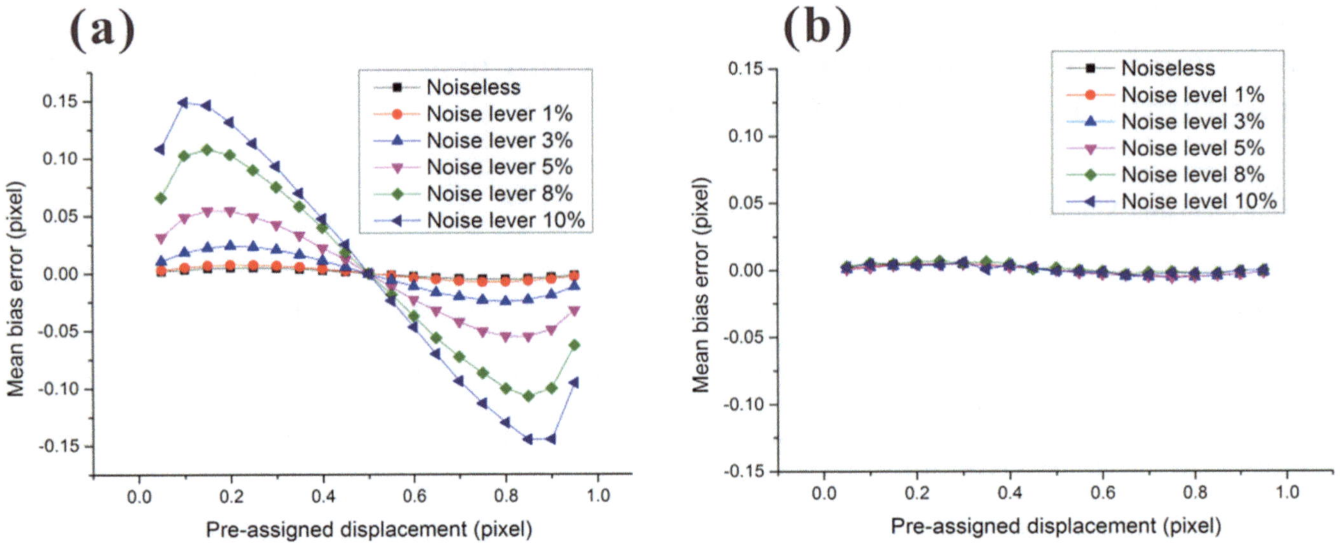

Fig. 18.1 Mean bias error of measured u displacements for FA-NR (**a**) and IC-GN (**b**) using the same bicubic spline interpolation

$$\begin{aligned} E\left(u'\right) &\cong u_0 - \frac{\sum_{i=-M}^{M}\left[\mathrm{h}\left(\mathrm{x}_i\right) f_x\right]}{\sum_{i=-M}^{M}\left(f_x'\right)^2} \\ Var\left(u'\right) &\cong \frac{\sigma_1^2+\sigma_2^2}{\sum_{i=-M}^{M}\left(f_x'\right)^2}, \sigma_{u'} = \sqrt{Var\left(u'\right)} \end{aligned} \tag{18.1}$$

18.3 Conclusions

In this paper, a theoretical model of the IC-GN algorithm was derived and compared with that of the FA-NR algorithm. The statistical error formulation of the IC-GN algorithm is different with that of FA-NR algorithm. The noise-induced bias error is eliminated. This new statistical error formulation should be used as a new statistical error formulation of digital image correlation.

Acknowledgements This work is supported by the National Natural Science Foundation of China (under Grant Nos. 11272089, 11327201 and 11532005).

References

1. Sutton, M.A., Orteu, J.J., Schreier, H.: Image Correlation for Shape, Motion and Deformation Measurements. Springer, New York (2009)
2. Pan, B., Qian, K., Xie, H., Asundi, A.: Two-dimensional digital image correlation for in-plane displacement and strain measurement: a review. Meas. Sci. Technol. **20**, 062001 (2009)
3. Lucas, B.D., Kanade, T.: An iterative image registration technique with an application to stereo vision[C]. IJCAI. **81**(1), 674–679 (1981)
4. Baker, S., Matthews, I.: Equivalence and efficiency of image alignment algorithms[C]//Computer Vision and Pattern Recognition, 2001. CVPR 2001. Proceedings of the 2001 IEEE Computer Society Conference on IEEE, 2001, 1: I-1090-I-1097, vol. 1
5. Baker, S., Matthews, I.: Lucas-kanade 20 years on: A unifying framework[J]. Int. J. Comp. Vision. **56**(3), 221–255 (2004)
6. Shao, X.X., Dai, X.J., He, X.Y.: Noise robustness and parallel computation of the inverse compositional Gauss–Newton algorithm in digital image correlation. Opt. Lasers Eng. **71**, 9–19 (2015)
7. Schreier, H.W., Braasch, J.R., Sutton, M.A.: Systematic errors in digital image correlation caused by intensity interpolation. Opt. Eng. **39**(52), 2915–2921 (2000)

Chapter 19
DIC in Machining Environment, Constraints and Benefits

Guillaume Rebergue, Benoît Blaysat, Helene Chanal, and Emmanuel Duc

Abstract The manufacturing of structural aluminum alloy parts used in aeronautics requires several steps of both forming processes and heat treatments before final machining. Some of the process steps will generate residual stresses and some will reduce residual stresses. For highly alloyed thick aluminum plates and closed-die forging parts, the material removal during the machining of complex geometries will release residual stresses and it can lead to part distortion.

This work aims at providing a reliable experimental technique measuring the part deformations during machining process to validate a robust predictive numerical method. The backbone of the technique relies on digital image correlation (DIC).

Indeed, the use of DIC as a measurement means allows a nondestructive control of part deformations during machining, which is interesting and difficult to achieve by other measurement techniques. Moreover, DIC allows a full field measurement in one shot that is faster than other available measurement tools. However, experimentation need to be realized to validate the use of DIC in a machining environment.

This work discusses more particularly the relevance, the reliability and the feasibility of DIC in the machining environment. The potential offered by the machining center for the calibration of the proposed Global-DIC procedure is also investigated.

Keywords DIC calibration • Global DIC • Milling • Residual stresses • Optical distortion

19.1 Introduction

In aeronautical context, aluminum alloys are subjected to numerous processing steps before being used to produce aircraft structural parts. These treatments, whether they are of thermal origin or whether plastic deformation, induce residual stresses in the material [1]. For highly alloyed thick aluminum plates and closed-die forging parts, the remaining residual stresses at final temper may induce geometrical defect on the machined part. Indeed, during the machining process of aluminum alloy parts dedicated to aeronautical applications, almost 90% of the material volume is removed. This causes a redistribution of residual stresses which leads to problems of deformation of the part during machining. This is a major issue for industry [2, 3] since this leads to difficulties and extra costs to manufacture conform part. Indeed, without special precautions, these deformations lead to overcut or undercut in some area of the part. That is why the aeronautical industry needs to improve their knowledge about the residual stress consequences.

On the one hand, important works have already been carried out on the determination of residual stresses [4–6] and their influence during the machining process [7, 8]. A specific numerical tool has been developed [9] to predict the behavior of the part during machining depending on the initial residual stress, the clamping and machining sequences [10].

On the other hand, the measurement of the displacements induced by the reorganization of residual stresses during machining becomes essential to improve the link simulation-experience and to validate the numerical prediction. Actually, such kind of measurement are generally performed by measuring the positions of specific points by probing [11, 12].

G. Rebergue (✉) • H. Chanal • E. Duc
Université Clermont Auvergne, SIGMA Clermont, Institut Pascal, BP 10448, F-63000, Clermont-Ferrand, France

CNRS, UMR 6602, IP, F-63178, Aubière, France
e-mail: guillaume.rebergue@sigma-clermont.fr

B. Blaysat
Clermont Université, Université Clermont Auvergne, Institut Pascal, UMR CNRS 6602, BP 10448, 63000, Clermont-Ferrand, France

CNRS, UMR 6602, IP, F-63178, Aubiére, France

M. Sutton, P.L. Reu (eds.), *International Digital Imaging Correlation Society*, Conference Proceedings of the Society for Experimental Mechanics Series, DOI 10.1007/978-3-319-51439-0_19

In a machining context, other methods of measurement exist but again, they only provide point-wise measurements of the workpiece or of the tool. To initiate a robust simulation-experience link, full field measurements are therefore needed.

The first objective of this work is to provide a reliable methodology dedicated to the measurement of the global deformation of the workpiece during a machining process. Such measurements will allow analyzing the influence of a given material removal on the deformation of the workpiece throughout a machining sequence. This will ensure to validate the modeling assumptions and validate more precisely the results of the simulation.

This method is based on the development of a digital image correlation (DIC) technique specific to the context of machining. Introduced by Sutton [13, 14] in the 1980s, DIC is now commonly used in every experimental mechanical laboratories thanks to the constant decreases of cost of cameras and computers but also thanks to the richness its outputs provide. Contrary to the usual displacement/train measurement techniques (strain gauges, position sensors, laser tracking or probing) that provide accurate measurement but point-wise, DIC offers full-field measurements. Moreover, thanks to a multitude of measured data, we know that DIC allows solving problems of identification and validation of finite element models [15].

19.2 Distortion Issues

First results are promising, nevertheless because of large size of the zone under study, optical distortions become a problem. This paper details issues caused by optical distortion and the opportunity to solve the problem. There are several types of optical distortions, here we will retain only the first source of distortion, the radial [16–18]. It appears when the light enters in the optical system by making a large angle with the optical axis. Thus the Gaussian approximation of the path of light rays is not satisfied and the light rays do not encounter the sensor to the position predicted by the model.

Distortions are here considered as mechanical deformation, consequently these deformations can be modeled by an apparent field of displacement between the captured image and the perfect projection of the surface studied [16–18]. These displacement fields are described by mathematical models that allow to express the displacement $\overrightarrow{u}\,(x_0)$ between the perfect image and our captured image for each pixel.

$$\begin{aligned} {\delta_x}^R(x,y) &= rx\left(x^2+y^2\right) \\ {\delta_y}^R(x,y) &= ry\left(x^2+y^2\right) \end{aligned} \tag{19.1}$$

With ${\delta_x}^R(x,y)$ and ${\delta_y}^R(x,y)$ the components of the displacement vector between the perfect image and the captured image at the coordinated (x,y) and r the coefficient of the amplitude of the distortion. (x,y) coordinates are defined from the coordinates of the optical center of the image (C_x, C_y).

19.2.1 Measuring Distortion Coefficients of the Optical System

It is necessary to determine now the distortion coefficients of the optical system. In this work context, it is possible to use the machine tool which manufactures the part to determine the influence of the optical distortion. Indeed, a machine tool can perform accurate translations, so the shape of the displacement field is well known and it is possible to identify and deduce the parasite displacement field $\Delta(x_0, u(x_0))$ generated by optical distortion issues which corrupt the measurement of the machining-tool movement.

For these problems of distortion, we have:

$$f\,(x_1) - g\,(x_1 + u\,(x_0) + \Delta\,(x_0, u\,(x_0))) \approx 0 \tag{19.2}$$

With x_0 the true position of the information in the speckle pattern and x_1 the position of this information in the image (Pixel values). $f(x_1)$ is the reference image as it is taken by the optical system, it includes thus the distortion defects. Similarly, $g(x_1)$ is the deformed image and also includes the distortion defects. Therefore note that the displacement field obtained via the algorithm is not the real displacement field $u(x_0)$ but the real displacement field corrupted by the distortion: $u(x_0) + \Delta(x_0, u(x_0))$.

During our calibration procedure, the real displacement field is known and is a constant translation. To give more freedom to the algorithm to approximate the displacement field, the displacement of speckle pattern stated:

$$\mathrm{u}\left(\mathrm{x}_0\right) = \mathrm{u}_\mathrm{x}\overrightarrow{e_x} + u_y\overrightarrow{e_y} = u \quad \forall x_0 \in RoI \tag{19.3}$$

We have:

$$\Delta\left(x_0, u\left(x_0\right)\right) = \delta^R\left(x_0 + u\right) - \delta^R\left(x_0\right) \tag{19.4}$$

After injection of Eq. (19.1) in (19.4), projection on the $\overrightarrow{e_x}$ axis and simplification we obtain:

$$\Delta\left(x_0\right) = A^*x^2 + B * x + K \qquad A, B, K \in \mathfrak{R} \tag{19.5}$$

With:

$$A = 3 * r^* u_x; \quad B = r^*\left(3 * {u_x}^2 - 6 * C_x * u_x\right) \tag{19.6}$$

u_x is the value of the real displacement field without distortion. It can be determined experimentally because the center of the image is not subjected to distortion. So it is possible to obtain coefficients describing the radial distortion by inverting the system.

$$r = \frac{A}{3 * u_x} \qquad C_x = \frac{3 * u_x^2 - \frac{B}{r}}{6 * u_x} \qquad C_x = \frac{3 * u_y^2 - \frac{B}{r}}{6 * u_y} \tag{19.7}$$

Thus, by performing a horizontal translational movement and another vertical with the machine-tool, we can determine coefficients to correct the radial distortion issue. For instance, considering only one horizontal translational movement: $r = 2.8 * 10^{-10}$ and $C_x = 3515$.

19.2.2 Correction of Distortion Phenomena

After modifications of images thanks to settings found previously, the displacement field will be calculated from corrected images and not from images deformed by distortion phenomena. Now we look at the variation of the displacement field for eight couples of images in function of RoI center position. In an ideal case, they should be close to zero because for an image pair, the displacement field is considered constant and uniform so there is no reason to have a different movement between the right and left of the RoI. However, the first figure shows that the image pairs with a RoI located on one end of the image undergoes a strong disturbances from the radial distortion. (One curve of point correspond to one image pair.) We can see on Fig. 19.1 that variation of displacement field decreases consequently with the correction.

19.3 Conclusion

The presented method ensure to determine coefficients of radial distortion for our optical system. These coefficient were determinate without optical sighting that is easier for industrial issue. Results are correct but they can be improve thanks to the vertical translation which gives more information. Moreover it may be worthwhile to incorporate other types of distortion in the method.

Thanking This work benefited from the support of the project SIMP-AERO No. ANR-15-CE10-0005-01 of the French National Research Agency (ANR).

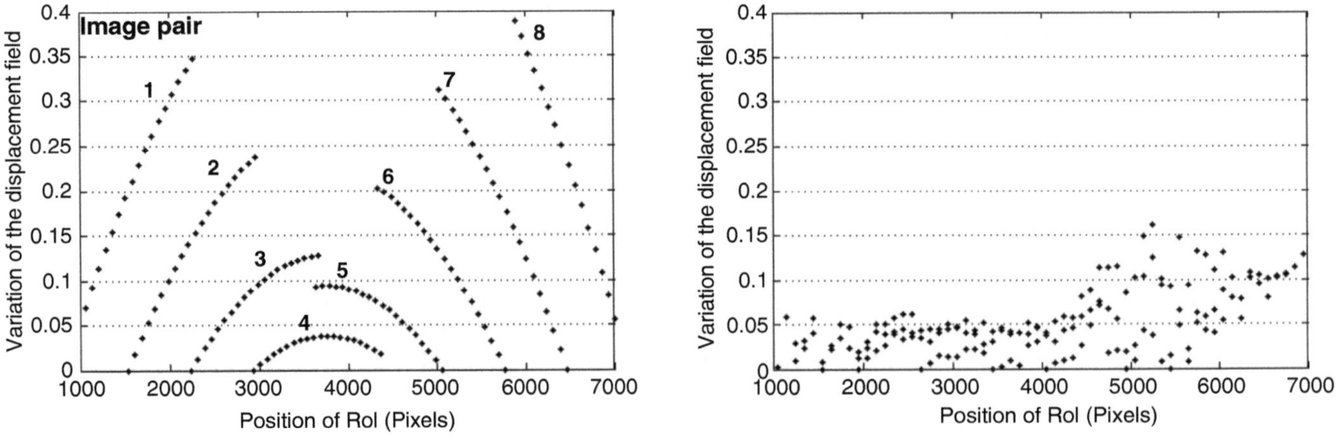

Fig. 19.1 Variation of the displacement field by couple of images. *Left*: without modification. *Right*: with modification

References

1. Das, S., Chandra, U.: Residual stress and distortion. In: Handbook of Aluminum Metallurgy: Processes and Equipment, vol. 1. Marcel Dekker, Inc., New York (1999)
2. Lequeu, P., Lassince, P., Warner, T., Raynaud, G.M.: Engineering for the future: weight saving and cost reduction initiatives. Aircr. Eng. Aerosp. Technol. **73**(2), 147–159 (2001)
3. Sim, W.-M.: Challenges of residual stress and part distortion in the civil airframe industry. Int. J. Microstruct. Mater. Prop. **5**(4–5), 446–455 (2010)
4. Jeanmart, P., Bouvaist, J.: Finite element calculation and measurement of thermal stresses in quenched plates of high-strength 7075 aluminium alloy. Mater. Sci. Technol. **1**(10), 765–769 (1985)
5. Tandon, R., Green, D.J.: Residual stress determination using strain gage measurements. J. Am. Ceram. Soc. **73**(9), 2628–2633 (1990)
6. Virkar, A.V.: Determination of residual stress profile using a strain gage technique. J. Am. Ceram. Soc. **73**(7), 2100–2102 (1990)
7. Wang, S.P., Padmanaban, S.: A new approach for fem simulation of nc machining processes. In: Materials Processing and Design: Modeling, Simulation and Applications-NUMIFORM 2004-Proceedings of the 8th International Conference on Numerical Methods in Industrial Forming Processes, vol. 712, no. 1, pp. 1371–1376 (2004)
8. Wei, Y., Wang, X.W.: Computer simulation and experimental study of machining deflection due to original residual stress of aerospace thin-walled parts. Int. J. Adv. Manuf. Technol. **33**(3), 260–265 (2007)
9. Cerutti, X., Mocellin, K.: Parallel finite element tool to predict distortion induced by initial residual stresses during machining of aeronautical parts. Int. J. Mater. Form. **8**, 255–268 (2015)
10. Cerutti, X., Mocellin, K., Hassini, S., Blaysat, B., Duc, E.: A methodology for machining quality improvement: towards the definition of machining process plans depending on the residual stress state of workpieces selected from aluminium alloy rolled plates. J. Manuf. Sci. Technol. (2016). doi:10.1016/j.cirpj.2016.07.004
11. Wang, Z., Chen, W., Zhang, Y., Chen, Z., Liu, Q.: Study on the machining distortion of thin-walled part caused by redistribution of residual stress. Chin. J. Aeronaut. **18**(2), 175–179 (2005)
12. Rai, J.K., Xirouchakis, P.: Finite element method based machining simulation environment for analyzing part errors induced during milling of thin-walled components. Int. J. Mach. Tools Manuf. **48**(6), 629–643 (2008)
13. Chu, T.C., Ranson, W.F., Sutton, M.A.: Applications of digital-image-correlation techniques to experimental mechanics. Exp. Mech. **25**(3), 232–244 (1985)
14. Sutton, M.A., Mingqi, C., Peters, W.H., Chao, Y.J., McNeill, S.R.: Application of an optimized digital correlation method to planar deformation analysis. Image Vis. Comput. **4**(3), 143–150 (1986)
15. Grédiac, M., Hild, F., Sutton, M.A.: Full-Field Measurements and Identification in Solid Mechanics. Wiley, Clermont Ferrand (2013). http://eu.wiley.com/WileyCDA/WileyTitle/productCd-1118578473.html
16. Brown, D.C.: Close-range camera calibration. Photogramm. Eng. **37**(8), 855–866 (1971)
17. Weng, J., Cohen, P., Herniou, M., et al.: Camera calibration with distortion models and accuracy evaluation. IEEE Trans. Pattern Anal. Mach. Intell. **14**(10), 965–980 (1992)
18. Sutton, M.A., Orteu, J.J., Schreier, H.: Image Correlation for Shape, Motion and Deformation Measurements: Basic Concepts, Theory and Applications. Springer Science & Business Media, New York (2009)

Chapter 20
Extracting High Frequency Operating Shapes from 3D DIC Measurements and Phased-Based Motion Magnified Images

Peyman Poozesh, Aral Sarrafi, Christopher Niezrecki, Zhu Mao, and Peter Avitabile

Abstract The three-dimensional digital image correlation (3D DIC) method in conjunction with a stereo-vision system can provide the full-field dynamic displacements of a structure with sub-pixel accuracy. However, stereo-photogrammetry systems are limited by camera resolution and intrinsic noise of the acquired images. Thus, in order to use optical sensing techniques to identify dynamic characteristics of a structure at high frequencies, the signal-to-noise ratio (SNR) in the sequence of images taken with a stereo-vision system needs to be improved. Within this paper phase-based video magnification, in conjunction with 3D DIC are used to visualize the high frequency operating shapes of a cantilever beam. The magnified sequence of images using motion magnification technique are post-processed using 3D DIC to quantify infinitesimal deformation that is not recognizable using only digital image correlation. The results obtained within this paper reveal the great potential of extracting 3D operating shapes of a high frequency structure using the motion magnification and stereo-photogrammetry techniques. Moreover, results of this paper indicate that using the motion magnification technique increases the SNR of the measurements, and could be used as a new approach to extract more information about the structure than previously possible compared to using 3D DIC alone.

Keywords Digital Image Correlation (DIC) • Video Magnification • High Speed Photogrammetry • Modal Analysis

20.1 Experimental Case Studies and Results

The experimental test case was designed to evaluate the feasibility of extracting full-field operating shapes of a structure using the motion magnification technique in conjunction with 3D DIC method. A pair of four-megapixel (2048 × 2048 pixel CMOS sensor) PHOTRON high-speed cameras equipped with 24-mm lenses were used for the measurement to extract high frequency operating deflection shapes of a small cantilevered beam subjected to an impulse excitation. The stereo-vision system, shown in Fig. 20.1, was placed 35 cm away from the test structure. A black and white speckled pattern was painted on the beam surface to monitor the 3D displacement of the patterned area. A proper pattern is one that has high contrast (large gray-scale variation) with appropriate size dots (~8 pixels in diameter) [1]. The stereo-vision system was calibrated using a small calibration panel (i.e. 380 × 320 mm^2) to identify the lens distortions and relative locations of the cameras with respect to each other. The calibration deviation (or calibration error) was determined was as 0.038 pixels (~0.008 mm for the measured field of view). According to the Aramis™ user manual [2], for a proper calibration, the calibration deviation needs to be less than or equal to 0.04 pixels.

20.1.1 Modal Parameter Extraction from Optically Measured Data

In the first case, a cantilevered beam shown in Fig. 20.1b was excited using an impact hammer and the full-field response of the beam was measured using the 3D DIC technique. The displacement noise floor of the measurement was within an acceptable range, approximately ±2 μm. The noise floor is essentially the lowest level of displacement that can be accurately measured using a conventional image sensing techniques (i.e. 3D DIC and 3DPT). A common acceptable displacement noise floor level for a typical DIC measurement in a laboratory setting is approximately 0.03 pixels (~5 μm) [2]. The distributed

P. Poozesh (✉) • A. Sarrafi • C. Niezrecki • Z. Mao • P. Avitabile
Structural Dynamics and Acoustic Systems Laboratory, One University Avenue, Lowell, MA 01854, USA
e-mail: Peyman_Poozesh@student.uml.edu

M. Sutton, P.L. Reu (eds.), *International Digital Imaging Correlation Society*, Conference Proceedings of the Society for Experimental Mechanics Series, DOI 10.1007/978-3-319-51439-0_20

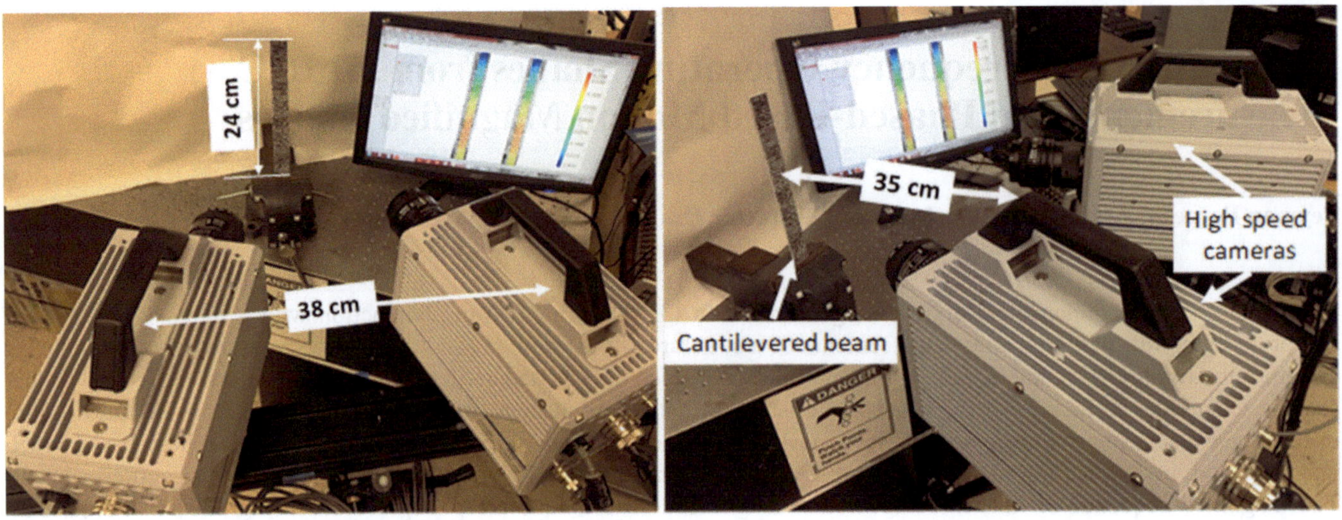

Fig. 20.1 Two photos of the test set-up showing the position of the cameras relative to each other and the cantilevered beam; a photo of test set-up shown from the front view (*left*) and a photo of test set-up shown from the an oblique angle view (*right*)

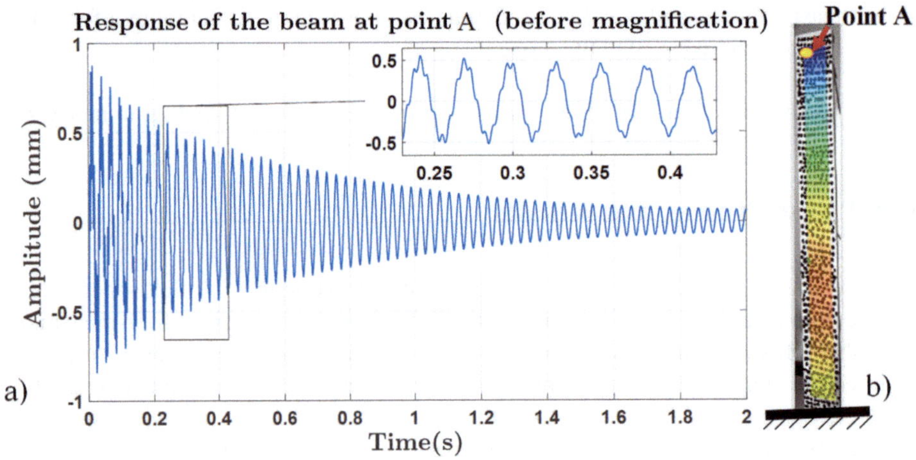

Fig. 20.2 The raw signal (i.e. before video magnification) extracted from optical measurement (3D DIC) displaying out-of-plane displacement of the beam subjected to an impulse excitation; (**a**) the out-of-plane displacement of the beam at point A, (**b**) the overlay of blade image and spatial out-of-plane displacement contour

displacement of the beam subjected to an impulse excitation was obtained using high-speed cameras and the 3D DIC method. A sampling frequency of 2500 Hz and shutter time of 1/3000 s were selected to avoid blurring and under/overexposure. The response of the beam at point A and the full-field displacement of the beam is shown in Fig. 20.2. A Fast Fourier Transform (FFT) was performed on the optically measured response of the beam at point A (see Fig. 20.3). The peak appeared at the frequency of 705 Hz indicates that there might be a structural mode at that frequency, assuming visual aliasing did not occur. However, the SNR of the obtained signal around that frequency (i.e. ~705 Hz) is low compared with the first three structural modes of the beam (see blue curve in Fig. 20.3, before magnification).

Applying phased-based motion magnification [3] to the source video in the frequency band of 700–709 Hz and then using 3D DIC technique to quantify the magnified motions in the captured images improves the SNR over that frequency band (see Fig. 20.3). The red curve in Fig. 20.3 shows the FFT of the response of the structure at Point A after magnification within a frequency band of 700–709 Hz. The operating shapes of the beam after the magnification clearly shows the first torsion mode of the beam associated with that frequency (i.e. 705 Hz). The peak picking technique was used to extract full-field operating mode shapes of the beam. As seen in Fig. 20.3, the video magnification technique including phased-based motion estimation and video reconstruction method improves the SNR at the frequency band of 700–709 Hz. Figure 20.3 shows that the proposed method can measure the vibrational response of structures at high frequencies by improving the SNR that would otherwise remain undetected or difficult to observe using an only conventional image-sensing techniques. It is worth

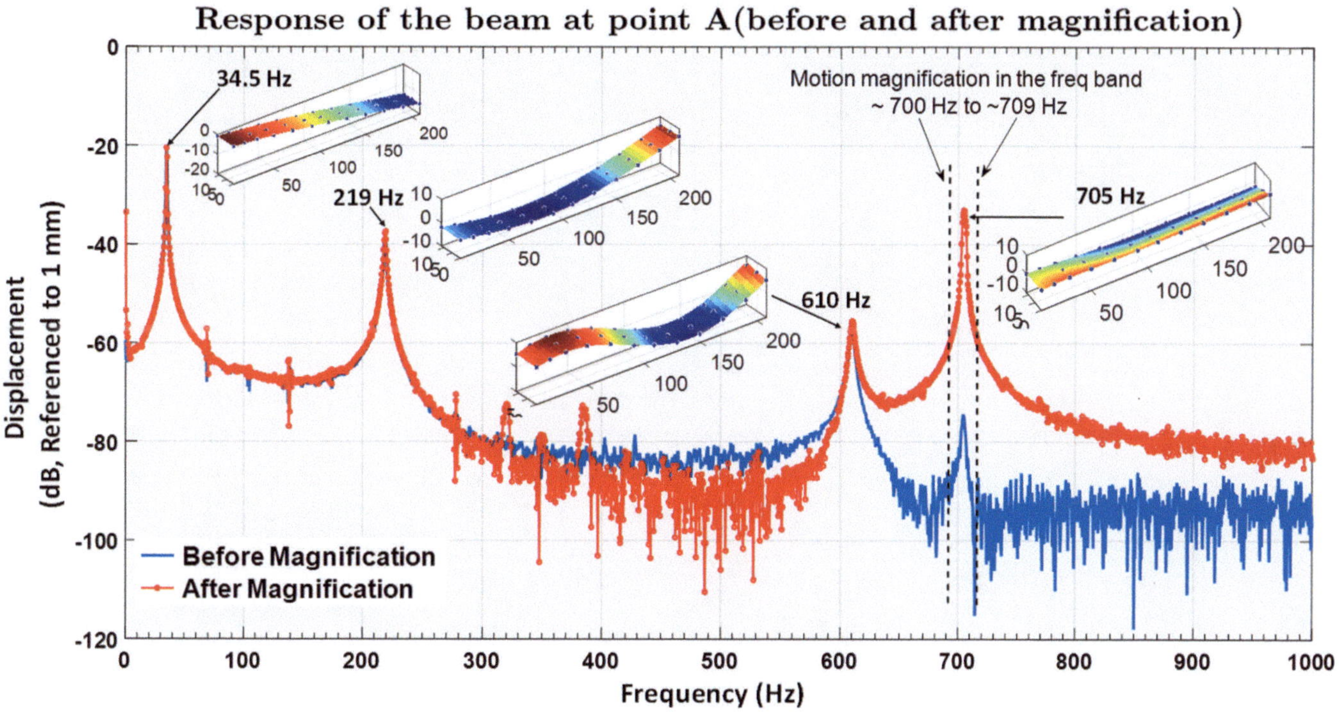

Fig. 20.3 Full-field 3D optically measured operating shapes of the beam and comparison of the response of the beam at Point A in the frequency domain before and after magnification

mentioning that using the magnification algorithm the pixels with high SNR gets amplified more than pixels with low SNR, therefore selecting a large frequency band enhances the pronounced and dominant structural modes in the frequency domain as seen in the blue curve in Fig. 20.3, and the modes with small responses remain undetected.

20.2 Conclusion

This paper proposes a novel methodology that makes it possible to extract the in-plane and out-of-plane small vibrational response of structures by improving the SNR and obtaining the frequency and operating shapes that would otherwise be invisible to conventional camera-based measurement methods (i.e. 3D DIC and 3DPT). The proposed technique employs the phased-based motion magnification technique and stereo-photogrammetry method and paves the way for successful use of image sensing techniques to replace conventional point-wise and contacting measurement sensors.

References

1. Poozesh, P., Baqersad, J., Niezrecki, C., Avitabile, P., Harvey, E., Yarala, R.: Large-area photogrammetry based testing of wind turbine blades. Mech. Syst. Signal Process. **86**, 96 (2016). doi:10.1016/j.ymssp.2016.07.021
2. ARAMIS v6.3, GOM mbH, Braunschweig, Germany 2011
3. Wadhwa, N.: Revealing and analyzing imperceptible deviations in images and videos, p. 198. Department of Applied Mathematics, Doctor of Philosophy, Massachusetts Institute of Technology (2016)

Chapter 21
2D DIC-Based Inverse Procedures for the Plastic Identification of Sheet Metals in High Strain Rate Tests

G. Chiappini, E. Mancini, M. Rossi, and M. Sasso

Abstract In this work, dynamic tension tests have been conducted by an SHB on sheet metals in order to characterize the plastic behaviour of the materials.

Pictures of the samples were acquired during the test by means of a fast camera and were post-processed by a global approach 2D DIC software. On the one hand, this permitted to validate the strain measurement by the classical SHB theory formulas; on the other hand, strain distribution maps were computed with the approximation of small out-of-plane displacements. Different analytical methods have been adopted to exploit quantitatively these strain maps and to extract the parameters of a strain hardening constitutive model: the most simple method uses only the data extracted from the necking region and accounts for the cross section reduction; the most complex is represented by the VFM which uses the information provided by all the pixels in the acquired images.

Keywords Hopkinson Bar • High strain rate • DIC • Inverse Methods • FastCam

21.1 Introduction

Nowadays Split Hopkinson Pressure Bar (SHPB) has been vastly used in order to determine the stress-strain curves at strain rates in the range of 10^2 to 10^4 s^{-1}. The Hopkinson Pressure Bar was first suggested by Hopkinson as an apparatus to measure stress pulse propagation in a long metal and it was extended by Davies and Kolsky. See [1] for a review of the method. More recently, SHPB is focused not only on compression tests but also on tensile and torsion tests for different classes of materials such as metals, polymers, ceramics, composites and foams [2].

Sheet metals have also been tested at high strain rate. In order to determine the constitutive behaviour of the material, and especially its strain rate sensitivity, several tests are needed, where the size of the specimen or the intensity of the impact are varied thus to obtain different strain rate values [3].

In this work, the direct Split Hopkinson Tension-Compression Bar (SHTCB) developed in [4] has been used to perform one dynamic test on a sheet metal sample. The strain analysis carried out through DIC highlights that within the same sample, different points experience different strain rate levels; indeed the deformation is not uniform and is characterized by a strong localization in the necking region. Therefore, in principle only one test could be used to assess the strain rate sensitivity of the material. Here the C coefficient of the Johnson-Cook model has been used as index of strain rate sensitivity, and it has been evaluated according to three different inverse procedures.

21.2 Experimental Results

A tensile test has been conducted and engineering stress and strain data have been collected. The specimen geometry and adopted gripping system, based on two dowel pins, are shown in Fig. 21.1.

G. Chiappini • M. Rossi • M. Sasso (✉)
Università Politecnica delle Marche, Ancona, Italy
e-mail: m.sasso@univpm.it

E. Mancini
Università degli Studi eCampus, Novedrate (CO), Italy

M. Sutton, P.L. Reu (eds.), *International Digital Imaging Correlation Society*, Conference Proceedings of the Society for Experimental Mechanics Series, DOI 10.1007/978-3-319-51439-0_21

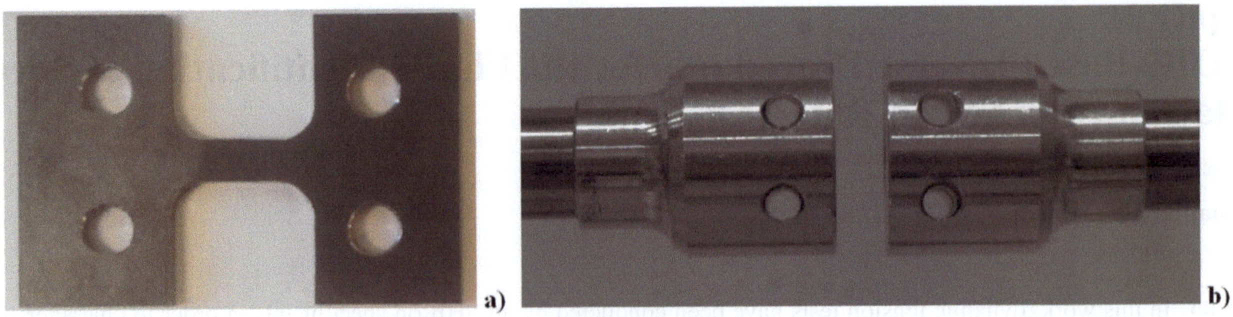

Fig. 21.1 Pictures of the sample (**a**) and of the fasteners (**b**)

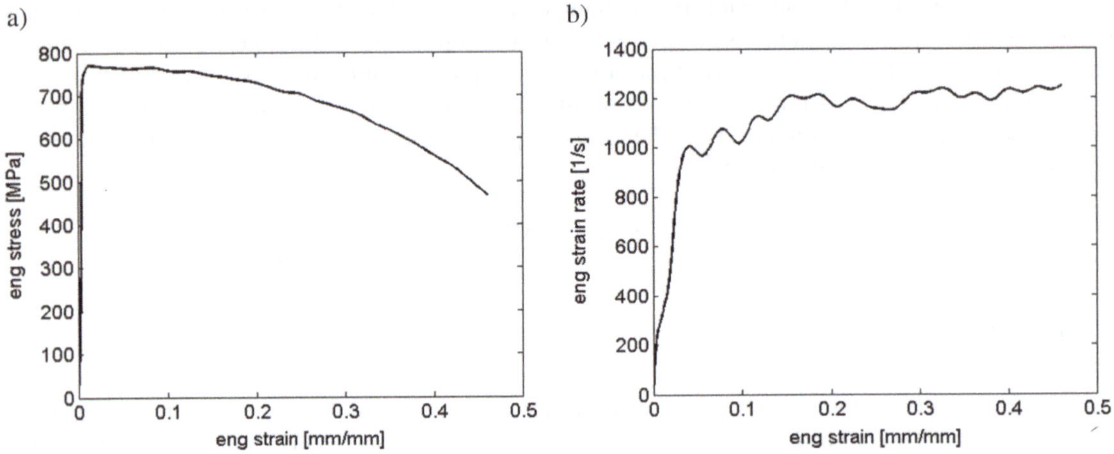

Fig. 21.2 Engineering stress (**a**) and strain rate (**b**) curves

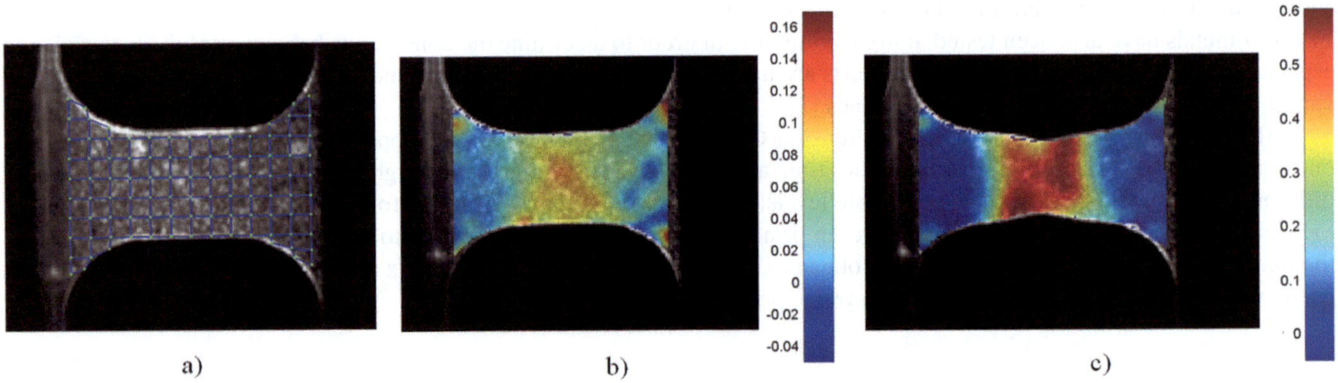

Fig. 21.3 (**a**) DIC grid, (**b**) strain ε_X at an intermediate instant, (**c**) strain ε_X at the final instant

The usual formulas coming from the SHB theory have been used to extract the engineering strain, strain rate and stress. The obtained curves are shown in Fig. 21.2. A fast camera has been used to acquire images during the test and to assess the actual deformation within the gauge length.

It is possible to note that, incidentally, the strain rate is almost nicely constant, exception made for the very first part of the test. Nevertheless, the stress curve is far from the typical shape, since relevant area reduction starts at low strain level, and the identification of real plastic behaviour is not possible. This is confirmed by the strain maps of Fig. 21.3.

21.3 Application of Inverse Methods

21.3.1 Evaluation at Neck

The first method used to correct the engineering curve into a true curve consists in evaluating the (average) strain in X and Y direction in the necking cross section; accepting the incompressibility hypothesis, the true stress S can be estimated from the engineering stress σ:

$$S(t) = \sigma(t) \cdot \exp\left[-\overline{\varepsilon_Y}(t)\right] \cdot \exp\left[-\overline{\varepsilon_Z}(t)\right] = \sigma(t) \cdot \exp\left[-\overline{\varepsilon_X}(t)\right] \tag{21.1}$$

21.3.2 Multiple Sections

The second method consisted in applying the previous equation to all the node columns of the DIC grid shown in Fig. 21.3a. It is found that each columns is characterized by an average strain rate, with the maximum located obviously in the central column. The effective strain rate varies from 200 to more than 1200 1/s. This permitted to correlate the effective stress computed by (21.1) to the strain rate and to evaluate the strain rate sensitivity.

Figure 21.4a compares the engineering curve with the true one coming from Eq. (21.1). Figure 21.4b shows, with thin lines, the reconstructed true stress-true strain for each column of the DIC grid nodes; the bold lines correspond to the best fitting Johnson-Cook model, where the sensitivity coefficient C is found to be quite small (i.e. 0.015).

21.3.3 Strain Map

The third adopted method consisted on the other hand in using all the data coming from DIC analysis, including the strain in directions Y and XY, thus to account for the multi-axiality of the stress distribution. This required to build up a numerical procedure, where the tentative constitutive law is coupled to a plasticity flow rule (the standard von Mises was used here) so that the (planar) stress components can be computed. For the equilibrium, neglecting inertial effects, each vertical section should have the same global force; so a cost function has been defined:

$$err = \sum_t \sum_c \left[\Pi_{11,c}(t) - \sigma(t)\right] \tag{21.2}$$

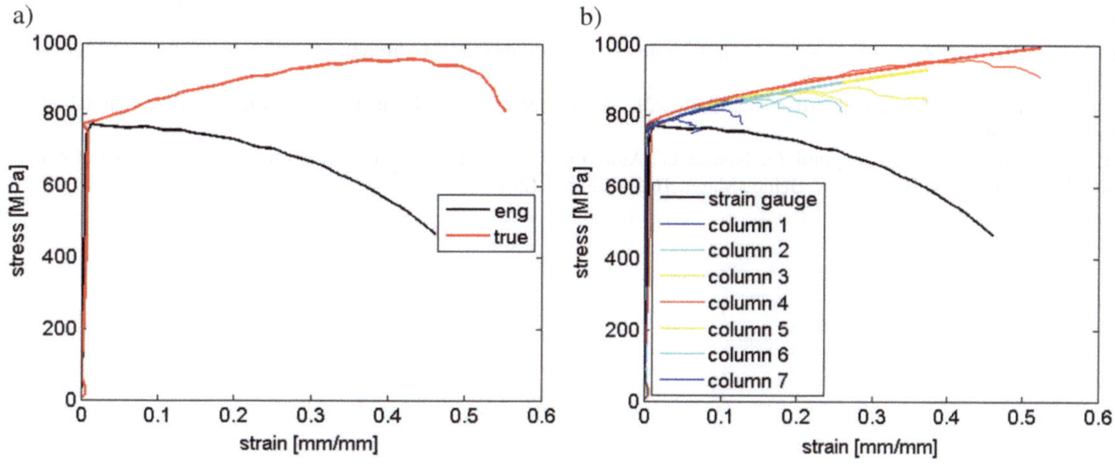

Fig. 21.4 Reconstructed true stress at the necking section (**a**) and for all the vertical columns of nodes (**b**)

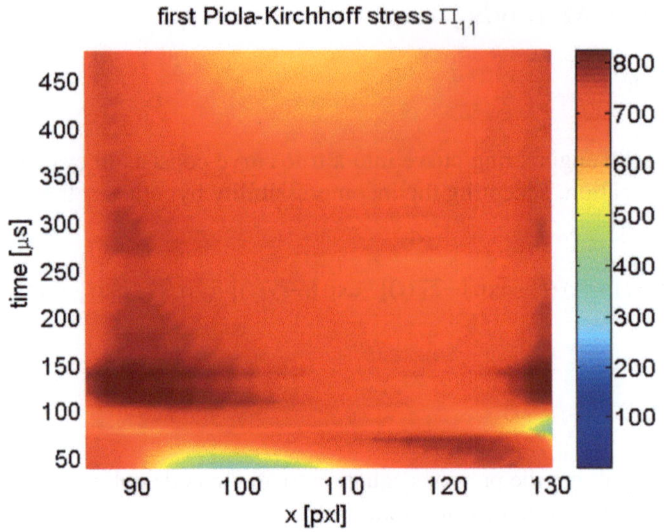

Fig. 21.5 Estimated evolution of the First Piola-Kirchhoff stress

where $\Pi_{11,c}(t)$ denotes the average of first component of the Piola-Kirchhoff stress tensor computed for all the pixels of the *c*th column at instant *t*. The evaluated engineering stress for each column versus time is shown in the contour map of Fig. 21.5. Even in this case the detected strain rate sensitivity is low, with a C coefficient of 0.017.

21.4 Conclusions

The work used the results from the DIC measurement carried out during an Hopkinson Bar test to convert engineering stress data into true ones. Different methods were adopted to identify the material strain rate sensitivity from one single test, exploiting the heterogeneous strain rate distribution within the sample.

References

1. Chen, W., Song, B.: Split Hopkinson (Kolsky) Bar, design, testing and applications. In: Mechanical Engineering Series, pp. 11–17. Springer, New York (2011)
2. Campana, F., Mancini, E., Pilone, D., Sasso, M.: Strain rate and density-dependent strength of AlSi7 alloy foams. Mater. Sci. Eng. A. **651**, 657–667 (2016)
3. Sasso, M., Fardmoshiri, M., Mancini, E., Rossi, M., Cortese, L.: High speed imaging for material parameters calibration at high strain rate. Eur. Phys. J. Spec. Top. **224**, 1–14 (2015)
4. Mancini, E., Sasso, M., Rossi, M., Chiappini, G., Newaz, G., Amodio, D.: Design of an innovative system for wave generation in direct tension–Compression Split Hopkinson Bar. J. Dyn. Behav. Mater. **1**(2), 201–213 (2015)

Chapter 22
Direct Measurement of R Value for Aluminum Alloy Sheet Metal Using Digital Image Correlation

Xin Xie, Junrui Li, Boyang Zhang, Bernard Sia, and Lianxiang Yang

Abstract Plastic Strain Ratio, also called the R value, is a key parameter to evaluate the formability of sheet metal which is widely used in the yield functions and constitutive equations. To measure the R value during a tensile test, the strain along the width direction and thinning direction have to be measured simultaneously. Conventionally, due to the difficulties of measuring the thinning strain, the R value is usually calculated from the principle strain 1 and 2 under zero plastic volume change assumption. This paper shows a novel testing setup using a 3D digital image correlation which can measure the strain in width and thinning strain simultaneously, thus the R value can be measured directly without volume constant assumption. The test plan, experiment setup and results are shown in detail in this article.

Keywords Plastic strain ratio • R value • Direct measurement • Digital image correlation • Zero plastic volume change assumption

22.1 Introduction

Plastic Strain Ratio, also called the R value, is a key parameter to evaluate the formability of sheet metal. For the emergency need of light weight vehicle design, an accurate measurement of R value is urgent needed by the automobile industry and stamping industry [1]. Practically, to measure the R value is more challenge than other material properties, such as the Young's modulus and Poison's Ratio. This is because, to measure the R value (the ratio of the strain in width and in thickness directions), the strain in width and in thickness directions need to be measured simultaneously during a standard tensile test. However, in sheet metal forming tests, due to the extreme small dimension in the thickness direction, an accurate and full-field measurement of thinning strain for sheet metal cannot be achieved by conventional strain measurement methods, such as strain gauge or extensometer [2, 3].

In recent years, a new full field strain measurement method which makes this task possible, called digital image correlation method (DIC), has been rapidly developed. DIC is a speckle pattern tracking based full-field, non-contact and robust optical measurement method which has the capability to measure full field strain distribution history [4, 5]. Combining a standard tensile test with DIC strain measurement enables a high accuracy measurement of R value. Currently, most R value measurement using DIC method is not a direct measurement due to the difficulties in measuring thinning strain and width strain simultaneously. In the current measurement procedure, usually the principle strain 1 (p_1) and principle strain 2 (p_2) of a standard dog-bone sheet metal coupon is measured first during a tensile test. Then, the thinning strain is calculated using zero plastic volume change assumption, stated as $p_1 + p_2 + p_3 = 0$. And finally, the R value is calculated as $R = p_2/p_3$. However, recently, Xie et al. reported a fact that in the deep plastic zone (near break), some materials such as aluminum alloy does not always follow the zero plastic volume change assumption [6]. In this case, an error caused by the non-conservation of plastic volume are included in all of the existing R value measurement using the conventional measurement method.

This article reports a novel digital image correlation application to directly measure R value instead of calculating the R value based on the zero plastic volume change assumption. In this research, with an optimized digital image correlation setup, the optimized DIC system can measure the major strain, minor strain and thinning strain simultaneously during a

X. Xie (✉)
Department of Mechanical Engineering, Lawrence Technological University, Southfield, MI 48075, USA
e-mail: xxie@ltu.edu

J. Li • B. Zhang • B. Sia • L. Yang
Optical Laboratory, Department of Mechanical Engineering, Oakland University, Rochester, MI 48309, USA

M. Sutton, P.L. Reu (eds.), *International Digital Imaging Correlation Society*, Conference Proceedings of the Society for Experimental Mechanics Series, DOI 10.1007/978-3-319-51439-0_22

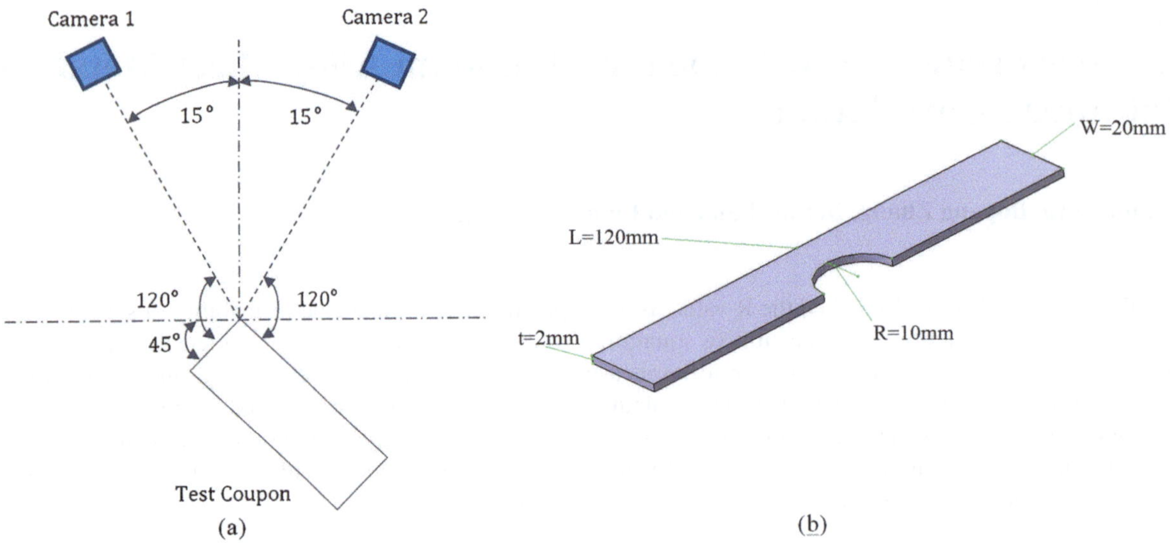

Fig. 22.1 (**a**) Schematic of the measurement setup; (**b**) a photo of testing setup

single tensile test. The R value of AA6000 aluminum alloy are measured. The R value directly measured from the novel DIC setup is then compared with the one calculated using the zero plastic volume change assumption. Experiment plan, optimization of measurement setup, experimental results and data analysis are shown in detail.

22.2 Experiment Setup

Figure 22.1a shows the schematic of the testing setup. Figure 22.1b shows a photo of the testing setup. An optimized 3D DIC system (Dantec Dynamics Q-400) is set to facing the corner of the testing sample, so that it can observe the top surface and thinning surface simultaneously. Then the strain distribution history on these two surfaces can be measured simultaneously during the loading. The notched testing coupon is tensiled by a MST tensile test machine at the speed of 5 mm/min. The acquisition rate of the cameras is set at 10 fps. A pair of 75–108 mm (F1.8–F22) zooming lenses are used to provide a small enough field of view. A high power LED (7500 lumens) is used to ensure a sufficient illumination.

In this research, a 120 mm × 20 mm AA6000 aluminum alloy sheet metal (2 mm thickness) with a 10 mm notch is used as the testing coupon. Figure 22.1 shows the sample shape configuration. The longitudinal dimension of the test coupon is 120 mm, the width of the test coupon is 20 mm, and an R = 10 mm notch is made in the middle of one edge to limit the failure location. The original AA6000 material comes in the size of 80 mm × 480 mm with thickness of 2 mm. The original metal sheet is cut into eight small sheets with a size of 40 mm × 120 mm. Then the R = 10 mm hole is punched using a flat punch tool from Uni-Punch at a punch clearance of 10%. Finally, the 40 mm × 120 mm sheet metal with 10 mm punch hole is cut through the center to form a pair of test coupons as shown in Fig. 22.2.

22.3 Results and Data Analysis

From the DIC measurement, a 3D contour along with the strain distribution on both the top surface and the thinning surfaces of the testing coupon during the tensile test can be measured. Figure 22.3a, b shows the measured strain distribution on both the top surface and the thinning surfaces of the testing coupon during the tensile test. Then the average strain value inside four 2 mm × 2 mm square areas were picked to check the volume conservation assumption, namely area A–D. From area A in the principle strain 1 map (true strain), the major strain inside the necking band can be calculated. Area A in the principle strain 2 map (true strain) also can be used to calculate the minor strain inside the necking band. From area B in the principle strain 2 map (true strain), the thinning strain inside the necking band can be calculated (the strain direction along the thinning direction as shown in Fig. 22.3b). In the same sense, the major strain outside the necking band can be calculated from area

Fig. 22.2 Schematic of the testing coupon

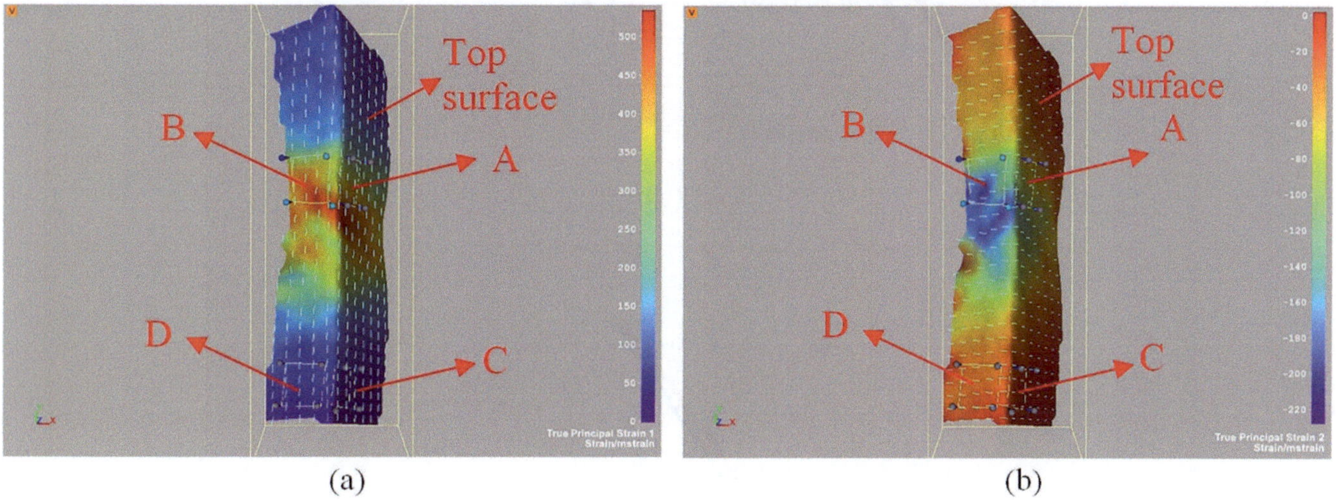

Fig. 22.3 Full field strain distribution measured by DIC system: (**a**) distribution of principal strain 1; (**b**) distribution of principle strain 2

C in the principle strain 1 map (true strain), the minor strain outside the necking band can be calculated from area C in the principle strain 2 map (true strain), and the thinning strain outside the necking band can be calculated from area D in the principle 2 strain 2 map (true strain).

Then two algorithms are used to evaluate the R value. One is using the directly measured p2 and p3 value to calculate the R value (in this paper this is called the directly measured R value for short). The other way is to calculate R value under zero plastic volume change assumption, where assume $p1 + p2 + p3 = 0$. In this approach, the thinning strain is calculated from measured p1 and p2. Then the R value is calculated from the measured p2 value and calculated p3 value under the zero plastic volume change assumption (in this paper this is called the calculated measured R value for short). Figure 22.3a, b shows the comparison of measured R value and calculated R value at the locations inside/outside of the necking band. In Fig. 22.4a, b, the blue line is directly plot from the measured strain data while red line shows the calculated data based on $p1 + p2 + p3 = 0$. From the comparison of Fig. 22.4a, b, it can be found that, at the location inside the necking band, the measured R value have a big difference between the directly measured value. While at the location outside the necking band, the directly measured R value is same with the calculated one.

Figure 22.5 shows a plot of volume strain history ($p1 + p2 + p3$ vs step). From Fig. 22.5 it can be found that at the location outside of the necking band AA6000 approximately follows the $p1 + p2 + p3 = 0$ assumption while at the location inside the necking band AA6000 does not follow the $p1 + p2 + p3 = 0$ assumption. This explained why the measured R

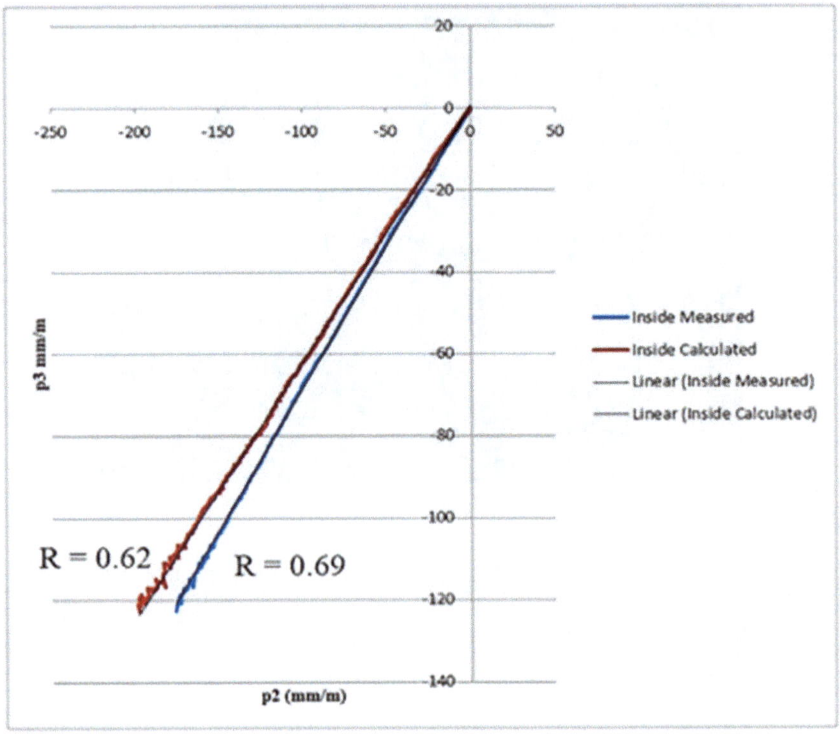

(a)

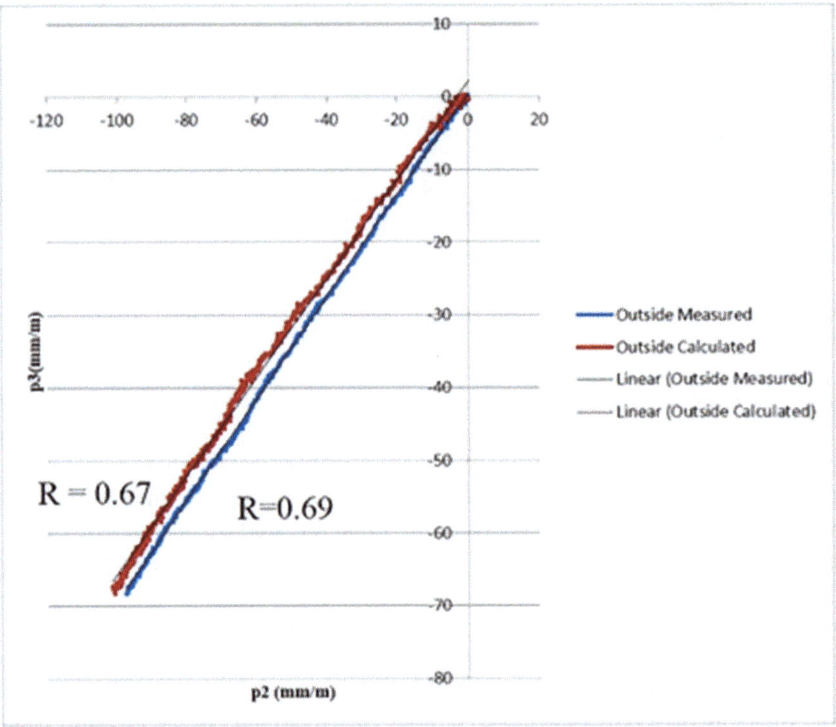

(b)

Fig. 22.4 (**a**) A comparison of measured and calculated R value inside necking band; (**b**) a comparison of measured and calculated R value outside necking band

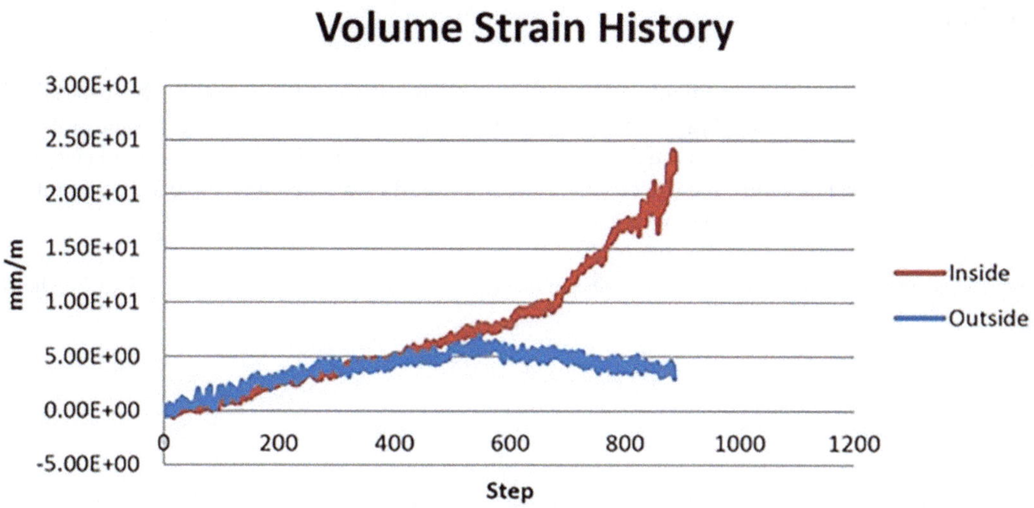

Fig. 22.5 A plot of volume strain history

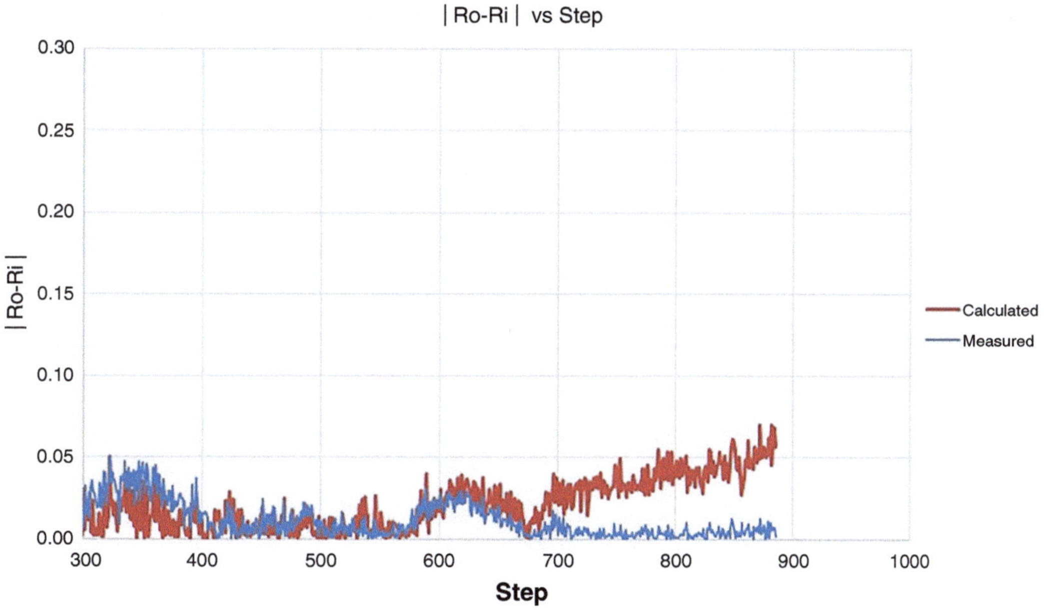

Fig. 22.6 A plot of $|R_o - R_i|$ vs Step

value and the calculated R value only have a difference at the location inside the necking band but don't have a significant difference outside the necking band, because non-conservation of plastic volume brings the calculated R value an extra error which make the calculated value slightly smaller than the measured value.

Figure 22.6 shows a plot of $|R_o - R_i|$ history, where R_o is the R value measured inside the necking band and R_i is the R value measured outside the necking band. The blue line in Fig. 22.6 shows the $|R_o - R_i|$ value get from the direct measurement method. The red line in Fig. 22.6 shows the $|R_o - R_i|$ value get from the calculated method. From the comparison of the two lines, it can be found, before step 680 (where volume conservation assumption valid inside and outside of the necking area, see Fig. 22.5), the calculated data and directly measured data overlaps well. However, after step 680 (where volume conservation assumption only valid outside of the necking area, see Fig. 22.5), the calculated data and directly measured data then have a significant difference of 0.05, which is the exactly difference shown in Fig. 22.4.

22.4 Conclusion

This article presents the first direct measurement of R value using DIC method for AA6000 sheet metal. The experiment shows that at the location outside the necking band (where the zero plastic volume change assumption does valid), the directly measured R value data meet well with the calculated data. However, inside the necking band (where the zero plastic volume change assumption does not valid), the directly measured data and calculated data have a significant difference which caused by the volume non-conservation of the material.

Acknowledgement The authors would like to express their sincere thanks to Dr. Changqing Du from FCA US LLC, who provided valuable criticisms and suggestions during this research.

References

1. Stickels, C.A., Mould, P.R.: The use of Young's modulus for predicting the plastic-strain ratio of low-carbon steel sheets. Metall. Mater. Trans. **1**(5), 1303–1312 (1970)
2. Pan, B., Qian, K., Xie, H., Asundi, A.: Two-dimensional digital image correlation for in-plane displacement and strain measurement: a review. Meas. Sci. Technol. **20**(6), 062001 (2009)
3. Chen, X., Xie, X., Sun, J., Yang, L.: Full field strain measurement of punch-stretch tests using digital image correlation. SAE Int. J. Mater. Manuf. **5**, 345–351 (2012)
4. Sutton, M.A.: Digital image correlation for shape and deformation measurements. In: Springer Handbook of Experimental Solid Mechanics, pp. 565–600. Springer, US (2008)
5. Chen, F., Chen, X., Xie, X., Feng, X., Yang, L.: Full-field 3D measurement using multi-camera digital image correlation system. Opt. Lasers Eng. **51**(9), 1044–1052 (2013)
6. Xie, X., Li, J., Sia, B., Bai, T., Siebert, T., Yang, L.: An experimental validation of volume conservation assumption for aluminum alloy sheet metal using digital image correlation method. J. Strain Anal. Eng. Des. **52**(1), 24–29 (2017)

Chapter 23
Failure Process of Recycled Concrete Aggregate Mortars Based on Digital Image Correlation

K.L. Apedo, S. Braymand, F. Hoerd, F. Feugeas, and C. Fond

Abstract This study addresses the failure process of mortars made of three different sands. Prismatic 40 × 40 × 160 mm^3 specimens were tested less than three points bending after 28 days of curing. Compressive tests were conducted on the two fragments of each mortar resulted from above flexural tests. The three different materials were: standard mortars performed with 0/2 mm natural rolled sand, mortars performed with 0/2 mm crushed sand and mortars performed with 0/2 mm fine recycled concrete aggregate (FRCA).The main aim of this experimental campaign was to study the influence on the mechanical properties of the substitution of natural sand (standardized or crushed gravel) with recycled sand when granular skeletons are identical. Crack propagation, displacement and strain fields, mechanical properties and other model parameters were obtained using 2D-DIC.The findings in these investigations could also be useful to understand the failure process and to improve the mechanical properties of mortars performed of FRCA by optimizing the granular skeleton.

Keywords Failure process • Uniaxial compression • Digital image correlation (DIC) • Crack propagation • Fine recycled concrete aggregate (FRCA)

23.1 Introduction

Fine recycled concrete aggregates (FRCA) arise mainly from demolition wastes. These aggregates, also known as recycled sands, are essentially composed of crushed natural aggregates and attached primary fine mortar [1]. Therefore they induce a significant water demand which makes them difficult to be recycled into mortars and concretes. Despite this difficulty, a larger part of these materials could be recycled into mortars and concretes to decrease the amounts of wastes which have to be disposed in landfill, and thus to preserve natural resources [1, 2]. It appears very important to understand the mechanical behaviour of mortars and concretes made of this kind of aggregates. In the literature, many studies have shown that the mechanical properties of mortars and concretes decreased as the replacement percentage of FRCA increased [3–6]. Some authors have demonstrated that it is possible to obtain similar properties when specific admixtures are used [1, 7]. When a lot of studies have made it possible to know the influence of the replacement percentage of FRCA on the mechanical properties of mortars and concretes, to our knowledge a very limited studies addressed the failure process (which characterizethe material's durability) of these materials and generally the traditional destructive testing techniques have been used. Nowadays, in the field of cement-based materials, several nondestructive testing techniques have been introduced according to their simplicity. Digital Image Correlation (DIC) is one of them, enabling to record a clear depiction of the surface displacement and strain fields and its transient changes according to stress redistribution which occurs after crack propagation. This study addresses the failure process of mortars made of three different sands namely natural rolled sand (NSR), natural crushed sand (NCR) and FRCA.

K.L. Apedo (✉) • S. Braymand • F. Hoerd • C. Fond
ICube, Université de Strasbourg, CNRS, 2 rue Boussingault, 67000 Strasbourg, France
e-mail: apedo@unistra.fr

F. Feugeas
ICube, INSA de Strasbourg, CNRS, 24 Bld de la Victoire, 67084 Strasbourg, France

M. Sutton, P.L. Reu (eds.), *International Digital Imaging Correlation Society*, Conference Proceedings of the Society for Experimental Mechanics Series, DOI 10.1007/978-3-319-51439-0_23

23.2 Materials and Methods

The cement used in the manufacture of the mortars was an Ordinary Portland Cement (CEM I 52.5N) in accordance with the standard NF EN 197-1. Three different mortar types were analyzed in the frame of this work, including standard mortars performed with 0/2 mm NRS, mortars performed with 0/2 mm NCS and mortars performed with 0/2 mm FRCA. The particle size distribution of fine aggregates was obtained as result of the sieve analysis and was fitted to obtain almost identical granular skeletons. The physical properties of the aggregates were obtained according to the standard NF EN 1097-6.The aim of the experimental campaign was to study the influence on the mechanical properties of the substitution of natural sand (standardized or crushed gravel) with recycled sand when granular skeletons are identical. Therefore, the mix proportioning of FRCA was not done considering a target compressive strength but maintaining almost the same mix proportioning of that of the standard mortars. The following mortar types were considered: mortars manufactured with NRS (NRSM) also known as Standard mortars, NCS mortars (NCSM) and FRCA mortars (FRCAM).

Following standard, the total water-to-cement ratio was equal to 0.5 for the standard mortar. The efficient water-to-cement ratio was identical then equal to 0.482. Due to different absorption of the three kinds of aggregates, the total water-to-cement ratio were 0.5, 0.5 and 0.68 respectively. A small quantity of superplasticizer (SP) MC-PowerFlow 3140 was used to improve the workability of the fresh mortars NCSM and FRCAM [8]. The mortars were molded in prismatic $40 \times 40 \times 160$ mm^3 casts and compacted using a vibrating table.

The specimens were demolded 24 h later and then put into water to insure 100 % of relative humidity under a temperature of 20 ± 1°C until the date of testing. The specimens were removed from water several hours before testing and a speckle pattern of black spots was applied on the specimen's surface after drying in order to facilitate the DIC post treatment using VIC-2D software.

Compressive and flexural strength tests were carried out according to EuropeanStandards NF-EN 196-1. Three-point flexural tests were performed using an Instron 3384 machine, with a capacity of 150 kN. Compressive strength tests were done on the two fragments of each mortar resulted from above flexural tests, using the same machine. The flexural strengths were determined by taking the average of three test results whereas the compressive strengths were determined by taking the average of six test results.

During the compression loading process, the whole front surface of each specimen was captured with a camera of the DIC system. The image size was 81.5×54.1 mm, which was larger than the specimen's front surface (i.e., 80×40 mm and an area of interest of 40×40 mm). A digital camera with a resolution of 4288×2848 pixels (256 gray levels) and an advanced image grabber was used. The image analysis software VIC-2D was then used to calculate the displacements and strain distribution over the specimen's front surface. All the Specimens under uniaxial compression could be regarded as a state of plane stress, so the two-dimensional DIC technique was capable of examining graphically the cracking propagation on the specimen's front surface [9]. Two white-light lamps were used as the light source.

23.3 Results

Mechanical properties are given in Table 23.1. Even if compressive strength test results show that the FRCAM strength is 6.31% lower than the reference NRSM specimen, this discrepancy can be neglected considering the standard deviations. However, NRSM and NCSM present almost the same strengths. Tensile strength test results show that the FRCAM strength is 39.05% lower than the reference NRSM and NCSM is 26.86% greater than the reference. In term of the elastic moduli, FRCAM presents poor value compared with NRSM (51.05% lower) and is then more deformable compared with the reference. The mortar cracks related to different loading stages and the final failure pattern for all the specimens are shown in Figs. 23.1, 23.2 and 23.3. During the tests of all the specimens, it was found that the crack developing process and the failure pattern for FRCAM were almost similar to those for NRSM and NCSM, even if the strain fields were different from

Table 23.1 Mechanical properties and model parameters of mortars after 28 days of curing

Mortar	Strengths			Results obtained by DIC			
	f_t (MPa)	f_c (MPa)	f_c/f_t	E (GPa)	ν	$\varepsilon_{d0}\times10^4$	$\varepsilon_{max}\times10^3$
NRSM	5.25 ± 0.49	55.08 ± 2.55	10.49	30.85	0.23	1.70	1.00
NCSM	6.66 ± 0.49	55.50 ± 2.30	8.33	32.47	0.22	2.11	1.40
FRCAM	3.20 ± 0.49	51.60 ± 1.52	16.13	15.10	0.23	2.28	1.69

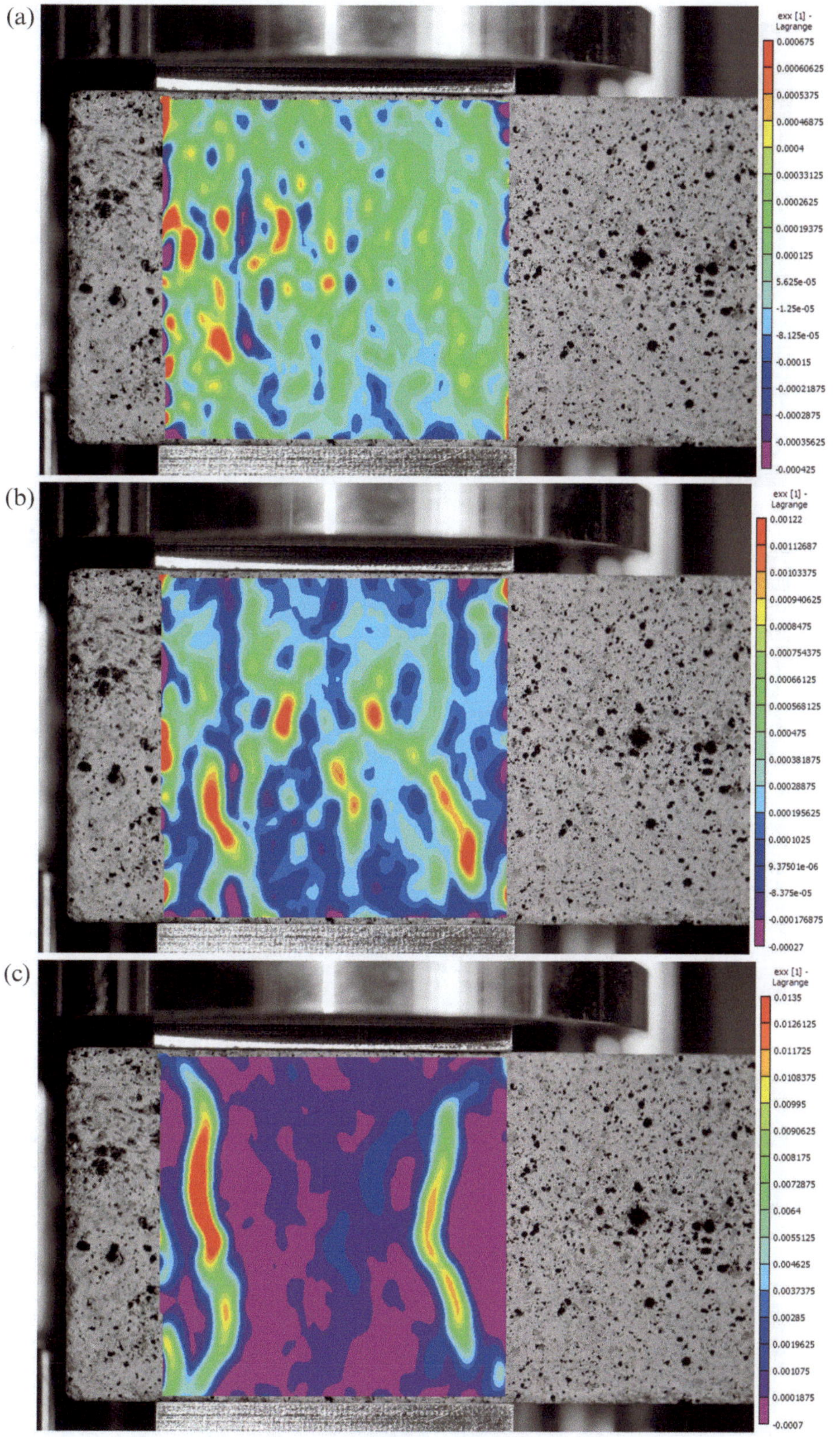

Fig. 23.1 Lateral Green-Lagrange strain contour maps related to different loading stages for NRSM under uniaxial compression. (**a**) 50% of peak load. (**b**) 75% of peak load. (**c**) 100% of peak load

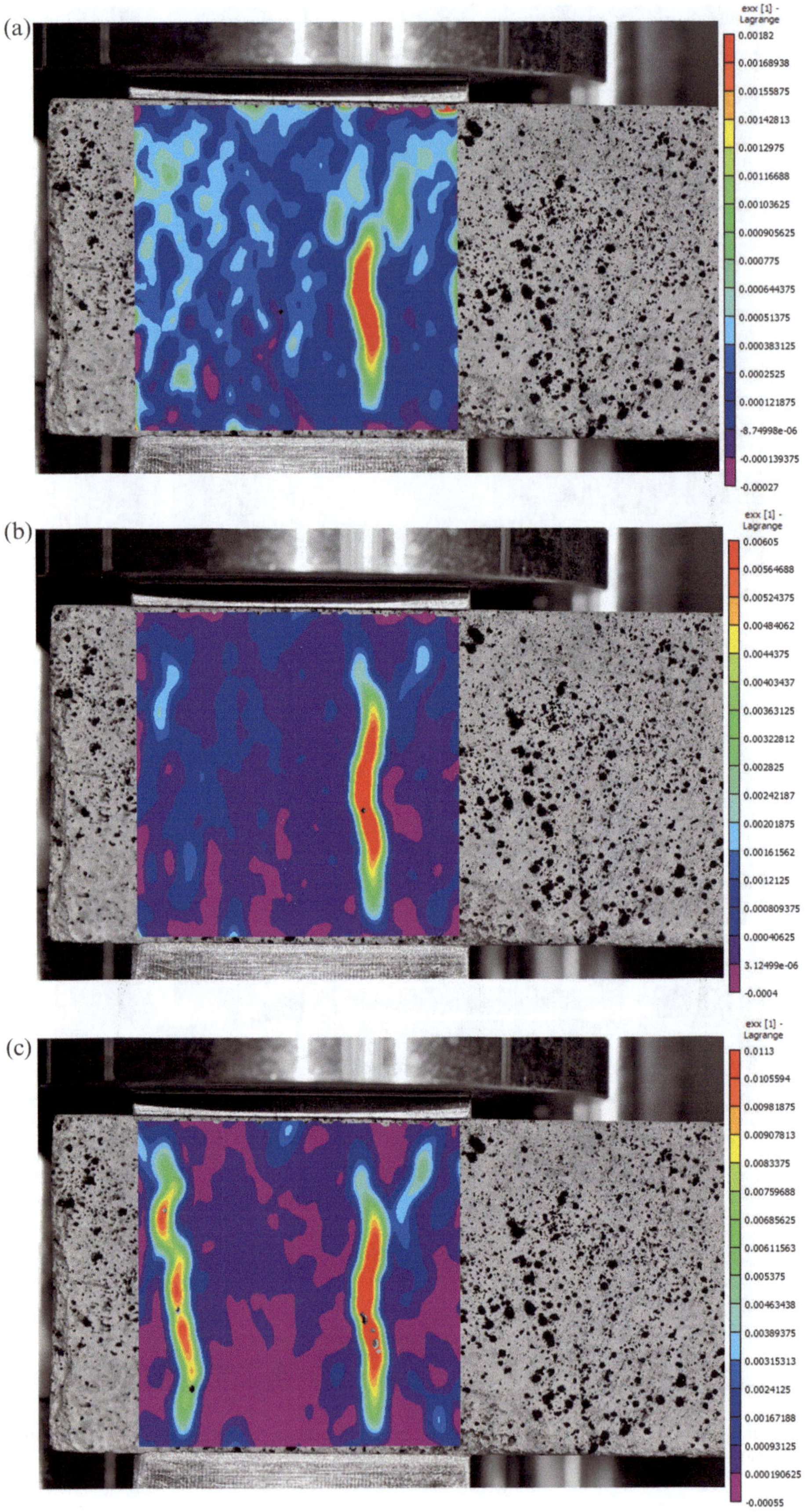

Fig. 23.2 Lateral Green-Lagrange strain contour maps related to different loading stages for NCSM under uniaxial compression. (**a**) 50% of peak load. (**b**) 75% of peak load. (**c**) 100% of peak load

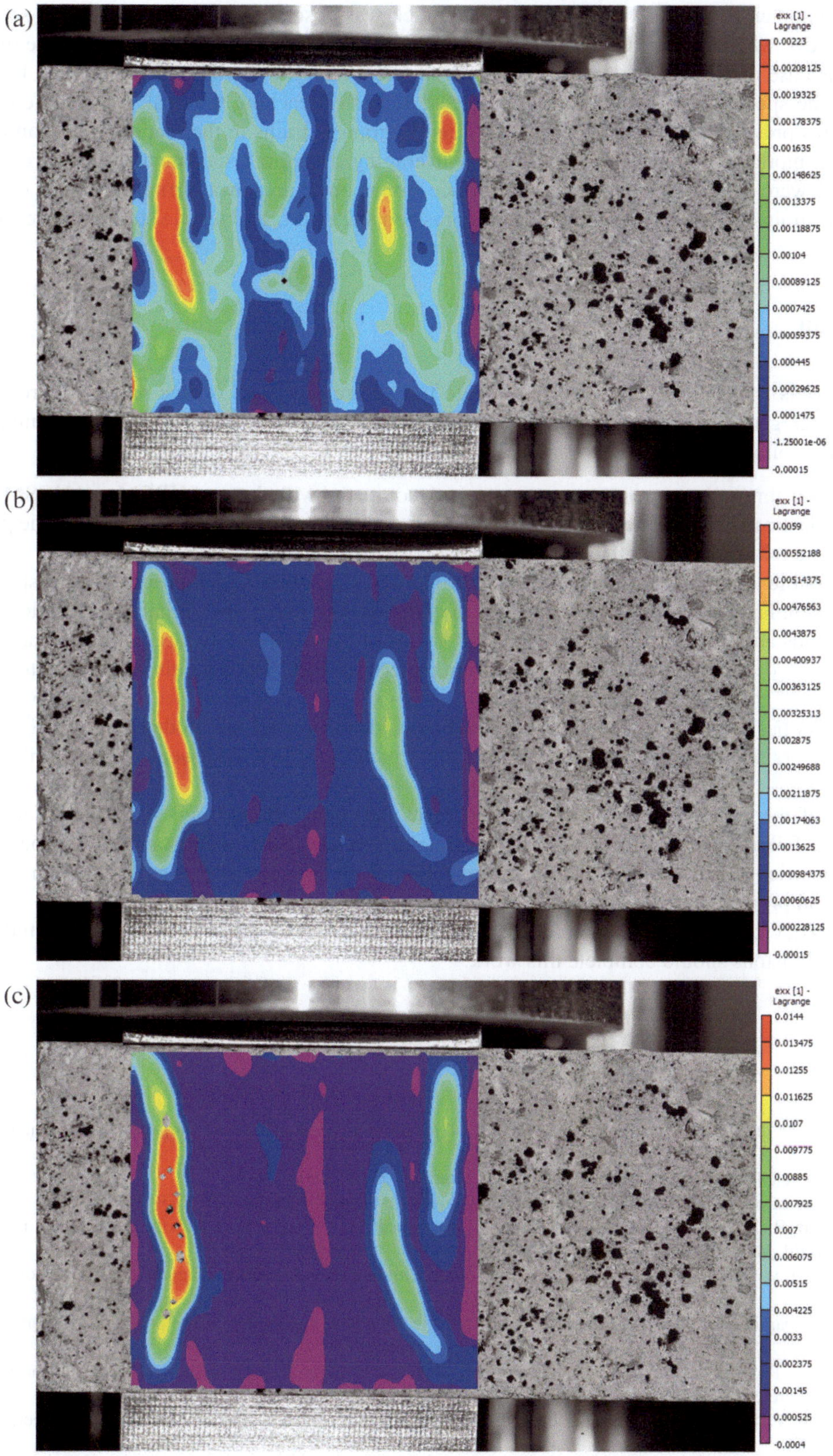

Fig. 23.3 Lateral Green-Lagrange strain contour maps related to different loading stages for FRCAM under uniaxial compression. (**a**) 50% of peak load. (**b**) 75% of peak load. (**c**) 100% of peak load

a type of mortar to another. In the early loading stage, no crack was observed on the surface of the specimens. In the case of NCSM and FRCAM, when the applied load reached 50% of the peak load, cracks occurred on the mid-height surface of the specimen. In the case of NRSM, the first crack appeared later. With the increase of the applied load, the initial cracks propagated toward the specimen's top and bottom surfaces; and after reaching the peak load, some discontinuous cracks coalesced. The strains presented in Table 23.1 are the Green-Lagrange strains in the loading direction (vertical direction in this study). For each mortar type, they were obtained by averaging over the 3 specimens tested in compression. For each specimen, the strains were obtained by averaging the surface strains on an area away from the loaded edges. ε_{max} and ε_{d0} are the Green-Lagrange strains which correspond to the compression strength f_c and tensile strength f_t, respectively.

23.4 Conclusion

Cement-based materials were manufactured to investigate the failure processes of fine natural aggregate mortars and fine recycled concrete aggregates mortars under uniaxial compression when using almost the same granular skeletons. In the light of this study, the following conclusions can be drawn:

(1) FRCAM presented almost the same compressive strength compared with those of NRSM and NCSM. FRCAM presented however lower elastic moduli, and tensile strength compared with those of NRSM and NCSM. The peak strain of NCSM and FRCAM were relatively higher than that of NRSM.
(2) DIC made it possible to identify certain mechanical properties of mortars. Crack propagation, displacements and strains were also provided. From a mortar type to another, in compression, DIC showed that the evolution of the displacement and the strain fields are different for almost the same values of the compressive strengths. DIC also make it possible to know that the crack developing process and the failure pattern for FRCAM were almost similar to those for NRSM and NCSM.
(3) The global cracks basically developed along the loading direction for all the modeled cement-based materials. Under compression, the crack developing process and the failure pattern for FRCAM were almost similar to those for NRSM and NCSM.
(4) Understanding of the differences in the failure processes between FRCAM and standardized mortars could provide insights into the failure mechanism of recycled aggregate mortars. Such understanding can be used to improve the mechanical properties of FRCAM through the optimization of the granular skeleton. The influence of granular skeleton on the mechanical properties of mortars could also be considered.
(5) Following these first results, no significant failure processes differences were observed. A depth study of the strain fields, especially their divisions should complete this work.

References

1. Deodonne K., Studies of the physicochemical characteristics of recycled aggregate concrete and their environmental impact, PhD Thesis, Strasbourg University, 2015
2. Oikonomou, N.D.: Recycled concrete aggregates. Cem. Concr. Compos. **27**, 315–318 (2005)
3. Khatib, J.M.: Properties of concrete incorporating fine recycled aggregate. Cem. Concr. Res. **35**, 763–769 (2005)
4. Shi-Cong, K., Chi-Sun, P.: Properties of concrete prepared with crushed fine stone, furnace bottom ash and fine recycled aggregate as fine aggregates. Constr. Build. Mater. **23**, 2877–2886 (2009)
5. Evangelista, L., de Brito, J.: Mechanical behaviour of concrete made with fine recycled concrete aggregates. Cem. Concr. Compos. **29**, 397–401 (2007)
6. Zhao, Z., Remond, S., Damidot, D., Xu, W.: Influence of fine recycled concrete aggregates on the propertiesof mortars. Constr. Build. Mater. **81**, 179–186 (2015)
7. de Brito, J., Ferreira, J., Pacheco, J., Soares, D., Guerreiro, M.: Structural, material, mechanical and durability properties and behaviour of recycled aggregates concrete. J. Build. Eng. **6**, 1–16 (2016)
8. Braymand, S., François, P., Feugeas, F., Fond, C.: Rheological properties of recycled aggregate concrete using superplasticizers. J. Civil Eng. Arch. **9**, 591–597 (2015)
9. Shen, B., Paulino, G.H.: Direct extraction of cohesive fracture properties from digital image correlation: hybrid inverse technique. Exp. Mech. **51**, 143–163 (2011)

Chapter 24
Identification of a 3D Anisotropic Yield Surface Using a Multi-DIC Setup

K. Denys, S. Coppieters, and D. Debruyne

Abstract Two synchronised stereo-DIC setups, referred to as a multi-DIC setup, have been used to acquire the displacement and strain fields over the front and thickness surface of a 10 mm thick high strength steel (S690QL). In a first step, using Finite Element Model Updating, the strain hardening behaviour has been identified beyond the point of maximum uniform elongation up to strains of 70%. As a result of the multi-DIC setup, full-field surface strain information is available in all three material directions: the lateral, longitudinal and through thickness direction. Consequently, a 3D anisotropic yield surface could be identified. Since the parameters of a yield surface are only identifiable if the sensitivity to these parameters is substantial, a new type of specimen has been designed having a perforation over the front and thickness surface. This double perforated specimen has then successfully been used to identify a 3D anisotropic yield surface.

Keywords Multi-DIC • FEMU • Thick HSS • S690QL • 3D anisotropy

24.1 Introduction

The use of thick High Strength Steels (HSS) enables a lighter and slender structure without loss in strength. Consequently, the use of HSS has increased noticeably in the recent years in the construction industry. Unfortunately, the increase of the yield strength does not coincide with an increase in ductility; hence brittle fracture and delamination can occur in HSS structures [1]. Moreover, the increased yield strength of HSS exhibits a proportionally lower fatigue resistance [2]. Therefore a profound understanding of the plastic material behaviour up to fracture, including the strain hardening and anisotropy, is mandatory [3, 4].

The material under investigation is 10 mm thick S690QL. Tests on this HSS have shown that true plastic strains in excess of 1 can be reached while it only has a maximum uniform strain of 0.05. Usually, the strain hardening behaviour is extrapolated beyond the point of maximum uniform elongation, however it is well known that this can lead to very different and unsafe results [5]. Therefore, an inverse method, in this work Finite Element Model Updating (FEMU) [6, 7], has been used to identify the strain hardening behaviour up to fracture. The FEMU procedure minimises the difference between the computed and experimentally measured forces and surface strains while updating the material parameters in an iterative procedure.

To measure the surface strains used in the FEMU approach, Digital Image Correlation (DIC) has been employed. Moreover, multiple synchronised stereo-DIC setups, denoted as multi-DIC, have been used to acquire the surface strains over the front and thickness surface. Consequently, surface information is available in all three material directions, making a 3D anisotropic yield criterion identifiable, whereof a 3D Hill 48 yield criterion [8] has been considered and identified in this work.

K. Denys (✉) • S. Coppieters • D. Debruyne
Department of Materials Engineering, KU Leuven, Campus Gent, Gebroeders De Smetstraat 1, 9000 Gent, Belgium
e-mail: kristof.denys@kuleuven.be

M. Sutton, P.L. Reu (eds.), *International Digital Imaging Correlation Society*, Conference Proceedings of the Society for Experimental Mechanics Series, DOI 10.1007/978-3-319-51439-0_24

24.2 Methodology

Since full-field surface information is available in the lateral, longitudinal and through thickness material direction, the 3D Hill 48 yield surface is identifiable:

$$\sigma_{eq}^2 = F(\sigma_{22} - \sigma_{33})^2 + G(\sigma_{33} - \sigma_{11})^2 + H(\sigma_{11} - \sigma_{22})^2 + 2L\sigma_{23}^2 + 2M\sigma_{31}^2 + 2N\sigma_{12}^2 \tag{24.1}$$

where 1 is the rolling direction (RD), 2 the transverse direction (TD) and 3 the through thickness direction. From Eq. (24.1) it can be inferred that the parameter L can be identified along the TD, and parameter M along the RD, respectively. Using a specimen along the 45 degree (45D) direction, both parameters can be identified simultaneously. Because it is assumed that the sum of G and H equals 1 [7], the parameter G should not be included in the identification procedure and can easily be calculated afterwards.

Further, the parameters L, M and N are most accurate when identified on a heterogeneous deforming zone. Therefore, a double perforated (DP) specimen, see Fig. 24.1, where a perforation is present at the front and thickness surface, has been machined and employed to identify the Hill 48 yield criterion.

24.3 Experiments

24.3.1 *Strain Hardening Behaviour*

First, the strain hardening behaviour has been identified on a DP specimen in the RD using FEMU. A tensile test has been conducted with a constant crosshead speed of 1 mm/min on a tensile equipment with a load capacity of 250 kN equipped with wedge grips. During the tensile test, the surface displacements and strains over the front and thickness surfaces (shown in Fig. 24.1) have been acquired using two synchronised stereo-DIC setups. The obtained displacements near the grip zones of the wedges have been applied in a displacement-driven FE simulation used in the FEMU procedure to identify a p-model [5]. A p-model has been identified as phenomenological strain hardening law because it was most adequate to follow the real strain hardening behaviour for the investigated HSS grade [6]. The identified strain hardening behaviour can be seen

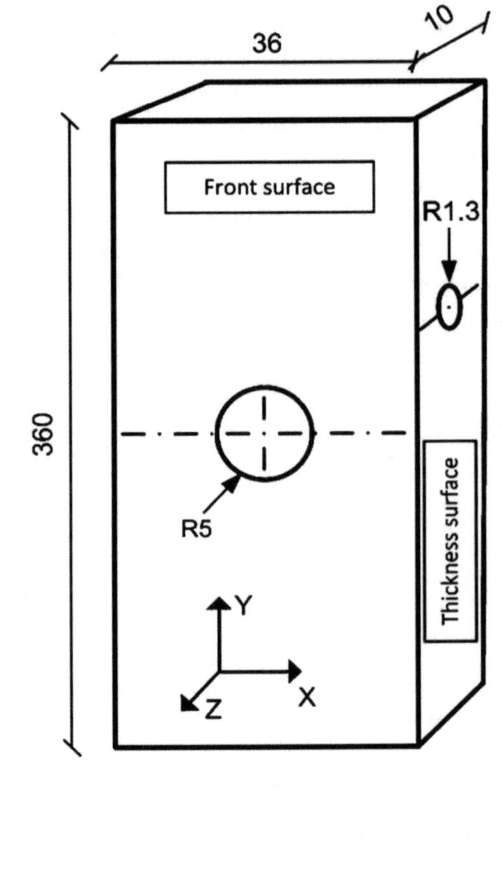

Fig. 24.1 Double perforated specimen

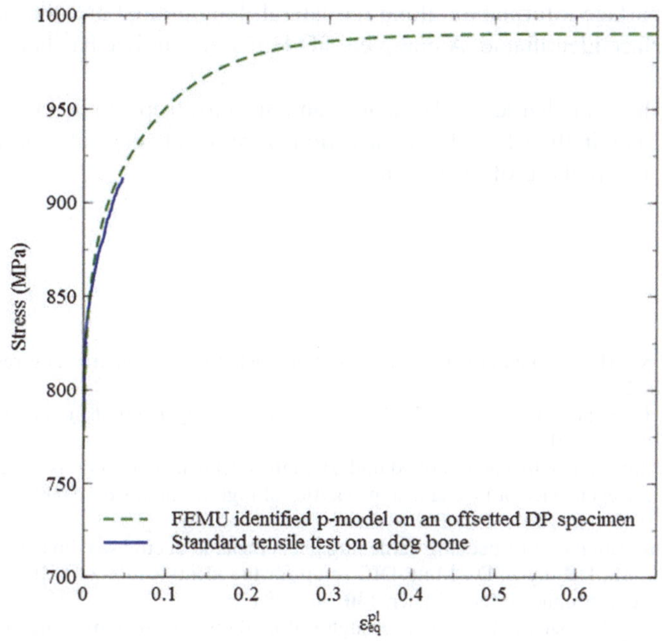

Fig. 24.2 Strain hardening behaviour of S690QL

Table 24.1 Identified Hill 48 yield criterion

Direction	F	G	H	N	M	L
RD	0.4817	0.4996	0.5004	1.5197	1.6040	–
TD	0.4927	0.4744	0.5256	1.4025	–	1.4454
45D	0.4791	0.4700	0.5300	1.4115	1.8772	1.2585

in Fig. 24.2 as the dashed green line, while the solid blue line represents the strain hardening behaviour obtained from a standard tensile test. A good agreement can be found in the pre-necking region. Further, it can be seen that an extrapolation of the standard tensile test will lead to a different result than the identified p-model and would therefore be unsafe.

24.3.2 3D Anisotropic Yield Surface

The next step is to identify a 3D anisotropic yield criterion, in this work the Hill 48 yield criterion (see Eq. (24.1)). Because no convergence could be reached when identifying all 5 parameters simultaneously starting from their isotropic values [6], the parameters F, H and N have first been identified using the surface strain field over the front surface. Meanwhile, the parameters L and M are kept constant at their isotropic value of 1.5. The identified parameters F, H and N have then been used as initial parameters to identify all 5 Hill 48 parameters simultaneously together with the strain field over the thickness surface in the case of a specimen along the 45D direction. For a specimen along the RD, the parameters F, H, N and M have been identified and F, H, N and L for a specimen along the TD. The results can be found in Table 24.1. The difference between the identified parameters can be explained by the difference in identified parameters.

24.4 Conclusion

A multi-DIC setup has been employed to acquire measurements over multiple surfaces of a specimen to identify more parameters simultaneously. In this work the strain hardening behaviour has been identified using FEMU on a DP specimen up to strains of 0.7. Because multi-DIC has been used, measurements were available over the front and thickness surface, resulting in available full-field surface information along the lateral, longitudinal and through thickness material direction, making the 3D Hill 48 yield surface identifiable. A complete 3D Hill yield surface has been identified on a DP specimen in the 45 degree direction.

There can be concluded that the strain hardening behaviour and 3D anisotropic yield criterion can be identified using only 2 specimens, namely a DP specimen in the RD and 45D direction. That results in a substantial reduction of the experimental work compared to the conventional methods of measurement.

References

1. Guo, W., Dong, H., Lu, M., Zao, X.: The coupled effects of thickness and delamination on cracking resistance of X70 pipeline steel. Int. J. Press. Vessel. Pip. **79**, 403–412 (2002)
2. de Jesus, A., Matos, R., Fontoura, B., Rebelo, C., de Silva, L., Veljkovic, M.: A comparison of the fatigue behaviour between S355 and S690 steel grades. J. Constr. Steel Res. **79**, 140–150 (2012)
3. Hmlinen, O.P., Bjirk, T.: Fretting fatigue phenomenon in bolted high strength steel plate connections. Steel Constr. **3**, 174–178 (2015)
4. Qiang, X., Bijlaard, F., Kolstein, H.: Dependence of mechanical properties of high strength steel S690 on elevated temperatures. Constr. Build. Mater. **30**, 73–79 (2012)
5. Coppieters, S., Kuwabara, T.: Identification of post-necking hardening phenomena in ductile steel metal. Exp. Mech. **54**, 1355–1371 (2014)
6. Denys, K., Coppieters, S., Seefeldt, M., Debruyne, D.: Multi-DIC setup for the identification of a 3D anisotropic yield surface of thick high strength steel using a double perforated specimen. Mech. Mater. **100**, 96–108 (2016)
7. Cooreman, S.: Identification of the plastic material behaviour through full-field displacement measurements and inverse methods. PhD thesis, University of Brussels (2008)
8. Hill, R.: A theory of the yield and plastic flow of anisotropic metals. Proc. R. Soc. Lond. A: Math. Phys. Eng. Sci. **193**, 281–297 (1948)

Chapter 25
Inertial Impact Method for the Through-Thickness Strength of Composites

Lloyd Fletcher and Fabrice Pierron

Abstract This work presents the design of a novel inertial impact test for the determination of the through-thickness tensile strength of composite materials at high strain rates. The principle behind this new test method is as follows: A rectangular plate specimen is impacted by a projectile that imparts a short compressive stress wave. This compressive wave will then travel through the specimen and reflect off of the free end becoming a tensile wave. The strength of a laminate composite is significantly higher in compression than in tension thus, the test parameters (e.g. specimen or projectile dimensions, projectile speed) can be selected such that the reflected tensile wave will cause failure in the specimen without exceeding the compressive strength. Full-field measurements can then be taken throughout the test and stress-strain curves can be reconstructed using the Virtual Fields Method. The stress-strain curves can then be used to determine the tensile strength of the composite. A preliminary design of test has been optimised using explicit dynamics simulations. Future work will include experimental validation of the test principle.

Keywords High strain rate • Through-thickness • Tensile strength • Full-field measurement • Composite

25.1 Introduction

The increasing use of composites in industry has led to the design of structural components with thick cross sections that deviate significantly from standard thin composite plate theory (i.e. composite components that are subjected to complex three dimensional stress states). Despite the use of sections with increasingly thick cross sections, the through-thickness tensile strength for fibre composites is not well understood, especially for high strain rates. This is shown by the significant variation in the reported through-thickness tensile strength of fibre composites [1]. There are a multitude of reasons for the observed scatter for the through-thickness tensile strength of laminate composites, several of them originating from the difficulties in implementing a through-thickness material test for laminate composite materials.

In practice it is difficult to test the through-thickness tensile strength of laminate composites due to constraints on specimen size, specimen alignment and gripping the specimen. Most of the available experimental data for the high strain rate through-thickness strength of composites has been obtained using the Split Hopkinson Pressure Bar (SHPB) technique [2, 3]. Reported experimental data using the SHPB technique has significant scatter in the through-thickness strength values for laminate composites. Part of the reason for this stems from the assumptions made in processing the SHPB data, with any deviation from the assumptions of the SHPB method leading to unquantifiable sources of error. For example: the SHPB technique relies on the assumption of quasi-static equilibrium to neglect inertial effects; that is, the stress wave within the specimen needs to have adequate time to completely damp out before this assumption becomes valid. For relatively soft materials with low wave speeds (e.g. the through-thickness direction for a laminate composite) this assumption can be extremely difficult to satisfy as inertial effects become significant. Therefore, the overall aim of this work is to develop new inertial impact test methods that address the limitations of existing high strain rate testing methodologies. In order to do this ultra-high speed imaging technologies will be combined with inverse identification techniques.

L. Fletcher (✉) • F. Pierron
Engineering Materials Research Group, Faculty of Engineering and the Environment, University of Southampton, Southampton, SO17 1BJ, UK
e-mail: L.C.Fletcher@soton.ac.uk

M. Sutton, P.L. Reu (eds.), *International Digital Imaging Correlation Society*, Conference Proceedings of the Society for Experimental Mechanics Series, DOI 10.1007/978-3-319-51439-0_25

25.2 Formulation of the Test Principle

The basic principle behind the inertial impact test presented here follows from a combination of the ideas presented in [4, 5]. The general method is as follows: (1) impose a short compressive pressure pulse on one edge of a rectangular specimen by impacting it with a projectile; (2) the compressive stress wave will then travel through the specimen to the opposing free edge of the specimen; (3) the compressive wave reflects off of the free edge of the specimen and becomes a tensile wave travelling back to the struck edge of the specimen. As long as the compressive strength of the material is higher than the strength in tension the input pulse can be tuned (by selecting projectile material, velocity and geometry) to not exceed the compressive strength while the reflected tensile wave causes failure in the specimen.

In order to capture displacement and strain data during the inertial impact test ultra-high speed imaging is used in conjunction with a full-field measurement technique such as the grid method. The displacement and strain data are then processed using the Virtual Fields Method (VFM) to reconstruct the stress-strain curves and identify the strength of the material. For example: the application of a simple rigid body virtual field leads to the following equation for the reconstruction of the average axial stress in the specimen 'σ_{xx}' at an arbitrary position 'x' from the free end at time 't':

$$\overline{\sigma_{xx}(x,t)} = \rho x \overline{a(x,t)}. \tag{25.1}$$

where 'ρ' is the material density and '$\overline{a(x,t)}$' is the average acceleration over the specimen surface from the free edge to the point 'x'. The acceleration map for the specimen can be determined by taking the second temporal derivative of the displacement map.

The proposed experimental setup is shown in Fig. 25.1. The specimen length 'L_s' is set to 20 mm (representative of a thick composite section). In order to maximise the spatial resolution of the displacement measurements (camera: Shimadzu HPV-X) the specimen aspect ratio is set to be the same as the camera sensor of 1.6:1 giving a specimen height 'H_s' and projectile height 'H_p' of 12.5 mm. The parameters that can be selected in order to 'tune' the input compressive pulse include: the length of the projectile 'L_p', the projectile velocity 'V_p' and the material properties of the projectile.

25.3 Finite Element Design Optimisation

The initial design of the inertial impact test was carried out using explicit dynamics simulations in ANSYS LS-DYNA (v 16.2). All simulations used PLANE162 elements (2D, 4-node, reduced integration, plane stress assumption). A parametric sweep of beta damping coefficient, time step safety factor and element size was conducted. These parameters were then chosen to minimise the error in the stresses reconstructed from the acceleration using Eq. (25.1). The optimal simulation parameters are as follows: beta damping coefficient of 0.1; time step safety factor of 0.5; and element edge size of 0.125 mm.

The material properties for the specimen were set to be representative of a unidirectional carbon fibre laminate: $\rho = 1600$ kg/m^3, $E_x = 10$ GPa, $E_y = 135$ GPa, $G_{xy} = 4.5$ GPa, $\nu = 0.022$. Two different projectile materials were analysed: Aluminium, $\rho = 2700$ kg/m^3, $E_x = 70$ GPa, $\nu = 0.32$; and glass fibre rod $\rho = 1900$ kg/m^3, $E_x = 30$ GPa, $\nu = 0.25$. A parametric sweep of the projectile length 'l_p' and the projectile velocity 'v_p' was conducted in order to maximise the tensile stress while minimising the compressive stress experienced by the specimen during the first wave reflection. Figure 25.2 shows the results of the parametric sweep. From available experimental data the target tensile stress of the parametric sweep was set to 75 MPa (Note that the compressive strength of a typical carbon fibre is 300 MPa in the through-thickness direction).

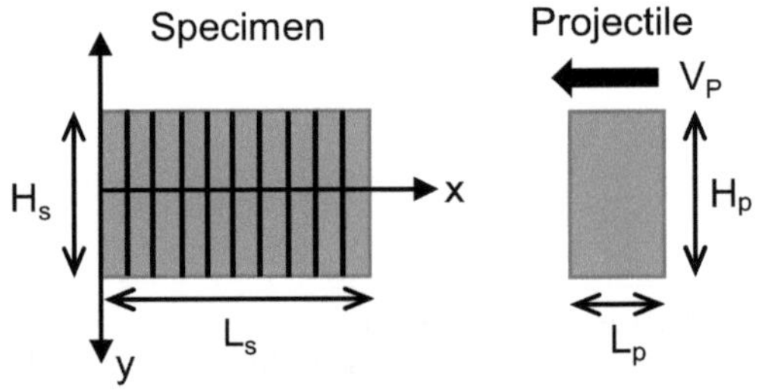

Fig. 25.1 Schematic of the proposed inertial impact test, the fibres in the specimen are aligned parallel to the y axis

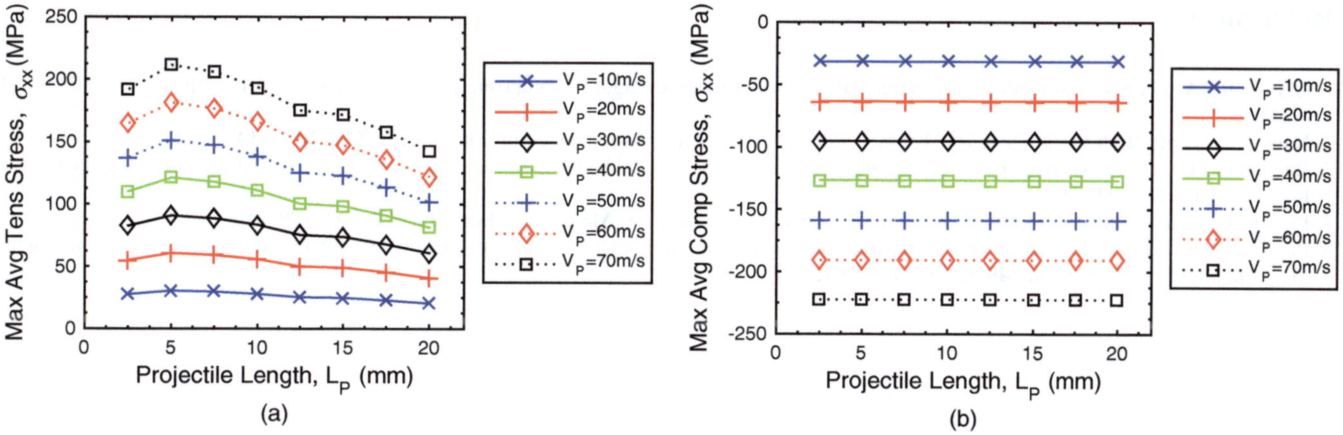

Fig. 25.2 Maximum average tensile stress (**a**) and compressive stress (**b**) plotted as a function of projectile length and velocity, projectile material is aluminium

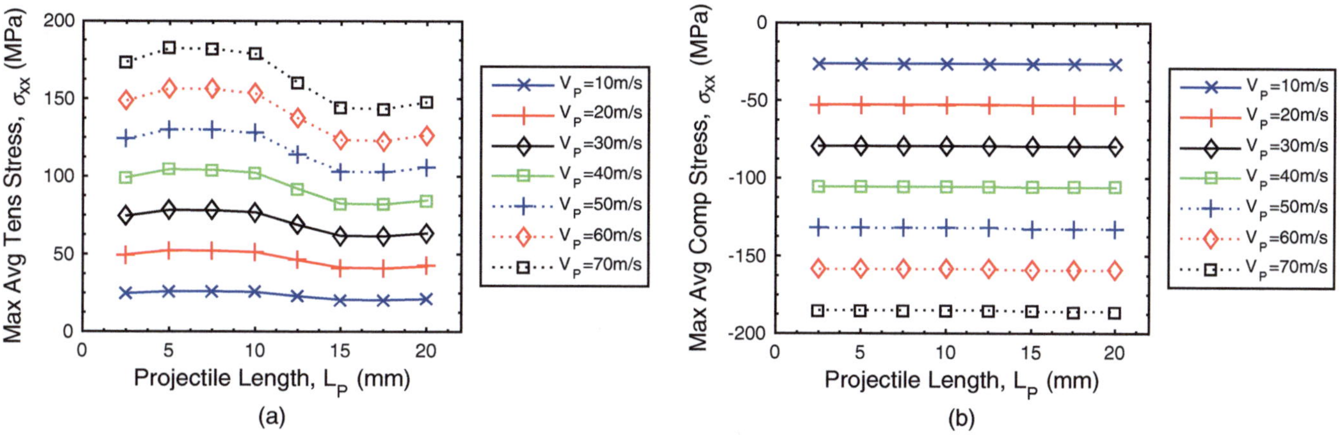

Fig. 25.3 Maximum average tensile stress (**a**) and compressive stress (**b**) plotted as a function of projectile length and velocity, projectile material is glass fibre rod

Based on the results shown in Figs. 25.2 and 25.3 the projectile material was chosen to be glass fibre rod with a length 'L_P' of 5 mm and a velocity 'V_P' of 30 m/s. This configuration gives a maximum input compressive stress of 79.2 MPa and a reflected tensile stress of 78.3 MPa. Note that glass fibre projectile was selected over the aluminium due to the lower input compressive stress relative to the output tensile stress.

25.4 Summary and Future Work

This work proposes a new inertial impact test for the through-thickness tensile strength of laminate composites. The proposed test uses a simple impact test to impart a compressive pulse on the specimen. The compressive stress wave then reflects off the free end of the specimen becoming a tensile wave, causing failure of the specimen. Data acquisition for the test is performed using full-field measurements with subsequent data processing performed using the VFM to identify the strength of the material. The preliminary results show the optimisation of the test configuration using finite element simulations. Based on the finite element modelling, the test configuration required to cause tensile yield in the specimen requires a projectile made from glass fibre rod with a length of 5 mm and impact velocity of 30 m/s. Future work will include the experimental implementation and validation of the proposed test method.

References

1. Olsson, R.: A survey of test methods for multiaxial and out-of-plane strength of composite laminates. Compos. Sci. Technol. **71**(6), 773–783 (2011)
2. Naik, N.K., et al.: High strain rate tensile behaviour of woven fabric E-glass/epoxy composite. Polym. Test. **29**(1), 14–22 (2010)
3. Ravikumar, G., et al.: Analytical and experimental studies on the mechanical behaviour of composites under high strain rate compressive loading. Mater. Des. **44**, 246–255 (2013)
4. Pierron, F., et al.: Beyond Hopkinson's Bar. Philos. Trans. R. Soc. Lond. A: Math. Phys. Eng. Sci. **372**, 2023 (2014)
5. Pierron, F., Forquin, P.: Ultra-high-speed full-field deformation measurements on concrete spalling specimens and stiffness identification with the virtual fields method. Strain. **48**(5), 388–405 (2012)

Chapter 26
Assessing the Metrological Performance of DIC Applied on RGB Images

Benoît Blaysat, Michel Grédiac, and Frédéric Sur

Abstract Thanks to the rise of digital reflex cameras, RGB CMOS sensors present nowadays interesting features, at relatively low prices. For instance, the Canon 5DSR sensor displays 50 million pixels. Nevertheless, such cameras being dedicated to photography and not to measurement, their high pixel density makes the associated sensor noise to be relatively high compared to the sensor noise of measurement cameras. Moreover, a Bayer filter mosaic is added in front of the pixels of such sensors in order to produce RGB images by interpolation. Embedded camera processors demosaice these data to elaborate full RGB pictures. Additional post-processing steps, including white balance and gamma correction, are also used in order to prevent any color shift and to give a pleasing-to-the-eye output.

These post-processing steps considerably transform the raw sensor data to produce the output images. The fundamental assumption of DIC, namely the conservation of brightness for a given material point, may thus be violated.

In this context, we propose in this study a consistent use of the separate channels of RGB sensors for the DIC minimization. The metrological performance of the DIC measurement using this approach is discussed and compared to that dealing with "classic" gray level images.

Keywords Digital image correlation • Full field measurement • Bayer filter • Metrological performance

26.1 Introduction

DIC is nowadays used in almost all experimental mechanical laboratories. It allows to recover the full displacement field of an imaged specimen surface; details are given in [1]. The main principle consists at back deforming the image (g) of the current state in order to match the image of the reference state (f). This process is driven by the minimization of the brightness difference over the matching function space U defined on the pixels that are contained in the Region of Interest (*RoI*):

$$u = \text{Argmin}_{u \in U} \left\{ \sum_{x_p \in RoI} \left(f\left(x_p\right) - g\left(x_p + u\left(x_p\right)\right)\right)^2 \right\} \tag{26.1}$$

This problem is now well-known, and a topical issue is to associate any measurement with its metrological properties, such as the measurement resolution, the spatial resolution or the bias, see [1–4].

Nevertheless, RGB camera sensors have been often used over the last few years thanks to their low cost but also to their high pixel density, see [5, 6], amongst others. Nevertheless, such sensors give data on three channels which are spatially shifted, as illustrated in Fig. 26.1. A regular speckle used to measure displacements is also shown in Fig. 26.2. The pattern corresponding to the Bayer filter is almost no noticeable on the large size image (see Fig. 26.2a), which is not the case any more in the close-up shown in Fig. 26.2b.

B. Blaysat (✉) • M. Grédiac
Clermont Université, Université Clermont Auvergne, Institut Pascal, UMR CNRS 6602, BP 10448, 63000 Clermont-Ferrand, France
e-mail: benoit.blaysat@univ-bpclermont.fr

F. Sur
Laboratoire Lorrain de Recherche en Informatique et ses Applications, UMR CNRS 7503, Université de Lorraine, CNRS, INRIA projet Magrit, Campus Scientifique, BP 239, 54506 Vandœuvre-lès-Nancy Cedex, France

M. Sutton, P.L. Reu (eds.), *International Digital Imaging Correlation Society*, Conference Proceedings of the Society for Experimental Mechanics Series, DOI 10.1007/978-3-319-51439-0_26

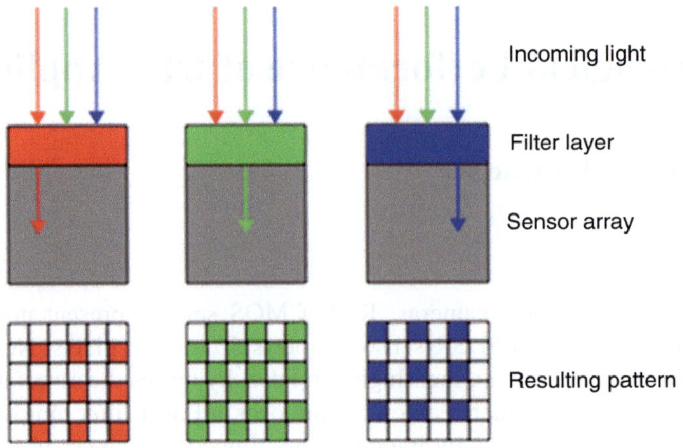

Fig. 26.1 Bayer arrangement of color filters on the pixel array of an image sensor [1]

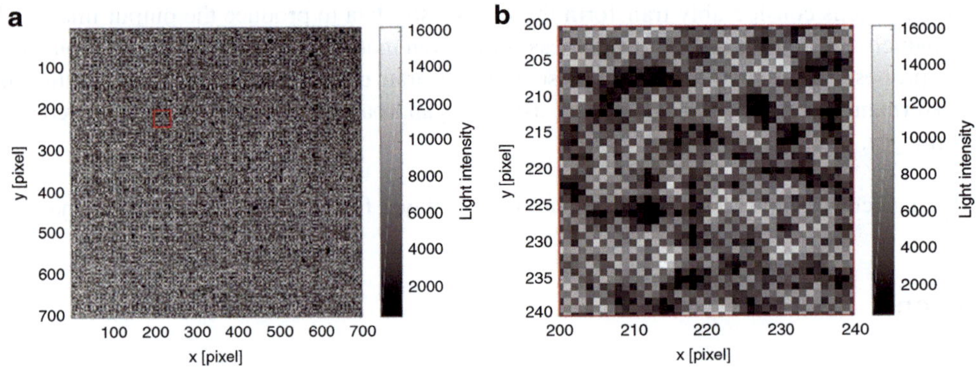

Fig. 26.2 Illustration of an imaged speckle with a RGB camera. (**a**) Large size image. (**b**) Close-up (*red box* of **a**). The Bayer filter pattern is clearly observable

Since DIC relies on the minimization of the brightness difference, cf. Eq. (26.1), a gray level image needs to be built from the RGB one. This requires two steps: data are first demosaiced in order to have a full RGB picture; then a post processing is necessary to elaborate a gray level picture.

We propose here to revisit the brightness conservation in order to avoid all image pre-processing for DIC purposes. This image processing makes the determination of the metrological performances more complicated, if not impossible when considering sensor noise propagation from gray level images to final displacement maps. Moreover, with such processing steps the gray level at any pixel is no more proportional to the brightness of the underlying physical point, which violates the fundamental brightness assumption.

26.2 Method

Recent studies highlight the artefacts of image demosaicing when images as input data of DIC. For instance, [7] averages pixel values with each 2×2 block to avoid Bayer filter effects. In [8], the authors investigate the effect of several demosaicing algorithms to determine the best approach. Here, we propose to avoid any image processing. Indeed, the signal-dependent noise in RGB sensors can be easily characterized, but any process applied to the raw information changes its characteristics, and thus complexifies noise assessment in DIC output. We propose therefore to define the brightness conservation channel per channel.

It has been shown in this study that using this approach for calculating the displacements enables us to control and thus correctly predict noise propagation from the RGB speckle images to final displacement maps. This is not the case with the classic procedures relying on images in which the gray level at any pixel is deduced from the raw information by using the aforementioned processing steps.

26.3 Conclusion

In this paper, we introduced a new definition of the brightness conservation, and thus DIC may now be directly applied to raw images captured by RGB sensors. No image processing is thus needed to format the image to DIC. This allows to properly investigate the metrological performance of color cameras as measurement tools to estimate displacement with DIC.

References

1. Schreier, H., Orteu, J.-J., Sutton, M.A.: Image Correlation for Shape, Motion and Deformation Measurements: Basic Concepts, Theory and Applications. Springer (2009). http://link.springer.com/book/10.1007%2F978-0-387-78747-3
2. Blaysat, B., Grédiac, M., Sur, F.: Effect of interpolation on noise propagation from images to DIC displacement maps. Int. J. Numer. Methods Eng. **108**(3), 213–232 (2016)
3. Blaysat, B., Grédiac, M., Sur, F.: On the propagation of camera sensor noise to displacement maps obtained by DIC - an experimental study. Exp. Mech. **56**(6), 919–944 (2016)
4. Grédiac, M., Blaysat, B., Sur, F.: A critical comparison of some metrological parameters characterizing local digital image correlation and grid method. Exp. Mech. In revision.
5. Ferreira, M.D.C., Venturini, W.S., Hild, F.: On the analysis of notched concrete beams: from measurement with digital image correlation to identification with boundary element method of a cohesive model. Eng. Fract. Mech. **78**(1), 71–84 (2011)
6. Shen, B., Paulino, G.H.: Direct extraction of cohesive fracture properties from digital image correlation: a hybrid inverse technique. Exp. Mech. **51**(2), 143–163 (2011)
7. Dufour, J.-E., Hild, F., Roux, S.: Integrated digital image correlation for the evaluation and correction of optical distortions. Opt. Lasers Eng. **56**, 121–133 (2014)
8. Forsey, A., Gungor, S.: Demosaicing images from colour cameras for digital image correlation. Opt. Lasers Eng. **86**, 20–28 (2016)

Chapter 27
Experimental Investigation of Compaction Wave Propagation in Cellular Polymers

Suraj Ravindran, Behrad Koohbor, and Addis Kidane

Abstract In the present study, an experimental setup has been developed to facilitate the in-situ characterization of wave propagation and full-field deformation response of polymeric foams subjected to high velocity direct impact. Full-field measurements are conducted using ultra high speed digital image correlation (DIC). Impact experiments are carried out over a range of impact velocities, while the formation and propagation of compaction wave across the length of the specimen, from the impact side to the support end of the specimen, is observed and quantified through a comprehensive deformation analysis. Full-field stress distribution obtained from experiment through non-parametric method along with the measured full-field strain are used to identify the energy dissipation mechanisms associated with wave propagation in the material.

Keywords Wave propagation • DIC • Polymer foam • Direct impact • High speed photography

27.1 Experimental

Closed-cell polyurethane foam specimens with initial density of 160 kg/m^3 are subjected to direct impact loading. Impact experiments are conducted using a gas gun shooting a polycarbonate projectile directly at the specimen. The specimen is fixed on the transmitted bar in a conventional Hopkinson bar apparatus. To observe and characterize the full-field deformation response in the specimen, an ultra-high speed camera (HPV-X2, Hadland Imaging, Inc.) is utilized. Figure 27.1 shows the schematic representation of the experimental setup used in this work. Images acquired during dynamic deformation are analyzed using commercial digital image correlation software Vic-2D (Correlated Solutions, Inc.).

In order to provide a detailed understanding on mechanisms associated with energy and wave dissipation in the utilized material, stress distribution in the specimen is investigated. Full-field axial stress is calculated from the inertia stress superimposed with the boundary-measured force, using a non-parametric approach introduced in [1] for the case of compressible cellular polymers. Once the distribution of stress in the material is obtained, Rankine-Hugoniot equations are used to study and identify the deformation and wave propagation mechanisms across the wave front (see Fig. 27.2).

27.2 Selected Results

Figure 27.3 shows the full-field distribution of axial strain across the length of the specimen at various times after the impact. Location of the compaction wave front is detected using the strain distribution in the material. As shown in the figure, the compaction front is visible starting from t = 12 μs and has reached half of the specimen at t = 42 μs. Shock front thickness is an important quantity in the determination of stress, strain and material velocity across the wave front and facilitates a fundamental understanding of the energy dissipation characteristics in the material [2]. Thickness of the shock front is quantified from the full-field strain along with distribution of axial velocity in the gauge area. Details of the analysis and results are discussed in the presentation.

S. Ravindran • B. Koohbor (✉) • A. Kidane
Department of Mechanical Engineering, University of South Carolina, 300 Main St, Columbia, SC 29208, USA
e-mail: koohbor@email.sc.edu

M. Sutton, P.L. Reu (eds.), *International Digital Imaging Correlation Society*, Conference Proceedings of the Society for Experimental Mechanics Series, DOI 10.1007/978-3-319-51439-0_27

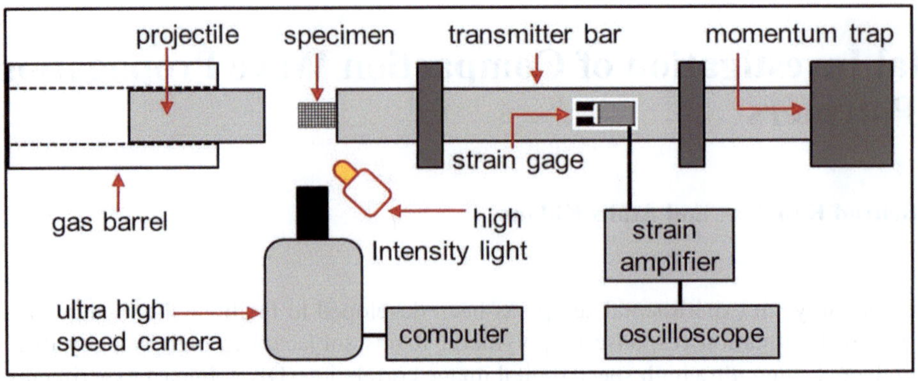

Fig. 27.1 Schematic view of the experimental setup

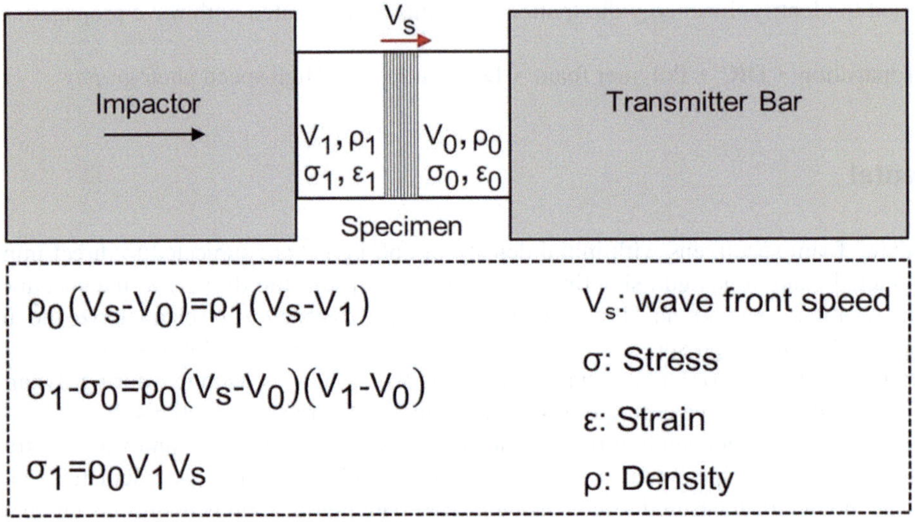

Fig. 27.2 Illustration of shock front in the specimen. Rankine-Hugoniot Equations are detailed in the figure

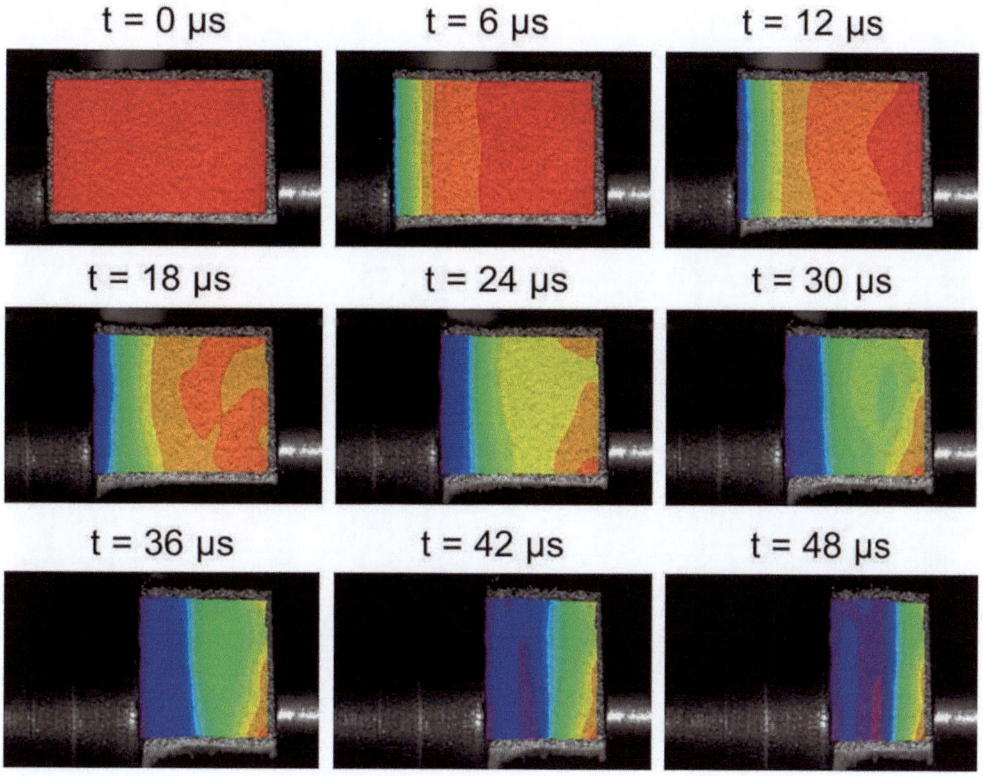

Fig. 27.3 Full-field axial strain distribution at different times after the impact

References

1. Koohbor, B., Kidane, A., Lu, W.-Y., Sutton, M.A.: Investigation of the dynamic stress-strain response of compressible polymeric foam using a non-parametric analysis. Int. J. Impact Eng. **81**, 170–182 (2016)
2. Gomez, L.R., Turner, A.M., Vitelli, V.: Uniform shock waves in disordered granular matter. Phys. Rev. E. **86**, 041302 (2012)

Chapter 28
Full-Scale Damage Detection of Railroad Crossties Using Digital Image Correlation

Alessandro Sabato and Christopher Niezrecki

Abstract Railroads experience different track failure mechanisms, which may cause problems ranging from service interruptions to derailments. With reference to the crossties-ballast system, one of the most acknowledged high-risk scenarios is the center binding condition, a situation in which the support of the ballast under the crossties is initially concentrated at the rail-seat rather than uniformly distributed. As a result, a firm support of the ballast is only provided at the center of the crosstie, while it cantilevers over its two ends. For this reason, novel track inspection technologies for detecting crossties' vertical deflection and displacement are continuously sought. In recent years, three-dimensional (3D) Digital Image Correlation (DIC) systems have proven to be an effective technique for extracting structural deformations and geometry profiles of large-sized civil structures. In this study, the feasibility of a novel railroad tracks examination system is proposed for crossties-ballast conditions assessment. It consists of four pairs of cameras installed on the underside of a rail car to detect the induced displacements. To validate the performance of the proposed system, different center binding conditions were simulated on a full-scale railroad mock-test. Experimental results showed that the 3D-DIC system was able to measure the relative tie deflections with a resolution of approximately 10^{-4} m, well below the deflections that these crossties experience in operation.

Keywords Ballast • Crosstie • Digital Image Correlation • Railroad • Structural health monitoring

28.1 Introduction

Statistics released from the Federal Railroad Administration (FRA), show that nearly 20,000 train accidents happened in the last decade and they accounted for more than \$2.5 billion USD in reportable costs [1]. One-out-of-three of those accidents were caused by track failures at different levels (e.g. rails, crossties, etc.). One of the most common failure mechanisms is generated by the different levels of abrasion experienced by the crosstie-ballast system at different locations. Because of the cyclic loading of the vehicles over time, the deformation and abrasion of the ballast is most severe under the rail-seat area of the crosstie. As the support of the ballast under the crossties becomes concentrated at the ties' middle-point rather than being uniformly distributed, the two far ends of the crosstie begin to cantilever producing the high-risk phenomenon known as center binding. Therefore, a method for efficiently assessing the state of crossties and ballast support and preventing train derailments is sought.

Current employed methods, such us human visual inspections and other techniques performed using traditional sensors (e.g. strain gages, accelerometers, ultrasound sensors, and ground penetrating radars), are extremely expensive, labor intensive, time consuming, and can be used for obtaining information at only a few discrete points. Those methods are not suitable for large scale railroad track monitoring [2]. For these reasons, the FRA continuously solicits proposals for developing automated track inspection technologies capable to detect defects at an earlier stage and prevent failures in service.

In recent years, interest has grown in performing structural health monitoring (SHM) by measuring structural deformations, full-field strains, and geometry profiles using three-dimensional (3D) Digital Image Correlation (DIC) systems [3, 4]. In this study, the feasibility of a novel railroad tracks examination system that detects crossties' vertical deflection and full field displacement using 3D-DIC techniques is described. If fully developed, the proposed system could expand the use of automated recording methods to provide more frequent and cost-effective measurements of tracks and to assess the conditions of the crosstie-ballast interface.

A. Sabato (✉) • C. Niezrecki
Department of Mechanical Engineering, University of Massachusetts Lowell, 1 University Avenue, Lowell, MA 01852, USA
e-mail: Alessandro_Sabato@uml.edu

M. Sutton, P.L. Reu (eds.), *International Digital Imaging Correlation Society*, Conference Proceedings of the Society for Experimental Mechanics Series, DOI 10.1007/978-3-319-51439-0_28

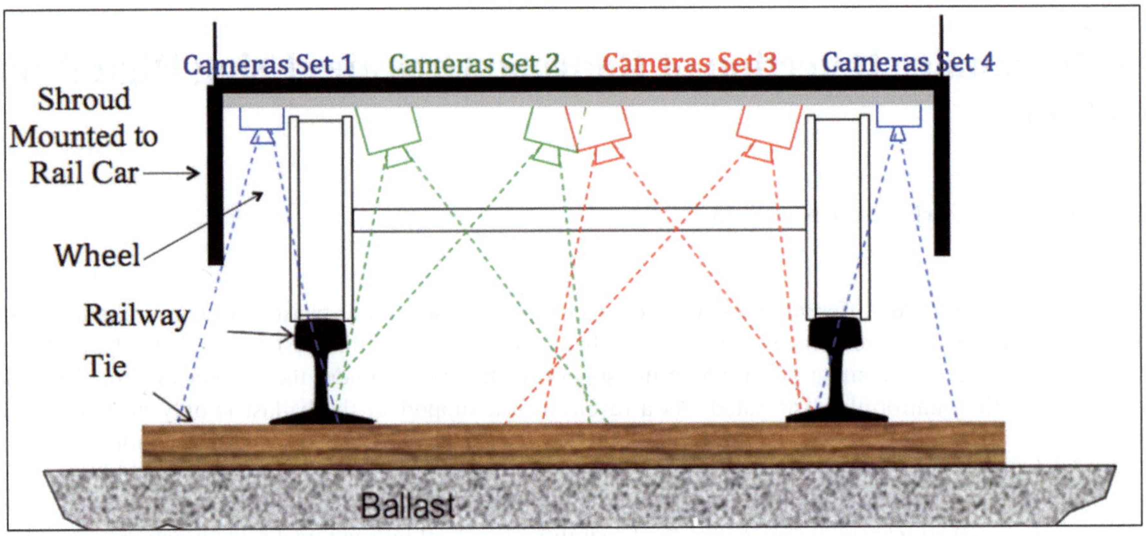

Fig. 28.1 Concept setup of the proposed stereo-photogrammetry system used to observe the whole crosstie deflection profiles attached to the underside of a rail car using four pairs of cameras sets (not to scale)

28.2 Methodology

In this work, the possibility to develop a 3D-DIC system to monitor the crossties' deflection profiles from a full-size rail car is investigated. In particular, the feasibility of using four sets of camera-pairs attached to the underside of a rail car to detect the induced displacements is discussed. The selected DIC system consists of a pair of 2 Megapixel FWX201 series digital cameras installed with 8.5 mm focal length lenses. Because of geometrical limitations due to the features of the train cars commonly used, the working distance of the cameras (i.e. 1 m) allowed for measuring a limited area of 0.81 × 0.67 m. Since these dimensions correspond to nearly 56% of the distance between the two rails, it is clear that a single stereo-photogrammetry system cannot cover the whole length of the crosstie. For this reason the field of view (FOV) of the full crosstie has been obtained by stitching together the images recorded using four pairs of cameras [5], in a setup similar to that shown in Fig. 28.1. In order to combine the individual sections into a single image, the recorded fields of view were laid out with several centimeters of overlap [5]. The SVIEW™ software, developed by the German company GOM, was then used for stitching the multiple fields of view together.

The 3D-DIC system proposed in this work has been evaluated through extensive experiments on a full-scale railroad model, which experiences displacement levels similar to what is experienced in the field when a railcar passes over a section of the track [6]. To measure the performance of the proposed system, different center binding conditions were simulated on the model and measurements were recorded for two different loading conditions applied to the crosstie being tested. Analyses of the results proved the capability of the 3D-DIC system to measure crossties' full-field vertical displacement and to highlight changes in their longitudinal profile.

28.3 Results

To demonstrate the feasibility of the 3D-DIC system described in the previous paragraph, two different loading conditions (i.e. zero and 510 N (114.65 lbf)) were applied to a crosstie for simulating the transit of a train. The mock crosstie is a wooden board that has geometrical length and width similar to a typical railway tie. The thickness of the board is only $\sim 2.0 \times 10^{-2}$ m and is designed to generate a transverse displacement similar to a crosstie in the field for the loading provided in the lab. The crosstie experienced three different support conditions to simulate different ballast configurations. In the first test, the crosstie was attached to the rails only to simulate an evenly supportive ballast; while in the second and third tests another support was placed below the crosstie; one-quarter and one-half of the tie's whole length respectively. These two conditions were used for simulating the ballast providing localized support in specific areas only, rather than being uniformly distributed. Analysis of the noise floor performed on the recorded images proved the systems' capacity to measure vertical displacement, Z, on the order of 10^{-4} m. Experimental results obtained by stitching together the fields of view recorded using the four sets of

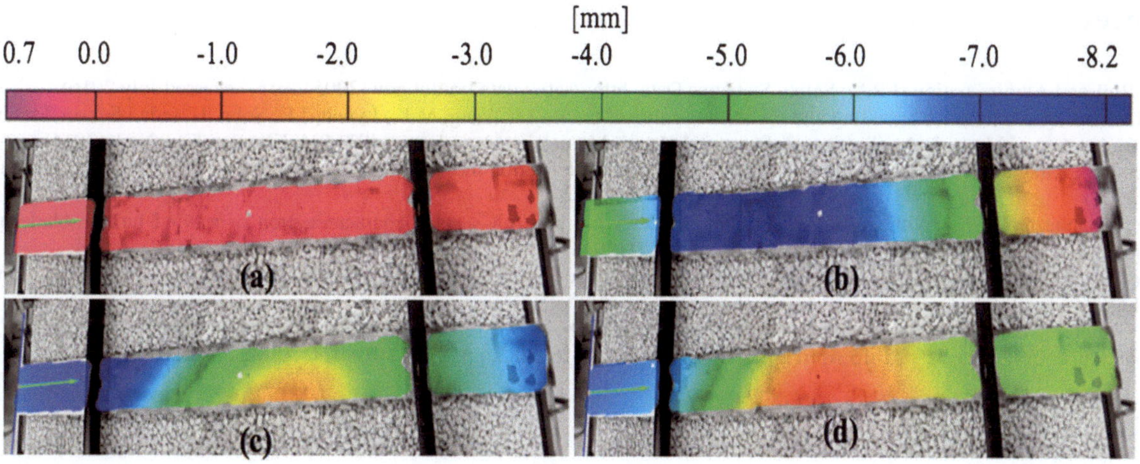

Fig. 28.2 Full-field vertical displacement Z for: (**a**) unloaded condition; (**b**) evenly supported crosstie and 510 N load applied; (**c**) crosstie supported 0.37 m from the right rail and 510 N load applied; and (**d**) crosstie 0.74 m from the right rail and 510 N load applied

cameras are shown in Fig. 28.2, where the displacement values plotted in false colors are superimposed on an image of the mock-test.

In particular, the full-field distribution measured in the vertical direction when the 510 N loads are applied to the rails is plotted in Fig. 28.2b–d for the three different simulated ballast conditions. Those results were compared to that measured of the unloaded (Fig. 28.2a) state. It is evident that when different supporting conditions occur, the vertical displacements of the crosstie are different. As the crosstie is evenly supported (Fig. 28.2b), a fairly uniform deflection is observed in the portion between the rails. The profile is parabolic with the lower vertex equal to -8.0×10^{-3} m in correspondence with the crosstie's middle point. On the other hand, it is observed that when the support is located in a specific area, as the load increases the portion of the tie close to the support experiences smaller displacements, while the vertical deflection is maximum at the tie's free edges. With reference to the results plotted in Fig. 28.2c, d, it is possible to observe that the area where the support is located experiences a displacement of nearly -2.0×10^{-3} m and -1.0×10^{-3} m for conditions (c) and (d) respectively, and a negative displacement in the range -5.0×10^{-3} m; -6.0×10^{-3} m at the free edges.

These results confirm that the crosstie behaves as two cantilever beams at its free edges, because a firm ballast support is only provided in the center. Furthermore, from an analysis of the data plotted in Fig. 28.2, the position of the support below the tie is easily detectable, highlighting any defect in the uniformity of the ballast. To finish, the crosstie's longitudinal profile is not shown here for the sake of brevity.

28.4 Conclusion

In this study, a novel 3D DIC system for the measurement of tie deflection profiles and ballast support assessment is proposed. The feasibility of the technique is verified through several experiments performed on a full-scale railroad track model. This operation is possible by just processing information from pictures, with no need to deploy a large array of sensors underneath the tie tested. It demonstrates that the 3D-DIC system can easily expedite the monitoring process and is able to extract information throughout the fields of view and not just at discrete points. Unlike most of the conventional SHM sensing systems (e.g. strain gages and accelerometers), which are only able to provide discrete sensing, the proposed imaging system allows for full-field displacement measurements over the entire length of the crosstie. Experimental results shown that the 3D-DIC system is able to detect out-of-plane vertical displacement with an accuracy of 10^{-4} m and characterize the differences in the crossties' deflection profiles as various ballast conditions are simulated.

Future development of the concept would consist of track inspections using a system of high-speed cameras mounted to a train moving at typical operating speeds (e.g. 60 mph), and in improving the measurement accuracy by taking into account the vibrations associated with the train motion.

References

1. Federal Railroad Administration (FRA): Safety Statistics Data. http://safetydata.fra.dot.gov (2005–2015, accessed July 2016)
2. Rizzo, P.: Sensing solutions for assessing and monitoring railroad tracks. In: Wang, M.L., Lynch, J.P., Sohn, H. (eds.) Sensor Technologies for Civil Infrastructures-Volume 2: Applications in Structural Health Monitoring, 1st edn, pp. 497–524. Elsevier, New York (2014)
3. Yoneyama, S., Kitagawa, A., Iwata, S., Tani, K., Kikuta, H.: Bridge deflection measurement using digital image correlation. Exp. Tech. **31**(1), 34–40 (2007)
4. LeBlanc, B., Niezrecki, C., Avitabile, P., et al.: Damage detection and full surface characterization of a wind turbine blade using three-dimensional digital image correlation. SHM. **12**(5–6), 430–439 (2013)
5. LeBlanc, B.: Non Destructive Inspection of Wind Turbine Blades with Three Dimensional Digital Image Correlation. University of Massachusetts Lowell (2011)
6. Yu, H.: Estimating deterioration in the concrete tie-ballast interface based on vertical tie deflection profile: a numerical Study. In: 2016 Joint Rail Conference, Columbia, SC, 12–15 April 2016, Paper No. JRC2016-5783, pp. V001T01A024. ASME, New York

Chapter 29
Leveraging Vision for Structural Identification: A Digital Image Correlation Based Approach

Mehrdad Shafiei Dizaji, Mohamad Alipour, and Devin K. Harris

Abstract This paper describes a case study on a large scale structural test specimen using 3D-DIC as an image-based metrology approach for structural identification (St-ID). For the identification process, a commercial FEA software package was interfaced with MATLAB to converge on the optimal unknown/uncertain system parameters of the experimental setup. The 3D-DIC results provided a rich full-field dataset that was used in the identification process, which was compared against ground-truth measurements derived from traditional physical in-place sensors typically used in St-ID. For the identification, a novel hybrid algorithm, incorporating a combination of a genetic algorithm and a gradient-based scheme was utilized for updating the FEA model and obtaining the optimal values of the selected parameters. Results demonstrated that deflection, strain, and rotation measurements derived from DIC mirrored those from the ground-truth sensors and allowed for convergence of the updating with a variety of measurement responses that are challenging to acquire in typical applications.

Keywords DIC • Optimization • St-ID • Image-base structural health monitoring (iSHM) • Finite element

29.1 Introduction

The concept of structural health monitoring (SHM) has shown promise as a strategy temporal condition assessment of the built environment. SHM provides a system performance evaluation strategy with the end goal of characterizing behavior and providing indications of damage and even forewarning of impending failure. Within a traditional SHM framework numerical models, typically finite element models (FEM), are commonly used to characterize the behavior of structural systems, but have difficulty in predicting the behavior of existing structures due to their approximate nature and uncertainties associated with boundary conditions and condition state. Structural identification (St-ID) describes a concept that correlates the response characteristics between a model and experiment, providing a basis for characterizing critical performance measures. This inverse problem aims to minimize differences between analytical and experimental results and is usually formulated as an optimization problem. Satisfactory correlation between the observed experimental behavior and the analytical results is critical, but equally essential is maintaining the physical significance of updated parameters. For this purpose, setting up of an objective function and selecting updating parameters are crucial steps in St-ID. The changes in these parameters are then determined iteratively and pushed to minimum by an optimization algorithm. St-Id aims to bridge the gap between the model and the real system by developing reliable estimates of the performance and vulnerability through improved simulations. This work describes an experimental case study from an ongoing investigation that leverages an image-based metrology approaches for St-ID.

29.2 Experimental Setup

The experimental program consisted of a series full-scale non-destructive tests on a structural steel beam (ASTM A992 W10x22) subjected to a flexural loading sequence within its elastic range ($\sigma_{\text{yield}} = 50$ ksi). The beam was tested in the Structures Laboratory at the University of Virginia and configured for strong-axis bending as a simple span with a concentrated load applied at midspan. The beam was instrumented with Bridge Diagnostic Inc. (BDI) sensors at both midspan and support locations to provide ground truth to the 3D-DIC measurements (Correlated Solutions *VIC-3D*) that were acquired

M. Shafiei Dizaji • M. Alipour • D.K. Harris (✉)
University of Virginia, 351 McCormick Road, Charlottesville, VA 22904, USA
e-mail: dharris@virginia.edu

M. Sutton, P.L. Reu (eds.), *International Digital Imaging Correlation Society*, Conference Proceedings of the Society for Experimental Mechanics Series, DOI 10.1007/978-3-319-51439-0_29

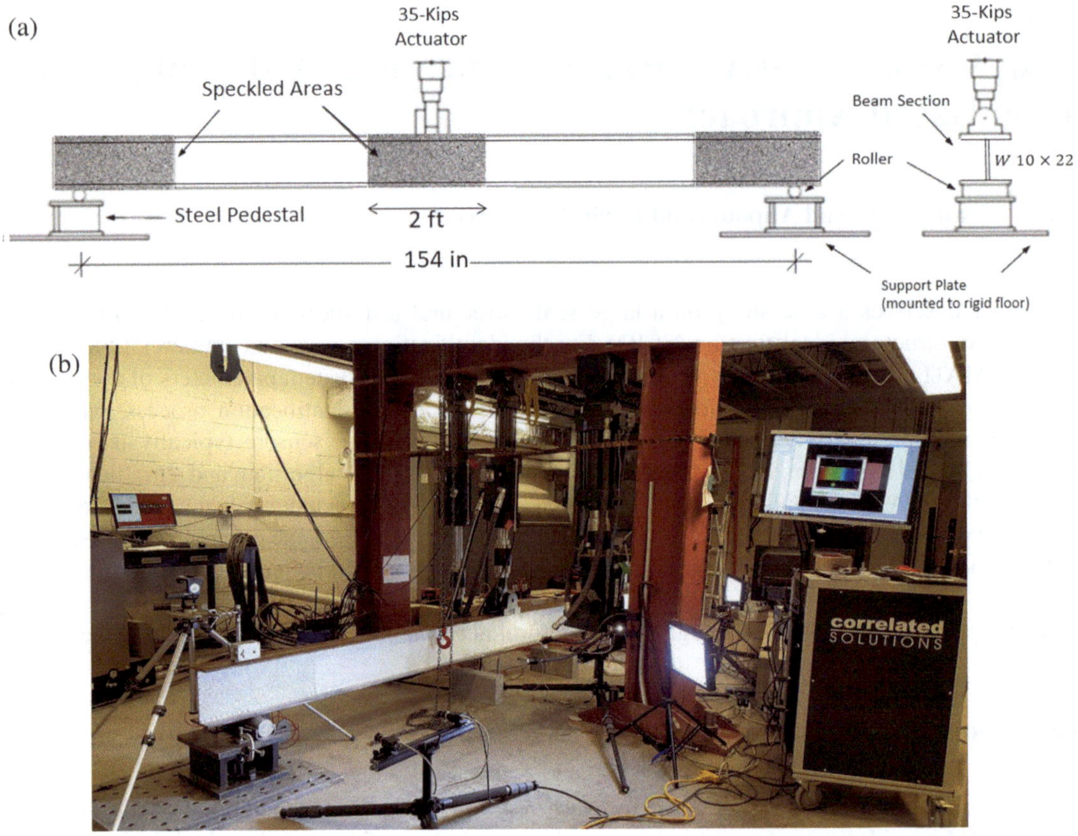

Fig. 29.1 (**a**) Experimental setup with elevation and x-section of the beam (**b**) six camera configuration (three independent systems configured)

over 24 inches long regions at the same locations. Three paired DIC camera (Point Grey Grasshopper 2.0 CCD with 5.0MP resolution) systems were used to evaluate the midspan (Schneider 8 mm lens) and two end span (Schneider 12 mm lenses) locations. After application of a flat while paint base coat to the regions of interest, the speckle pattern was applied manually with a permanent marker to create a random speckle pattern. Figure 29.1 illustrates the details of the experimental setup and instrumentation configuration used for the tests. For the experiments, the beam was step loaded in displacement-control in 0.05 in. increments (20 s per step) up to a peak displacement of 0.5 in. and then unloaded through the reverse sequence. During the loading, the ground truth sensors acquired data at 50 Hz while the DIC images were acquired at 2 Hz.

29.3 Numerical Model

As noted, St-ID requires the development of an initial numerical model that can be updated based on experimentally derived behavior results. In this investigation, the numerical model was developed in ABAQUS, a robust commercially available finite element software package. ABAQUS allowed for the development of an interface with MATLAB, a multi-paradigm numerical computing environment, which facilitated the iterative parameter optimization algorithm. A global view of the model of the steel beam has been shown in Fig. 29.2. The beam was modeled with solid elements. For modelling of the beam, particular attention was given to the supports to allow for realistic reproduction of the supports restraint conditions. In the models, a series of linear and rotational springs were used to create the necessary restraint.

The optimization algorithm developed in this investigation incorporated the features of a genetic algorithm and a gradient-based scheme to iterate on the unknown parameters. The Young's Modulus (E_s) was selected as an unknown parameter in the optimization scheme along with restraint stiffness at the support locations, which were determined to exhibit some rigid body movement in preliminary testing. While steel is generally classified as a macro-scale homogenous materials, iteration on E_s allowed for evaluation of the performance of the optimization scheme with multiple objectives.

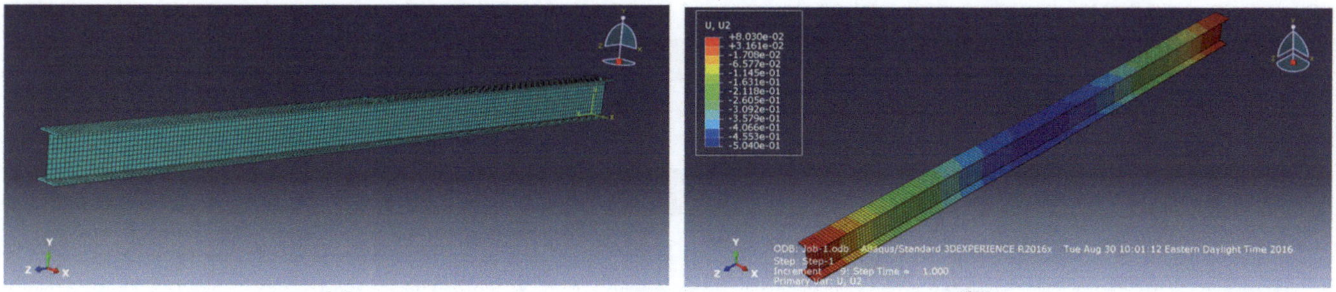

(a) (b)

Fig. 29.2 A global view of the numerical model of the steel beam (**a**) Mesh of the beam (**b**) Results with initial parameters

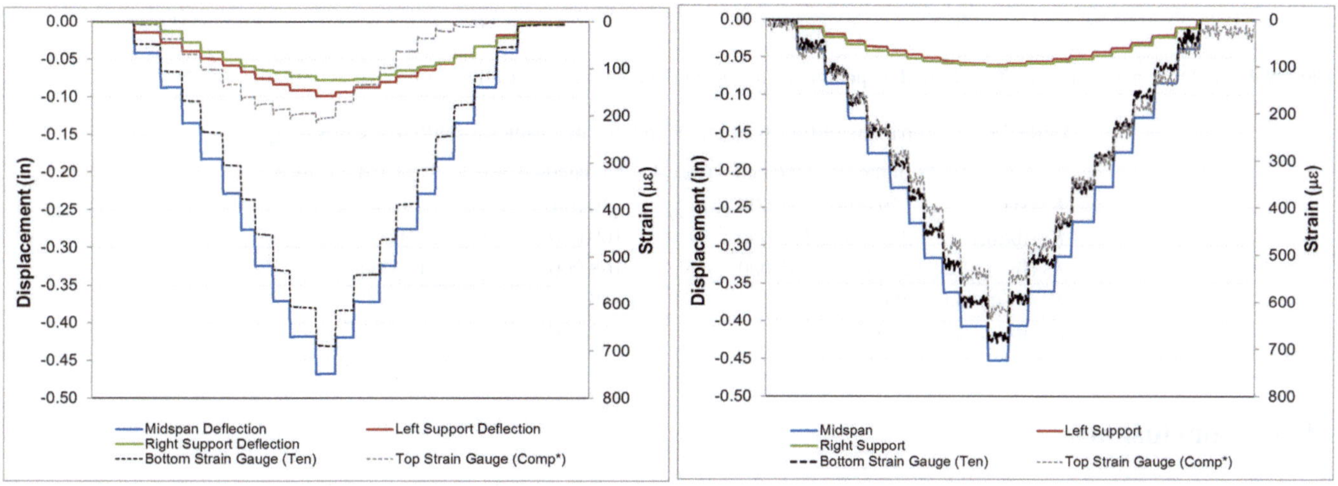

(a) (b)

Fig. 29.3 Comparison of results obtained from DIC and ground truth sensors (**a**) BDI sensors (**b**) 3D-DIC

29.4 Results

Results from the experimental program provided a basis for comparison of the 3D-DIC measurements with the in-place BDI sensors that are representative of those used in traditional structural testing. For the DIC results, the calibration error ranged from 0.04 to 0.05 for the three camera systems and the post-processing utilized a subset size of 49 and step size of 15. Figure 29.3 illustrates a comparison of the results of selected sensors for one of the experiments. It should be noted that the DIC measurements provided a supplemental benefit of measurement of load frame deformation, which would not have been realized with the selected in-place sensor configuration.

Leveraging the results from the DIC, the initial FE model was updated using the optimization algorithm to converge on predictions of the beam's Young's Modulus (E_s) and support stiffness parameters (K_1, K_2, K_3, K_4). As illustrated in Fig. 29.4a, three scenarios were selected for model updating with variations in the restraint conditions and objective function parameter. For all scenarios, the optimization was limited to 10 iterations to evaluate the impact of measurement parameters on optimization efficiency. The convergence results are shown in Fig. 29.4b, which demonstrate that convergence is achieved with an objective function based on a single displacement (Cases 1 and 2); however, convergence does not occur based on a single strain measurement (Case 3) (Table 29.1).

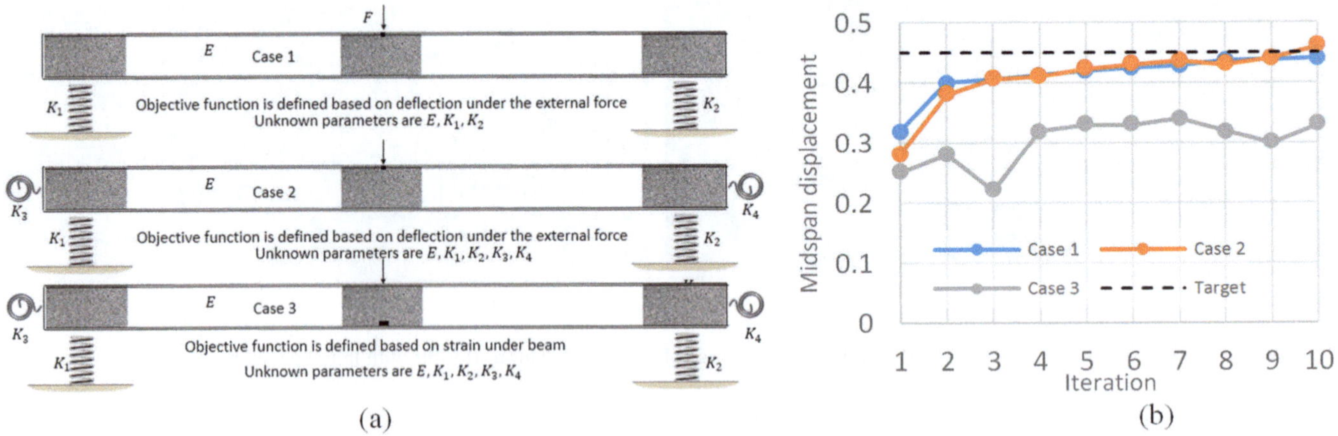

Fig. 29.4 St-ID Scenario (**a**) cases for FE model updating of the beam (**b**) convergence results of midspan deformation for the cases

Table 29.1 Comparison of results of initial and updated values for different cases

	Initial	Updated—Case 1	Updated—Case 2	Updated—Case 3
E (ksi)	29,000	31,000	30,011	29,110
K_1 (lb/in.)	250,000	180,000	160,000	220,000
K_2 (lb/in.)	250,000	185,000	168,000	204,000
K_3 (lb-in./rad)	500	–	10,000	25,000
K_4 (lb-in./rad)	500	–	12,000	24,444

29.5 Conclusions

This work describes the calibration and experimental validation of a numerical model of a steel beam based on displacement and strain parameters. Preliminary results suggest that image-based measurements such as 3D-DIC can be used as an alternative to physical in-place sensors for characterizing the response of large scale structural systems. Based on the cases evaluated in this work, DIC shows significant promise as a non-invasive tool for structural identification. However, these results also demonstrated the limitation of using a single parameter for the updating process. It was observed that while DIC measurements were able to mimic those responses derived from traditional sensors, limitations may exist with identification schemes that leverage local deformations (i.e. single strain measure) which may not be representative of global response (e.g. displacement, rotation, or vibration).

29.6 Future Work

The study described in this paper represents preliminary work in the area of image-based St-ID and is expected to serve as the foundation in a broader context for a non-contact assessment or image-based structural health monitoring (*iSHM*). In the near term, efforts will focus on multi-parameter optimization utilizing the full-field measurements derived from DIC. Future work in this domain aims to extend the outcomes of this investigation to increasingly complex structural systems such as heterogeneous structural members and full-scale structures in service.

Chapter 30
Uncertainty Quantification in the Evaluation of DIC-Based Dynamic Fracture Parameters

Logan Shannahan and Leslie Lamberson

Abstract A hybrid experimental-computational method is presented which takes full-field displacement data from DIC and evaluates dynamic stress intensity factors (SIFs) at fracture initiation and propagation. The technique relies on the principles of linear elastic fracture mechanics, and as such the input data must maintain a region of K-dominance ahead of the moving crack. This requires displacement fields taken outside local plastic and 3D effects at the crack tip, and mitigated from potential boundary effects, otherwise error can increase to nearly 20% of the actual material toughness. The region of useable DIC information and necessary speckle parameters directly influences the precision of the resulting dynamic fracture predictions; as does an accurate determination of elastic properties and location of the moving crack front within the speckle pattern. The sensitivity of these parameters is examined using both experimental and FEM-generated virtual displacement data. Results on brittle hard metals and polymers using nominal elastic properties have shown that misalignment of the methodology-determined crack tip location with the actual can contribute to error on the order of 1–2 MPa-m$^{1/2}$ per source.

Keywords Dynamic fracture • Mixed mode • Critical stress intensity factor • Full-field displacement data • Propagating crack

30.1 Introduction

More classical methods used for extracting crack tip energetics in dynamic fracture experiments have been based on optical techniques such as caustics and photoelasticity. While very useful, these techniques require a reflective or transparent, or birefringent, material, respectively. Photoelasticity can be extended to a greater number of materials by applying a birefringent coating, and Der et al. have demonstrated that the SIF measured in the coating accurately reflects the SIF in the material, though sufficiently thick coatings can still cause difficulty in accurately tracking the crack tip [1]. Photoelasticity does have the advantage of locating the crack tip by default, as the technique generates bands of isochromatic fringes around the crack tip.

However, modern advances in spatial and temporal resolution of high speed cameras have allowed the possibility of capturing full-field DIC displacement data around a propagating crack. While displacements alone do not give fracture information, asymptotic expressions for the displacements around a crack tip as a function of the SIFs can be used to extract SIFs from the displacement data. This technique has the key advantage of requiring only a suitable camera and applied speckle pattern, but has been explored relatively less when compared to more traditional dynamic fracture methods [2–4]. As a result, the metrology of using DIC for dynamic SIF extraction is the focus of this investigation. The goal of this study is to describe the impact of each parameter on the final SIF value using DIC, and quantify its impact on the dynamic material toughness measured.

30.2 Method

The technique used in this study is comprised of several interrelated methods. In all cases, an over-deterministic least squares analysis is used to fit the asymptotic expressions for displacement in terms of SIFs to the collected full-field displacement data. For the case of a stationary crack, subjected to dynamic loading, but prior to initiation of crack growth, the crack tip

L. Shannahan (✉) • L. Lamberson
Drexel University, 3141 Chestnut St, Philadelphia, PA 19104, USA
e-mail: logan.shannahan@gmail.com

M. Sutton, P.L. Reu (eds.), *International Digital Imaging Correlation Society*, Conference Proceedings of the Society for Experimental Mechanics Series, DOI 10.1007/978-3-319-51439-0_30

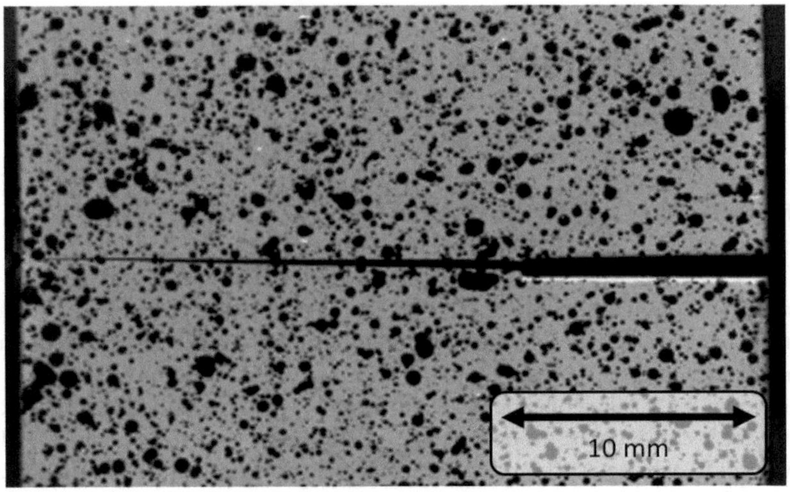

Fig. 30.1 A tungsten carbide (WC) sample is shown after impact at 4 m/s. The pre-notch can be seen at right, and the crack has propagated from the sharpened tip through the entire sample

location is known to an accuracy dependent on the resolution of the high speed imagery. The crack tip location is a necessary input, and as such allows for a linear version of the calculation, allowing the asymptotic expressions to be directly fit to the displacement data. When the crack has begun to propagate however, it is necessary to treat the crack tip location as an unknown in addition to the SIFs, and a nonlinear formulation in which an iterative numerical solution is used to solve for both SIFs and the crack tip location simultaneously is required.

An additional variation is the choice of asymptotic expression. Several exist, for cases ranging from simple quasi-static Mode I (opening) fracture, to dynamic Mode I and Mode II (sliding) fracture. In this study, two different sets of expressions are compared. The first are the expressions for a dynamically loaded mixed-mode (Mode I and Mode II) stationary crack in an anisotropic material [2]. This allows for complex loading conditions, but neglects the consideration of inertial effects once the crack has begun to propagate. As such, the expressions for a dynamically propagating Mode I crack in an isotropic were also used [5]. All expressions have the form $u = K_I f^I(r,\theta) + K_{II} f^{II}(r, \theta)$, in which u is the displacements, K_I and K_{II} are the Mode I and Mode II SIFs respectively, and f^I and f^{II} are functions of the location of the crack tip and the material properties in question. For the case of the dynamically loaded propagating crack tip, K^{II} and f^{II} are neglected and f^I is also a function of crack tip velocity.

In order to study the sensitivity of these expressions and analytical methods to various sources of error such as noise in the DIC displacement data, or inaccuracies in the elastic properties of the material, both virtual and experimental data have been used. The virtual data was generated using Abaqus Explicit to simulate a dynamic impact on a notched rectangular sample 60 by 30 by 4 mm, with a 10 mm notch located in the center of one 60 mm edge. PMMA, a brittle polymer, was used as a model material for this simulation. The nodal displacements were used as a noise-free idealized substitute for the full-field DIC displacement data. The shape and impact loading profile were chosen to match experiments carried out on tungsten carbide (WC) samples on a modified Kolsky bar. These samples were held in place with clay to provide as free a boundary condition as possible and struck at 4 m/s by a rounded tip steel rode with a diameter of 12.7 mm. One surface of each sample was coated by a layer of solid white paint and a speckle pattern of black paint to allow for DIC. High speed imagery was recorded at one million frames per second using a Shimadzu HPV-X high speed camera. The imagery was analyzed using MatchID to obtain full-field DIC displacement data both prior to and during crack growth (Fig. 30.1).

30.3 Results and Discussion

Both of sets of expressions used in the analysis are derived from the principles of linear elastic fracture mechanics, and as such the input data must maintain a region of K-dominance ahead of the moving crack. This requires displacement fields taken outside local plastic and 3D effects at the crack tip, as well as isolated from potential boundary effects near the surface of the sample. To study the limits of the region of K-dominance, a sensitivity study was run on a set of experimental data. Any given region of data from inside the zone of K-dominance should give the same SIF, with a small error resulting from noise in the DIC displacement data. However, error in the calculated SIF should increase rapidly if data was taken from the plastic zone or regions in which boundary effects dominated. The results of the sensitivity study matched this expectation, with error in the calculated SIF reaching nearly 20% of the actual material toughness when data from the plastic zone or boundaries was included in the analysis.

The presence of the speckle pattern increases the already substantial difficulty in determining the precise location of the crack tip. However, the analysis is dependent on an accurate value for this location. Choosing (or analytically determining, in the case of the nonlinear form of the analysis) a location ahead of the crack tip will result in a decrease in the calculated Mode I SIF compared to the actual material toughness. The magnitude of this decrease is dependent on the magnitude of the misalignment, but can be substantial, reaching several MPa-$m^{1/2}$. The converse effect holds with even greater magnitude if the nominal location of the crack tip is behind the actual location. In the worst cases, calculated SIFs can be greater than actual material toughness by a factor of ten or more. However, in these cases the degree of misalignment is sufficient that it can be recognized by eye if the nominal location of the crack tip is compared to the high speed imagery or full-field displacement data. In the region in which the crack tip cannot be accurately determined by sight, misalignment results in error of 1–2 MPa-$m^{1/2}$. Similar trends hold for Mode II SIFs, if the chosen crack tip location is above or below the actual crack.

It is also important to ensure that the elastic properties used match the material in question. A sensitivity analysis found that a change of 100 GPa in the elastic modulus of the WC sample (nominal elastic modulus 650 GPa) resulted in a corresponding change of approximately 1 MPa-$m^{1/2}$ in the calculated SIF result, or a change of approximately 5%. However, in the nonlinear version of the analysis in which the location of the crack tip is not specified, a discrepancy in the elastic properties also reduces the accuracy of the calculation of the location of the crack tip. The resultant misalignment results in further error in the calculated SIF, as detailed above.

In the case of the expressions for a dynamically loaded Mode I propagating crack, the analysis is also sensitive to a proper determination of crack speed during fracture. A sensitivity analysis run on the virtual PMMA data found that a decrease in crack speed of 500 m/s from an original crack speed of 740 m/s (determined from tracking crack propagation in the virtual data) resulted in a corresponding decrease in the measured Mode I SIF of 60%. Just as with the material elastic properties, inaccuracy in the crack speed also results in inaccuracies in determining the crack tip location.

30.4 Conclusion

The use of full-field DIC displacement data to extract SIFs expands the capability and utility of DIC as an experimental technique, particularly in the field of dynamic behavior of materials. In order to properly employ this method, it is necessary to understand the accuracy and potential limitations by quantifying the sensitivity of the results as a function of various potential sources of error, both computationally and empirically derived. Results of this study suggest that an improper choice of data in regards to K-dominant region ahead of the crack, and inaccuracies in the location of the crack tip have the largest effect on results, while error in material properties can also cause substantial deviation from the "true" material toughness. As the quality and resolution of cameras increases, the accuracy of this technique can also increase with continued careful implementation.

References

1. Der, V.K., Barker, D.B., Holloway, D.C.: A split birefringent coating technique to determine dynamic stress intensity factors. Mech. Res. Commun. **5**(6), 313–318 (1978)
2. Yoneyama, S., Morimoto, Y., Takashi, M.: Automatic evaluation of mixed-mode stress intensity factors utilizing digital image correlation. Strain. **42**(1), 21–29 (2006)
3. Lee, D., Tippur, H., Kirugulige, M., Bogert, P.: Experimental study of dynamic crack growth in unidirectional graphite/epoxy composites using digital image correlation method and high-speed photography. J. Compos. Mater. **43**, 2081–2108 (2009)
4. Shannahan, L., Weerasooriya, T., Gunnarsson, A., Sanborn, B., Lamberson, L.: Rate-dependent fracture modes in human femoral cortical bone. Int. J. Fract. **194**(2), 81–92 (2015)
5. Ravi-Chandar, K.: Dynamic Fracture. Elsevier (2004)

Chapter 31
Behavior Investigation of CFRP-Steel Composite Members Using Digital Image Correlation

Yuntong Dai, Haitao Wang, Gang Wu, Jianxiao Wan, Shuangyin Cao, Fujun Yang, and Xiaoyuan He

Abstract The two types of carbon fiber reinforced polymer (CFRP) applications used in steel structures are bond-critical application and contact-critical application, and the major failure modes for these two applications are debonding failure and buckling failure, respectively. Conventional electrometric techniques may not provide precise results because of the limitations associated with single-point contact measurements. Therefore, the nondestructive full-field measurement technique may be a suitable replacement for conventional methods. In this study, the digital image correlation (DIC) technique was adopted to investigate the bond behavior and buckling behavior of CFRP-steel composite members. The CFRP-to-steel bonded joint and the CFRP-strengthened square hollow section (SHS) steel column were tested to examine the suitability of the DIC technique. The stereo-DIC technique was utilized to measure continuous deformation. The bond-slip relationship of the CFRP-to-steel interface was derived using the DIC data. Additionally, a multi-camera DIC system consisting of four stereo-DIC subsystems was applied to the compressive test of CFRP-strengthened SHS steel column. The buckling location and CFRP delamination of the column were identified. The experimental results confirm that the stereo-DIC technique can provide effective measurements for investigating the behaviors of CFRP-steel composite members.

Keywords CFRP • Steel • Bond behavior • Buckling behavior • Digital image correlation

31.1 Introduction

The use of FRP in steel structures can be divided into two categories, namely, the bond-critical application and the contact-critical application [1]. In CFRP-strengthened steel members, the externally-bonded CFRP strengthening for improving the flexural and fatigue behaviors belongs to the first category, whereas the CFRP-confined steel columns can be assigned to the second category. The preceding analysis [2] concluded that some limitations exist in the conventional electrometric techniques that are applied to evaluate the bond and buckling behaviors of CFRP-steel composite members. A full-field measurement technique is an effective solution to this problem. This paper reports the effectiveness of the DIC technique in investigating the bond and buckling behaviors of CFRP-steel composite members. A study of the bond behavior was conducted using the CFRP-to-steel bonded joints via the stereo-DIC technique, whereas a study of the buckling behavior was implemented using the CFRP-strengthened SHS steel columns with a multi-camera DIC system.

31.2 Results

The single-shear test was combined with the stereo-DIC technique to investigate the bond behavior of the CFRP-to-steel interface. The continuous displacements of the CFRP plate and steel were directly obtained. Figure 31.1 presents the gradual debonding process of the CFRP-to-steel interface. The interfacial shear stress can be derived via differential operation of the continuous strain distribution. Figure 31.2 shows the bond-slip relationship of the CFRP-to-steel interface. The buckling behavior of the CFRP-strengthened SHS steel column was examined with the proposed multi-camera DIC system. Figure 31.3 shows the displacement contours during the loading, on which the CFRP delamination is directly reflected. Inward and outward buckling were observed on all four sides at the column ends.

Y. Dai • H. Wang • G. Wu • J. Wan • S. Cao • F. Yang • X. He (✉)
School of Civil Engineering, Southeast University, Nanjing 210096, China
e-mail: mmhxy@seu.edu.cn

M. Sutton, P.L. Reu (eds.), *International Digital Imaging Correlation Society*, Conference Proceedings of the Society for Experimental Mechanics Series, DOI 10.1007/978-3-319-51439-0_31

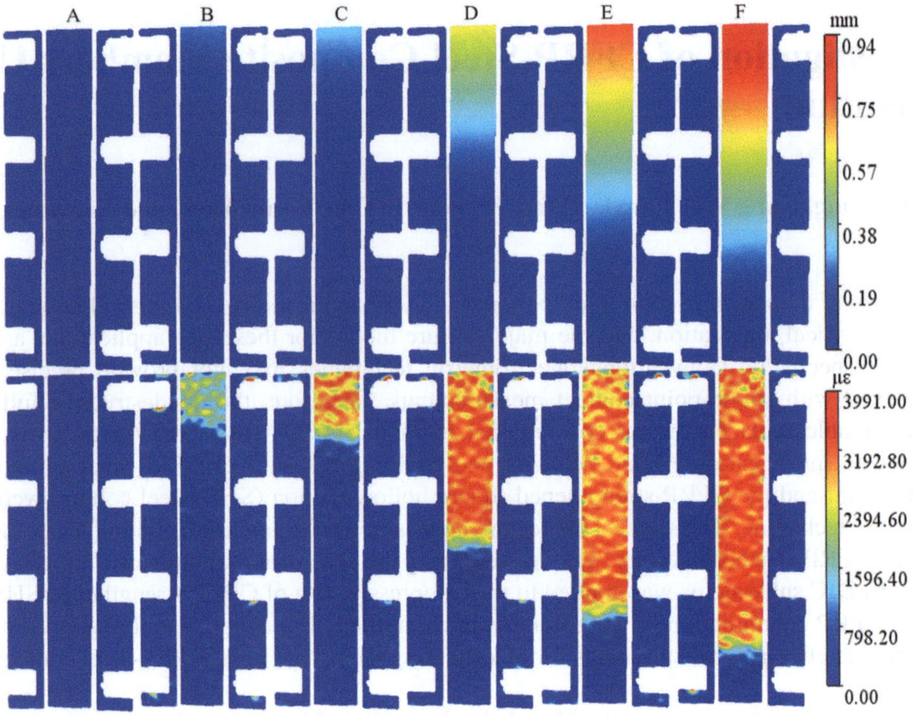

Fig. 31.1 Deformation fields on CFRP-to-steel interface

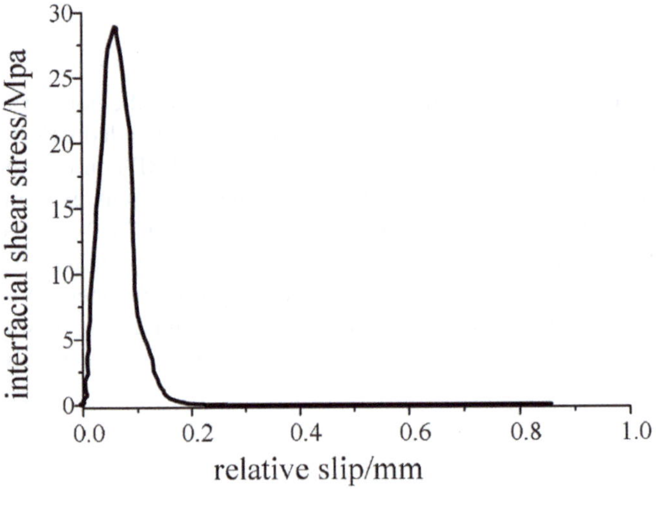

Fig. 31.2 Bond-slip relationship of CFRP-to-steel interface

31.3 Conclusions

(1) The bond behavior between the CFRP plate and the steel was tested using the stereo-DIC technique. Continuous CFRP and steel displacements can be obtained; thus, the continuous relative slip can be directly measured using this method. The bond-slip relationship of the CFRP plate-to-steel interface can be easily obtained from the DIC data.

(2) The first insight into the buckling behavior of a CFRP-strengthened SHS steel column using a multi-camera DIC system was presented. The multi-camera DIC system consists of four stereo-DIC subsystems that can capture the exact process of deformation and buckling. The experimental study shows that the application of CFRP sheets delayed the occurrence of local buckling and increased the ultimate load capacity and elastic stiffness of the SHS steel column.

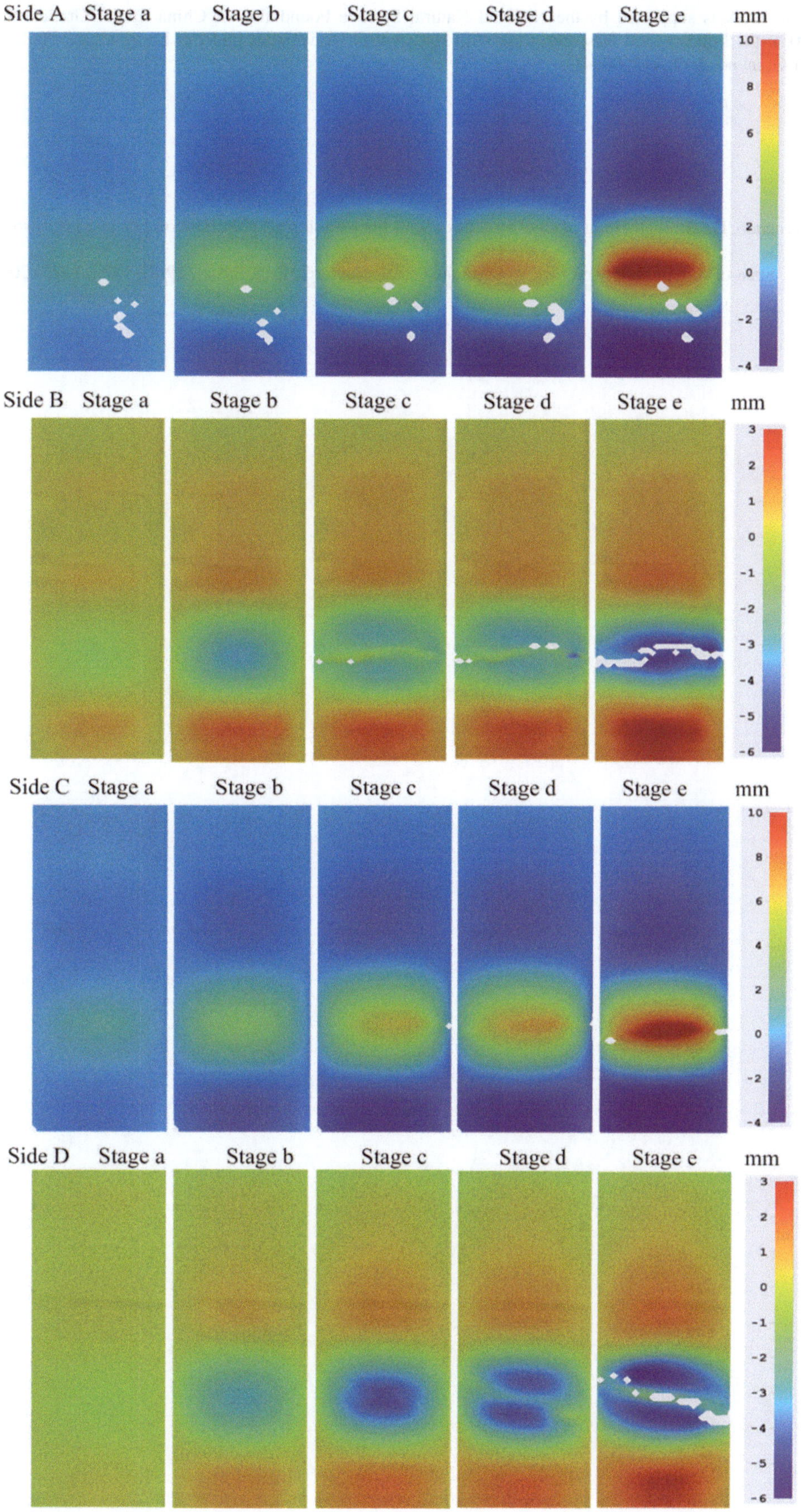

Fig. 31.3 Lateral displacement evolution of the CFRP-strengthened SHS steel column

Acknowledgements This work is supported by the National Natural Science Foundation of China (under Grant nos. 11272089, 11327201, 11332012 and 11532005); the Fundamental Research Funds for the Central Universities and the Postgraduate Student Innovative Project of Jiangsu Province (under Grant no. KYLX15_0094).

References

1. Teng, J.G., Yu, T., Fernando, D.: Strengthening of steel structures with fiber-reinforced polymer composites. J. Constr. Steel Res. **78**, 131–143 (2012)
2. Zhao, X.L., Zhang, L.: State-of-the-art review on FRP strengthened steel structures. Eng. Struct. **29**(8), 1808–1823 (2007)

Chapter 32
Analysis of Possible Registration of Inhomogeneous Deformation Fields in Composite Plates with Technological Defects

E.M. Spaskova and V.E. Wildemann

Abstract When using composite materials in industry there arises the need of structure state control to reveal damages and defects. The most frequent defects are discontinuity ones: defects of delamination, starved spot, nonpressing, cracks, air or gas inclusions. In most cases, such defects as discontinuity—are the defects which have technological background, as well as there is a possibility of their occurrence during operation. These defects must be eliminated by improving the process or their diagnostics during operation. A large number of foreign research works is devoted to defects diagnosis during operation using digital optical video system Vic-3D.

Keywords Composite material • Technological defects • Digital optical video system VIC-3D • Infrared thermal vision system Flir SC7600 • Tension and combined tension and torsion

32.1 Experimental Procedure

In this paper, there are obtained experimental data for laminated composite samples underwent tension and combined tension and torsion tests with pre-introduced technological defects corresponding to nonpressing and starved spot layers in a specified limited area [1–3]. Samples are rectangular plates with the size of 200 × 50 × 3 mm having square shape defects of various total surface areas of 400 and 800 mm^2. Mechanical tests on uniaxial tension and combined tension and torsion are conducted on the universal biaxial servohydraulic system Instron 8850. Digital optical video system Vic-3D allows to reveal an introduced defect and registrate strain fields on the surface of a sample. The mathematical apparatus of video system is presented in the work No 4. The inhomogeneous temperature distribution is obtained by using the infrared thermal vision system Flir SC7600 [4]. Installation of the equipment for the tests is shown in Fig. 32.1.

There are tested 3 defect-free samples under proportional tension with torsion to control the position and angle of twist with different ratios of specified displacement and angle of twist (u/φ). For test running there is chosen the u/φ ratio corresponding to speed longitudinal displacement of 2.5 mm/min and rotational speed of 90°/min. The average load value of defect-free fracture specimens is 47 kN. In accordance with the test program samples with defects (one per each type of defect) are loaded under specified u/φ ratio till the load reaches the level of 23.5 kN and the rotational moment—35 Nm; they are kept under this load for 15–30 s, then are unloaded to zero stress and rotational moment.

32.2 Result and Discussion

Video recording of deformation process is performed by the Q400 cameras with 4 megapixel resolution; main performance of correlation: local area (subset)—39 × 39 pixels, step—4 pixels. Shooting speed is 15 frames/s. As a result of tests, field of cross (ε_{xx}), shear (ε_{xy}) and longitudinal strain (ε_{yy}) at a certain point for samples with different total defect area are constructed: the time point of 16 s corresponds to the sample with the total defect area of 800 mm^2 under tension (Fig. 32.2), the time point of 15 s—to the sample with the total defect area of 400 mm^2 under tension with torsion (Fig. 32.3). Video system allows to fix the place of defect introduction. Defect under tension is less deformable, which is confirmed by field transverse strain ε_{xx} (Fig. 32.4).

E.M. Spaskova (✉) • V.E. Wildemann
Centre of Experimental Mechanics, Perm National Research Polytechnic University, 29 Komsomolsky prospekt, Perm, Russia
e-mail: cem.spaskova@mail.ru; wildemann@pstu.ru

M. Sutton, P.L. Reu (eds.), *International Digital Imaging Correlation Society*, Conference Proceedings of the Society for Experimental Mechanics Series, DOI 10.1007/978-3-319-51439-0_32

Fig. 32.1 The uniaxial tensile test and combined tension and torsion samples of laminated composite materials

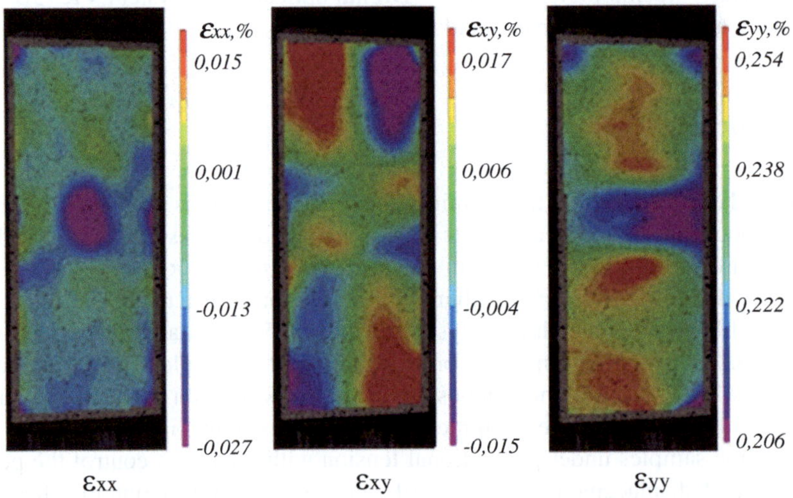

Fig. 32.2 Field of cross (ε_{xx}), shear (ε_{xy}) and longitudinal strain (ε_{yy}) under tension at time 16 s for a sample with a total area of 800 mm^2 defects

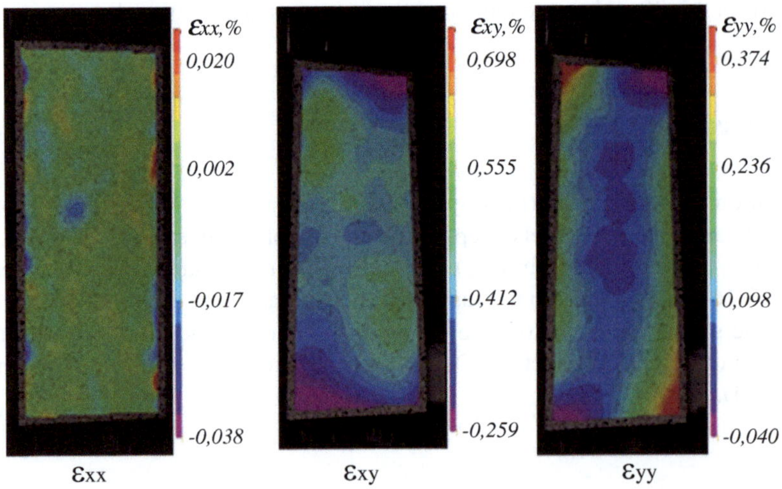

Fig. 32.3 Field of cross (ε_{xx}), shear (ε_{xy}) and longitudinal strain (ε_{yy}) under combined tension and torsion at time 15 s for a sample with a total area of 400 mm^2 defects

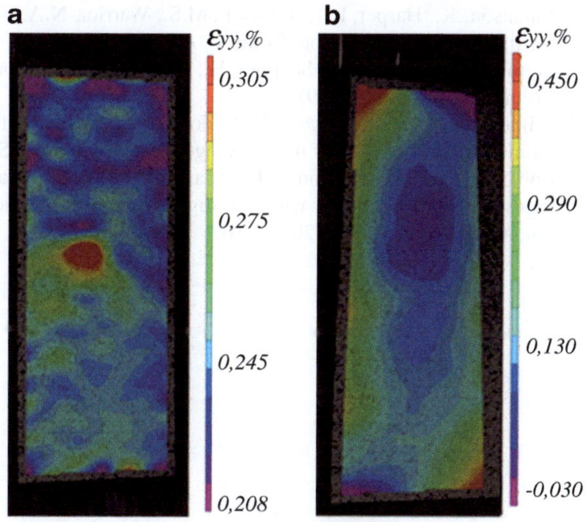

Fig. 32.4 Field of longitudinal strain (ε_{yy}) at time 15 s for a sample with a total area of 800 mm^2 defects under tension (**a**) under combined tension and torsion (**b**)

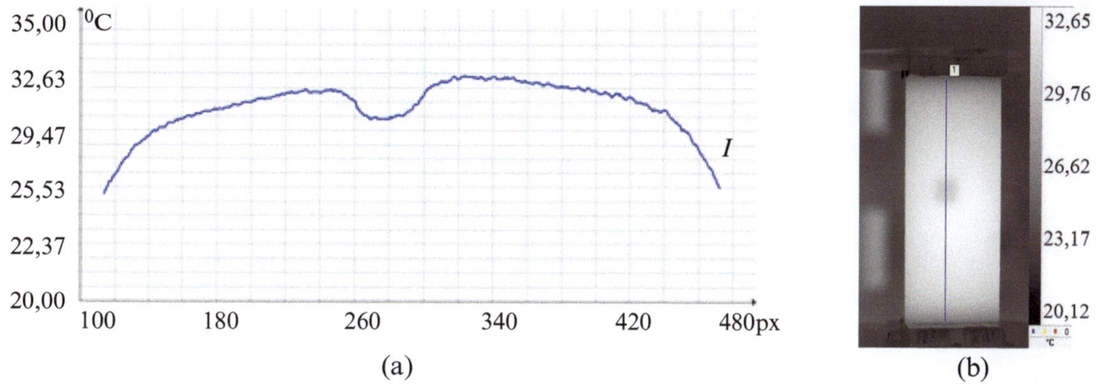

Fig. 32.5 Field temperature after turning off the light source (**b**), the temperature distribution on line 1 passing through the process defect (**a**)

There is registered nonuniform temperature distribution on sample surface in the defect zone after heating, in which the total area of defects is 800 mm^2, the light source makes up 150 W for 30 s. Figure 32.5a shows the temperature field after the shutdown of light source and temperature distribution lines (Fig. 32.5b). The obtained data confirm the presence of pre-introduced defects.

As a result of testing there are obtained the experimental data on samples being under tension and combined tension and torsion demonstrating the presence of inhomogeneous deformation fields, as well as strain changes in the vicinity of a defect registered by fiber optic sensors. The results obtained with the video system, and indications of fiber sensors are duly correlated with each other. In the course of the work there is investigated the evolution of inhomogeneous deformation fields and temperature fields in samples with pre-introduced defects, analyzed diagnostic procedure using advanced diagnostic facilities, which enables the detection of technological defects in samples of composite materials, and seems promising for monitoring and diagnosis of semi-natural structural elements, boundary determination for safe operation of composite structures [5, 6]. The work is carried out at PNRPU's Center of Experimental Mechanics with the joint participation of employees of PNRPU's Scientific-Educational Center of aviation composite technologies. The work is fulfilled with the support of the Russian Science Foundation (Project No. 15-19-00243) at Perm National Research Polytechnic University.

References

1. Panin, S.V., Burkov, M.B., Byakov, M.V., Lyubutin, P.S.: Combined method of study of deformation and destruction of samples of carbon-carbon composite material according to the acoustic emission, digital image correlation and strain measurement Siberia. Bull. Sci. Electron. Sci. J. **4**(5), 129–138 (2012)
2. Elhajjar, R.F., Shams, S.S.: A new method for limit point determination in composite materials containing defects using image correlation. Compos. Sci. Technol. **122**, 140–148 (2016)

3. Johanson, K., Harper, L.T., Johnson, M.S., Warrior, N.A.: Heterogeneity of discontinuous carbon fibre composites: damage initiation captured by digital image correlation. Compos. A Appl. Sci. Manuf. **68**(0), 304–312 (2015)
4. Sutton, M.A., Orteu, J.-J., Schreier, H.: Image correlation for shape, motion and deformation measurements, 364 p. University of South Carolina, Columbia, SC, USA (2009)
5. Wildemann, V.E., Spaskova, E.V., Shilova, A.I.: Research of the damage and failure processes of composite materials based on acoustic emission monitoring and method of digital image correlation. Solid State Phenom. **243**, 163–170 (2016)
6. Lobanov, D.S., Wildemann, V.E., Spaskova, E.M., Chikhachev, A.I.: Experimental investigation of the defects influence on the composites sandwich panels strength with use digital image correlation and infrared thermography methods. PNRPU Mech. Bull. **4**, 159–170 (2015). doi:10.15593/perm.mech/2015.4.10

Chapter 33
Regularization Techniques for Finite Element DIC

J.-F. Witz, J. Réthoré, and J. Hosdez

Abstract To compensate bad quality of speckles pattern or on increase the spacial resolution by decreasing the element size, finite based DIC often requires a additional regularization. Usually the regularization introduce a cut-off length below which the wave length of the measured displacement are filtered. While this might improve the measurement for low frequency displacement to be measure, discontinuities are artificially smoothed what may be detrimental for the analysis of cracks. To circumvent this drawback, a new regularization strategy using the median filter during the optimization process is proposed and compared with the other classical regularizations on perfect discontinuity in order to illustrate the effect of regularization on crack.

Keywords DIC • Median • Regularization • Filtering • Crack

33.1 DIC Method

33.1.1 Problem Formulation

Image Registration, also known as Digital Image Correlation in solid mechanics and Particle Image Velocimetry in fluids mechanics is used since the 1970s [1]. Some improvement has been made in order to take into account strain of the Zone of Interest in the 1980s by [2]. One of the major evolution of Image Registration has been the introduction of the Optical flow used in image registration by Lucas [3] in 1981.

Let x be the coordinates of the fixed image $I_F(x)$ and T the transformation of the moving image $I_M(x)$. The optical flow conservation is given in Eq. (33.1)

$$\mathcal{I}_{\mathcal{F}}(x) = \mathcal{I}_{\mathcal{M}}\left(T(x)\right). \tag{33.1}$$

Solving Eq. (33.1) can be achieved by solving the following problem:

$$\widehat{\mu} = \underset{\mu}{\operatorname{argmin}}\, C\left(T_\mu, \mathcal{I}_F, \mathcal{I}_M\right). \tag{33.2}$$

In Eq. (33.2), C is the metric,T_μ the transformation parametrized by μ defined in Eq. (33.3).

$$T_\mu(x) = x + u(x) = x + f\left(\mu, x\right). \tag{33.3}$$

The transformation used herein is the finite element quadrangle of the first order named $\phi(x)$ in Eq. (33.4) [4].

$$T_\mu(x) = T_{U_i}(x) = x + \{\phi(x)\}^T\{U_i\}. \tag{33.4}$$

J.-F. Witz • J. Hosdez (✉)
LML/Lam3, Cité scientifique, 59651, Villeneuve d'Ascq, France
e-mail: jean-francois.witz@ec-lille.fr; jerome.hosdez@ec-lille.fr

J. Réthoré
GeM, UMR 6183 CNRS, EC Nantes, Université de Nantes, Rue de la Noë, Nantes, France
e-mail: julien.rethore@ec-nantes.fr

M. Sutton, P.L. Reu (eds.), *International Digital Imaging Correlation Society*, Conference Proceedings of the Society for Experimental Mechanics Series, DOI 10.1007/978-3-319-51439-0_33

In the specific case of optical flow conservation the metric C used herein is the Sum of Squared differences (SSD) as defined in Eq. (33.4)

$$SSD(\mu,\mathcal{I}_{\mathcal{F}},\mathcal{I}_{\mathcal{M}}) = \frac{1}{|\Omega_{\mathcal{F}}|}\sum_{x_i\in\Omega_{\mathcal{F}}}\left(\mathcal{I}_{\mathcal{F}}(x_i) - \mathcal{I}_M\left(T_\mu(x_i)\right)\right)^2 \tag{33.5}$$

Even if the following developments have been made using the optical flow assumption, it is possible to relax it. One can use the Normalized Cross Correlation metric or even Mutual Information. An extended numbers of metrics can be found in the image registration software Elastix.

33.1.2 Resolution Process

Solving the optical flow is an ill-posed problem whose resolution process is iterative. In order to be able to solve such problem, one has first to reduce its complexity, this is done using a neighborhood (i.e. the size of the ZOI or element). The method used for solving such process is a Gauss-Newton gradient descent algorithm. It also requires to be able to evaluate the value of the mobile image at a non-integer coordinate, this is done here using a cubic interpolation. As solving this process consist in minimizing C as defined in Eq. (33.5) a gradient descent strategy is used. The gradient of C is given by Eq. (33.6).

$$\begin{aligned}\frac{\partial SSD(\mu,\mathcal{I}_{\mathcal{F}},\mathcal{I}_{\mathcal{M}})}{\partial\mu} &= \frac{2}{|\Omega_{\mathcal{F}}|}\sum\nolimits_{x_i\in\Omega_{\mathcal{F}}}\left(\mathcal{I}_{\mathcal{F}}(x_i) - \mathcal{I}_{\mathcal{M}}\left(T_\mu(x_i)\right)\right)\frac{\partial\mathcal{I}_{\mathcal{M}}(T_\mu(x_i))}{\partial\mu}\\ &= \frac{2}{|\Omega_{\mathcal{F}}|}\sum\nolimits_{x_i\in\Omega_{\mathcal{F}}}\left(\mathcal{I}_{\mathcal{F}}(x_i) - \mathcal{I}_{\mathcal{M}}\left(T_\mu(x_i)\right)\right)\left(\frac{\partial T_\mu(x_i)}{\partial\mu}\right)^t\frac{\partial\mathcal{I}_{\mathcal{M}}(T_\mu(x_i))}{\partial x}.\end{aligned} \tag{33.6}$$

The update rule is the following:

$$\mu_{k+1} = \mu_k + \alpha_k\frac{\partial SSD(\mu,\mathcal{I}_{\mathcal{F}},\mathcal{I}_{\mathcal{M}})}{\partial\mu}. \tag{33.7}$$

with α_k is equal to $\left[\frac{\partial C(\mu,\mathcal{I}_{\mathcal{F}},\mathcal{I}_{\mathcal{M}})}{\partial\mu}\frac{\partial C(\mu,\mathcal{I}_{\mathcal{F}},\mathcal{I}_{\mathcal{M}})^t}{\partial\mu}\right]^{-1}$, an Hessian approximation, in accordance with the Gauss-Newton algorithm.

In order to save time, the opposite transform is identified. This allows us to defined once for all the iterative process at this scale α_k on the fixed image.

33.2 Regularization

The DIC method is generally used for images that comes from experimental set-up with noises and biases that are difficult to quantify. In order to achieve good performances regularization are employed. The main objective of this work is to propose a new regularization and to compare it with the classical ones. The regularization is usually performed by adding a term to C. As in the Tikhonov regularization. The Tikhonov regularization consists in adding a penalization term in the metric as describe in Eq. (33.8)

$$\mathcal{M} = \mathcal{C} + \mathcal{P} \tag{33.8}$$

where C is the sum of squared differences and $\mathcal{P} = \frac{w_{\mathcal{P}}}{|\Omega_{\mathcal{F}}|}\sum_{x_i\in\Omega_{\mathcal{F}}}\nabla u(x_i)\nabla u(x_i)$ The above method is resolved replacing C by M. The gradient of the displacement is evaluated using the Q4P1 shape functions. The determination of the regularization length l_c is adjusted through the parameter w_P that balances the contribution of the regularization term P. More details about the relationship between w_P and the regularization length can be found in [5].

Cachier [6] claimed that using a filters during the iterations can be formulated as an energy minimization. Indeed, one can show that a Tikhonov regularization is equivalent to use a Gaussian filter during the iteration. Later on we choose to use this approach as it allows one to perform regularization that couldn't be implemented as a penalization due to the non-convexity of

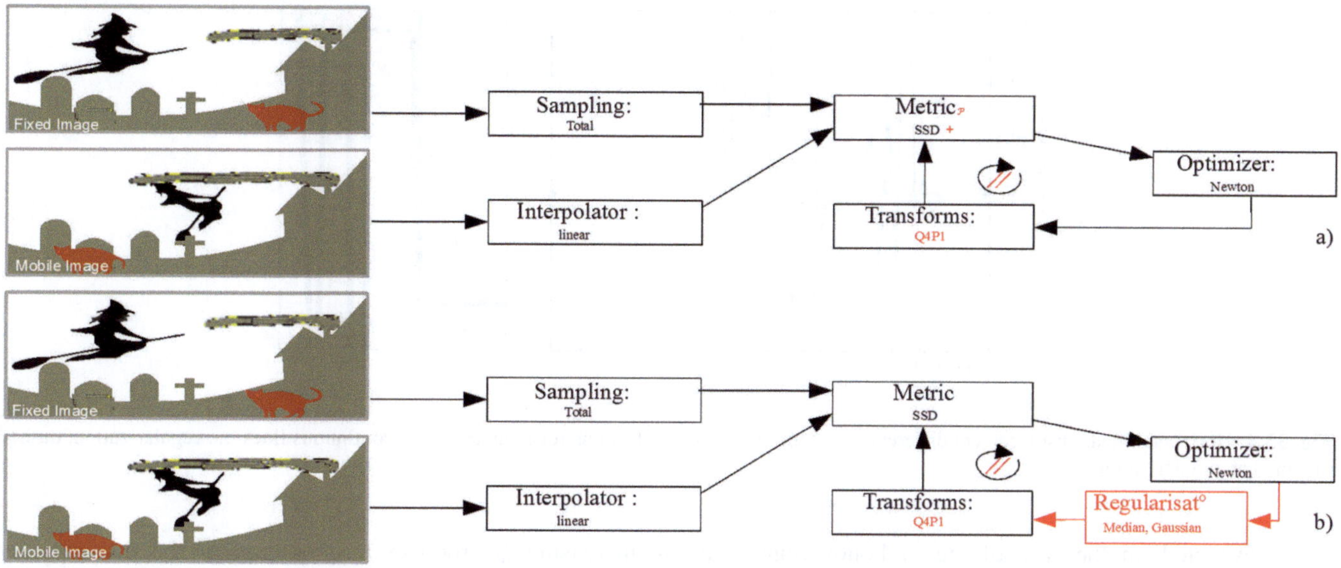

Fig. 33.1 (**a**) Classical FE-DIC. (**b**) FE-DIC with regularization in the optimization process

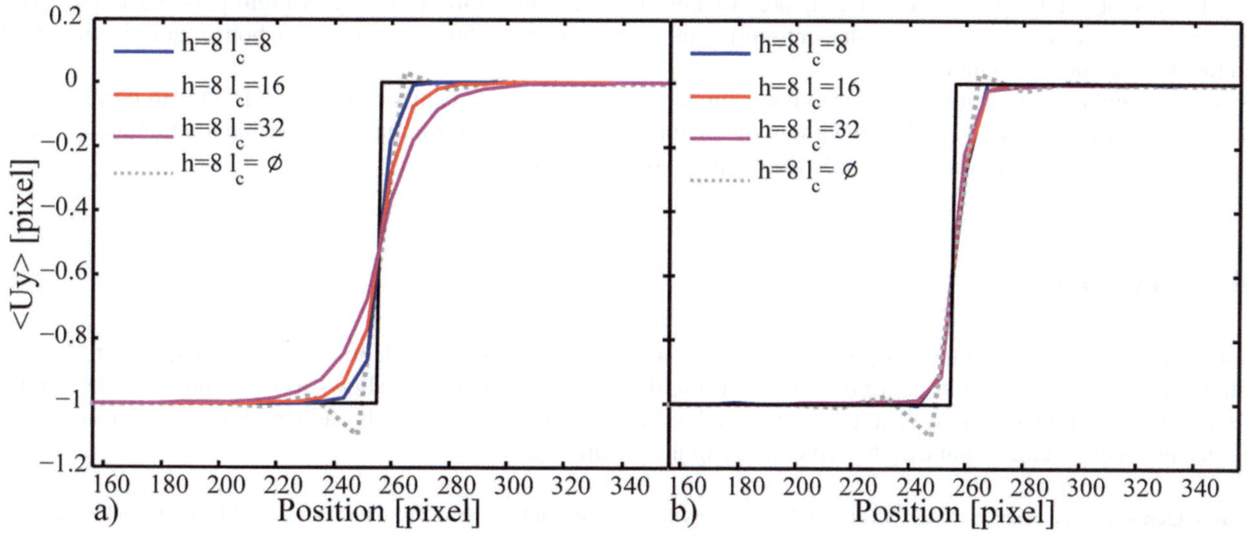

Fig. 33.2 Horizontal mean displacement, h is the element size, l_c the regularization length. (**a**) Tikhonov penalization. (**b**) Median regularization

the associated energy. The median filter is a good example of this kind of regularization. The global procedure is summarized in Fig. 33.1. At each iteration of the process the computed displacement field is filtered using a median filter avoiding the aberrant values to be kept during the iterative process. The regularization length is set by the number of neighboring nodes accounted for the estimation of the median estimate of the displacement. This filter has the ability to keep discontinuities whereas a classical Gaussian filter blur the fields. We have also noticed that *extrema* values of the identified parameters are preserved by the median filter whereas other filters lower the *extrema*.

33.3 Virtual Experiment and Results

In order to avoid the interpolation issues a horizontal one pixel displacement discontinuity is prescribed on a given speckle in the middle of the specimen. The measured displacement is then averaged along the vertical direction and compared to the prescribed displacement. First 8×8 pixel elements are used while the regularization length l_c is varied from 0 to 64 pixels. Figure 33.2 shows the impact of the regularization on the discontinuity.

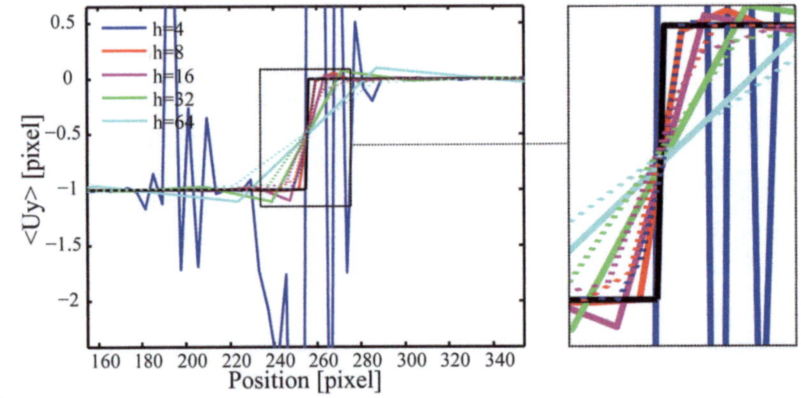

Fig. 33.3 Horizontal mean displacement different size with one element of radius for regularization, continuous lines: no regularization, dashed one median regularization

As expected, in the case of the Tikhonov regularization, increasing l_c produces a smoothing of the discontinuity over a length close to l_c on both side of the discontinuity. Per contra the median filter regularization preserves the discontinuous nature of the displacement, meaning that picking the displacement values on at the nodes of the elements cut by the discontinuity gives a robust estimates of the displacement jump. This results highlights the interest of such regularization process when there are discontinuities in an experiment. One can notes that both regularization methods avoid the displacement overshoot.

Next, the regularization length l_c is set to one element size while the element size is varied from 4 to 64 pixels. Figure 33.3 shows the results obtained with and without the median filtering. It is shown that it is possible to maintain convergence to a good estimate of the displacement while the element size is decreased.

33.4 Conclusion

A new procedure of regularization is proposed for the classical optical flow correlation with the possibility to be adapted to others metrics. The median as continuous filter during the optimization process allows to reject noise while it preserves discontinuities. Such method is well suited for the analysis of heterogeneous materials and structures which usually involve sharp displacement gradient that may be difficult to capture accurately.

Acknowledgements The support of ADEME through grant Cervifer and Région Pays de la Loire through grant MatSyMat, labelled by EMC2, are gratefully acknowledged.

References

1. Barnea, D.I., Silverman, H.: A class of algorithms for fast digital image registration computers. IEEE Trans. Comput. **C-21**, 179–186 (1972)
2. Sutton, M., Wolters, W., Peters, W., Ranson, W., McNeill, S.: Determination of displacements using an improved digital correlation method. Image Vis. Comput. **1**, 133–139 (1983)
3. Lucas, B.D., Kanade, T.: An iterative image registration technique with an application to stereo vision IJCAI81, pp. 674–679. ACM, New York (1981)
4. Besnard, G., Hild, F., Roux, S.: "Finite-element" displacement fields analysis from digital images: application to Portevin–Le Châtelier bands. Exp. Mech. **46**(6), 789–803 (2006)
5. Leclerc, H., et al.: Voxel-scale digital volume correlation. Exp. Mech. **51**(4), 479–490 (2011)
6. Cachier, P., Bardinet, E., Dormont, D., Pennec, X., Ayache, N.: Iconic feature based nonrigid registration: the PASHA algorithm. Comput. Vis. Image Understand. **89**, 272–298 (2003)

Chapter 34
Micro Speckle Stamping: High Contrast, No Basecoat, Repeatable, Well-Adhered

Andrew H. Cannon, Jacob D. Hochhalter, Geoffrey F. Bomarito, and Timothy Ruggles

Abstract Micro Speckle Stamping has been developed and tested whereby repeatable micro speckle patterns for DIC are applied with no basecoat. The speckle patterns are created on a stamp, and ink is applied to the stamp. The user then transfers the speckle pattern from the stamp to the specimen. Micro Speckle Stamping uses high optical contrast and high electrical contrast speckle materials and leaves no residue between speckles. This new method is more amenable to applying patterns to complex surface geometries and large surface areas and also allows investigations of the virgin surface with EDS and EBSD.

Keywords Digital image correlation • Micro speckle stamping • Designer patterns • Pattern • Optimization

Accurate surface strain measurements using Digital Image Correlation (DIC) depend on high-contrast patterns on the test specimen. Error in the strain measurement depends on the pattern applied, and repeatability of the pattern specimen-to-specimen is ideal. With Micro Speckle Stamping and lithography, repeatable micro patterns can be designed [1–15]. Lithography relies on low optical contrast speckle material and leaves residue between speckles. Some other micro speckling techniques suitable for 200 nm to 20 μm speckle base elements rely on base coats that can crack or debond during testing.

High contrast speckle patterns with no specimen base coat result from Micro Speckle Stamping as Fig. 34.1 shows. A high contrast ink is applied to a micro patterned stamp, and the inked stamp transfers its pattern directly to the specimen. The lack of base coat eliminates basecoat cracking as a potential problem during DIC experiments. Because the stamp does not alter the specimen surface between speckles, it is possible to investigate the virgin surface with Energy Dispersive X-ray Spectroscopy (EDS) and Electron Backscatter Diffraction (EBSD).

Variation in speckle patterning specimen-to-specimen causes statistical uncertainty in DIC results. Micro Speckle Stamping uses a designed template, so it creates repeatable speckle patterns specimen-to-specimen.

Speckles must adhere to the specimen for DIC to track specimen deformation accurately. Figure 34.2 shows the transfer printed pattern adheres well to the specimen in extreme conditions. A Micro Speckle Stamp transfer printed a 1 μm base element speckle pattern to an aluminum alloy single edge notch specimen. During tensile testing, a number of slip traces formed around the notch. No cracking of the speckle pattern resulted even on the slip band.

Speckles that result from Micro Speckle Stamping can create high quality, high resolution DIC results. Figure 34.3 shows typical DIC results from stamped 2 μm base element speckles.

Acknowledgements The authors thank Brian Wisner/Drexel for Fig. 34.1 SEM imaging and Prof. Michael Sangid/Purdue for DIC results

A.H. Cannon (✉)
1900 Engineering, LLC, Clemson, SC, USA

Department of Chemical and Biomolecular Engineering, Clemson University, Clemson, SC, USA
e-mail: andrew.cannon@1900engineering.com

J.D. Hochhalter • G.F. Bomarito
NASA Langley Research Center: Durability, Damage Tolerance, and Reliability Branch, Hampton, VA, USA

T. Ruggles
National Institute of Aerospace, Hampton, VA, USA

M. Sutton, P.L. Reu (eds.), *International Digital Imaging Correlation Society*, Conference Proceedings of the Society for Experimental Mechanics Series, DOI 10.1007/978-3-319-51439-0_34

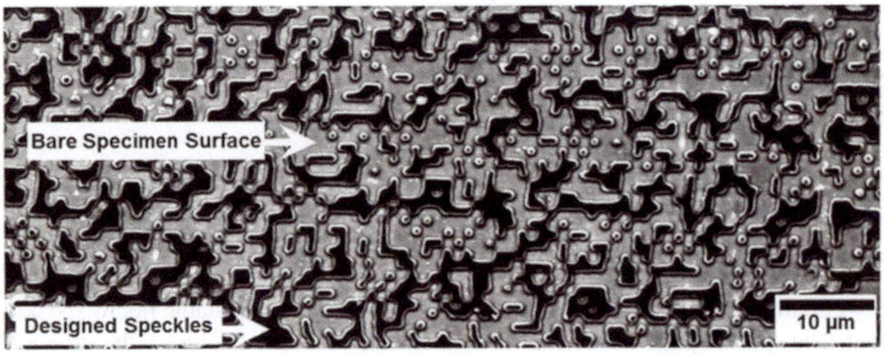

Fig. 34.1 High contrast speckle patterns result from Micro Speckle Stamping. No basecoat is necessary, allowing investigations of the virgin surface with EDS and EBSD

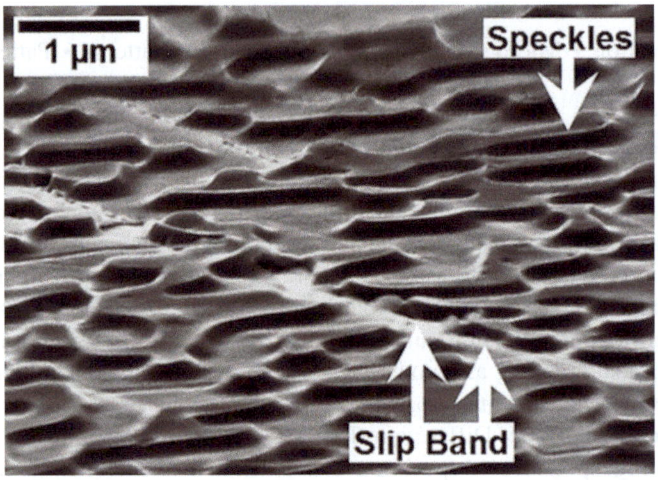

Fig. 34.2 The transfer printed speckle pattern adheres well to the specimen during mechanical shock. No cracking of the speckle pattern resulted on the slip band

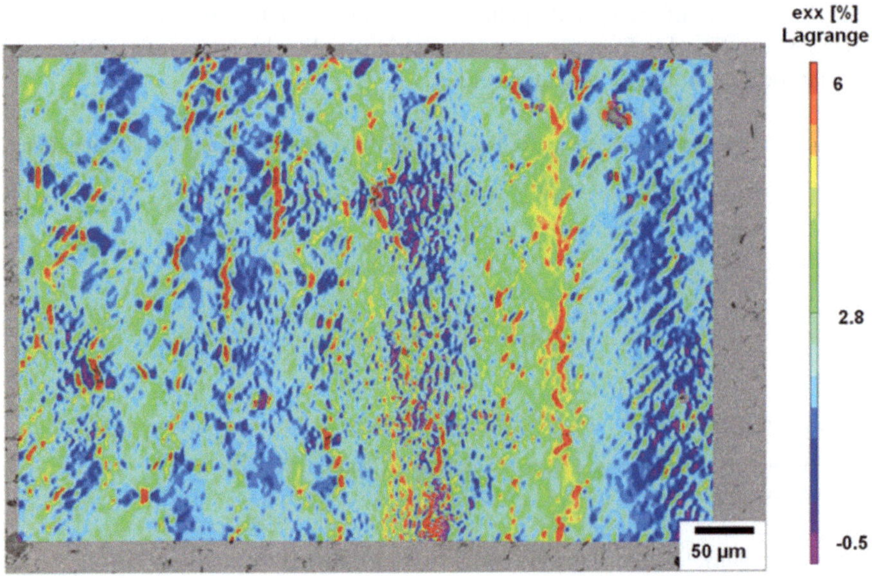

Fig. 34.3 Typical DIC results from Micro Speckle Stamping with 2 µm base elements

References

1. Cannon, A.H., Hochhalter, J.D., Mello, A.W., Bomarito, G.F., Sangid, M.D.: Microstamping for improved speckle patterns to enable digital image correlation. Microsc. Microanal. **21**(S3), 451–452 (2015)
2. Kammers, A.D., Daly, S.: Small-scale patterning methods for digital image correlation under scanning electron microscopy. Meas. Sci. Technol. **22**(12), 125501 (2011)
3. Pan, B., Xie, H., Wang, Z., Qian, K., Wang, Z.: Study on subset size selection in digital image correlation for speckle patterns. Opt. Express. **16**(10), 7037–7048 (2008)
4. Pan, B., Lu, Z., Xie, H.: Mean intensity gradient: an effective global parameter for quality assessment of the speckle patterns used in digital image correlation. Opt. Lasers Eng. **48**(4), 469–477 (2010)
5. Sutton, M.A., Orteu, J.J., Schreier, H.: Image Correlation for Shape, Motion and Deformation Measurements: Basic Concepts, Theory and Applications. Springer Science & Business Media (2009)
6. Wang, Y.Q., Sutton, M.A., Bruck, H.A., Schreier, H.W.: Quantitative error assessment in pattern matching: effects of intensity pattern noise, interpolation, strain and image contrast on motion measurements. Strain. **45**(2), 160–178 (2009)
7. Sun, Y., Pang, J.H.: Study of optimal subset size in digital image correlation of speckle pattern images. Opt. Lasers Eng. **45**(9), 967–974 (2007)
8. Hua, T., Xie, H., Wang, S., Hu, Z., Chen, P., Zhang, Q.: Evaluation of the quality of a speckle pattern in the digital image correlation method by mean subset uctuation. Opt. Laser Technol. **43**(1), 9–13 (2011)
9. Lecompte, D., Smits, A., Bossuyt, S., Sol, H., Vantomme, J., Van Hemelrijck, D., Habraken, A.M.: Quality assessment of speckle patterns for digital image correlation. Opt. Lasers Eng. **44**(11), 1132–1145 (2006)
10. Triconnet, K., Derrien, K., Hild, F., Baptiste, D.: Parameter choice for optimized digital image correlation. Opt. Lasers Eng. **47**(6), 728–737 (2009)
11. Stoilov, G., Kavardzhikov, V., Pashkouleva, D.: A comparative study of random patterns for digital image correlation. J. Theor. Appl. Mech. **42**(2), 55–66 (2012)
12. Crammond, G., Boyd, S.W., Dulieu-Barton, J.M.: Speckle pattern quality assessment for digital image correlation. Opt. Lasers Eng. **51**(12), 1368–1378 (2013)
13. Bossuyt, S.: Optimized patterns for digital image correlation. In: Imaging Methods for Novel Materials and Challenging Applications, vol. 3, pp. 239–248 (2013)
14. Bomarito, G.F., Hochhalter, J.D., Ruggles, T.J., Cannon, A.H.: Increasing accuracy and precision of digital image correlation through pattern optimization. Opt. Lasers Eng. **91**, 73–85 (2016)
15. Winiarski, B., Schajer, G.S., Withers, P.J.: Surface decoration for improving the accuracy of displacement measurements by digital image correlation in SEM. Exp. Mech. **52**(7), 793–804 (2012)

Chapter 35
Determination of Fracture Loci for Anisotropic AA6063-T6 Extrusions

Michael J. Nemcko, Clifford Butcher, and Michael J. Worswick

Abstract As alloy development increases in complexity so do the fracture criteria that describe these materials. For example, aluminum extrusions are processed such that the crystallographic texture and grain structure introduce anisotropy in the constitutive behavior of the material. Furthermore, the stress state has a significant impact on the failure strain. In this work, strain mapping based on the digital image correlation method is used to obtain an experimental failure criterion for AA6063 extrusions in the T6 condition. An experimental program is developed to determine the fracture strain in sample geometries which cover a wide range of stress states. In order to account for the anisotropic nature of the extrusions, samples are tested in the extrusion, 45 and transverse directions for each sample geometry. The fracture strains are extracted from the experiments and the finite element method is used to calculate the triaxiality for each testing condition. Fracture loci are developed for each direction based on a Modified Mohr-Coulomb expression. The results reveal the anisotropic nature of these aluminum extrusions and suggest that the material testing direction must be considered to accurately predict fracture.

Keywords Fracture • Triaxiality • Aluminum extrusions • Anisotropy • Modified Mohr-Coulomb

35.1 Introduction

The development of lightweight structural materials for automotive applications has continuously attracted significant research interest due to government regulations and the demand for enhanced fuel economy. However, as alloy design strategies become more complex, so do the material models used to evaluate material performance. The prediction of failure is essential for the design of lightweight structural materials. This work focuses on the development of a fracture criterion for anisotropic aluminum extrusions which are used as crash rails. In the context of ductile fracture, damage accumulation occurs by the nucleation, growth and coalescence of microvoids. Continuum-based damage models consider the macroscopic response of the material and are often calibrated phenomenologically. The Johnson–Cook model is commonly used to predict failure and incorporates the effects of strain rate and temperature [1]. Furthermore, the model highlights the significance of stress triaxiality (ratio of the mean stress over the effective stress) on the failure strain. More recent studies suggest that the third invariant of the stress tensor (Lode parameter) also has a significant impact on the failure limit [2, 3]. As a result, Bai and Wierzbicki developed the Modified Mohr-Coulomb (MMC) failure criterion by transforming the Mohr-Coulomb model into the space of stress triaxiality, Lode parameter and equivalent plastic strain [4]. This failure criterion has been applied to develop fracture loci for aluminum alloys and high strength steels [5, 6]. The Generalized Incremental Stress-State dependent MOdel (GISSMO) was recently developed by Daimler to predict ductile fracture within the finite element code LS-DYNA [7]. Its primary motivation is to improve numerical simulations which involve strain path dependent processes. Damage is accumulated in an incremental manner using a generalization of the Johnson–Cook model [4]. Once the damage parameter reaches a critical value, damage is introduced and element deletion occurs. The main input into the GISSMO model is tabulated data of the failure strain as a function of triaxiality which is obtained experimentally. Therefore, the objective of this work is to develop experimental fracture loci for AA6063 extrusions in the T6 condition using the digital image correlation method. A hybrid experimental-numerical approach is used to calibrate the MMC failure criterion over a wide range of stress states.

M.J. Nemcko (✉) • C. Butcher • M.J. Worswick
Department of Mechanical and Mechatronics Engineering, University of Waterloo, 200 University Avenue West,
Waterloo, ON, Canada N2L 3G1,
e-mail: mnemcko@uwaterloo.ca

M. Sutton, P.L. Reu (eds.), *International Digital Imaging Correlation Society*, Conference Proceedings
of the Society for Experimental Mechanics Series, DOI 10.1007/978-3-319-51439-0_35

35.2 Experimental Method

An extensive experimental program has been developed to determine the fracture loci of an anisotropic AA6063 aluminum extrusion in the T6 condition. Four sample geometries were investigated which cover a wide range of stress states. These samples include the mini shear, hole tension, regular notch and plane strain notch geometries shown in Fig. 35.1. The specimens were cut directly from the extrusion and the final sample thickness was roughly 1.85 mm. In order to characterize the anisotropy of the material, each sample geometry was tested in three directions such that the maximum principle loading direction coincides with the extrusion, 45 and transverse directions. A speckle pattern was applied to the sample surfaces with spray paint prior to tensile testing. The tensile tests were carried out using the MTS Criterion Model 45 tensile frame at strain rates of approximately 0.001/s.

Images were acquired with two Point Grey digital cameras with 180 mm (1:3.5) Tamron lenses in order to obtain three dimensional measurements. In addition, a 1.4X teleconverter was used to increase the magnification. The frame rate varied depending on the sample geometry; however, the rate was adjusted to obtain at least 500 images before fracture. Images were imported into the Vic-3D software package by Correlated Solutions to obtain the full field strain measurements [8].

35.3 Results

The strain at failure was determined by analysing the image before the first crack was observed on the sample surface. Figure 35.2 shows the strain contours of the maximum principle component at failure. The Virtual Strain Gage Length (VSGL) was calculated by:

$$VSGL = resolution \times filtersize \times stepsize$$

and set to 0.3 mm for all calculations. As a result, the filter and step size were adjusted accordingly for each camera set up. A circular area of interest with a diameter of $2VSGL$ was used to export the strain values in the region where the crack initiated. The local equivalent strain at failure was calculated assuming a Mises material by:

$$d\bar{\varepsilon} = \sqrt{\left[(d\varepsilon_2 - d\varepsilon_3)^2 + (d\varepsilon_3 - d\varepsilon_1)^2 + (d\varepsilon_1 - d\varepsilon_2)^2\right]/3}$$

where $d\varepsilon_i$, $i = 1,2,3$ represents the increments of the principle strain components.

In order to calibrate the MMC model the Lode parameter and triaxiality values are required in the region of interest. Due to the uncertainties of obtaining these parameters experimentally, finite element simulations were performed using LS-DYNA

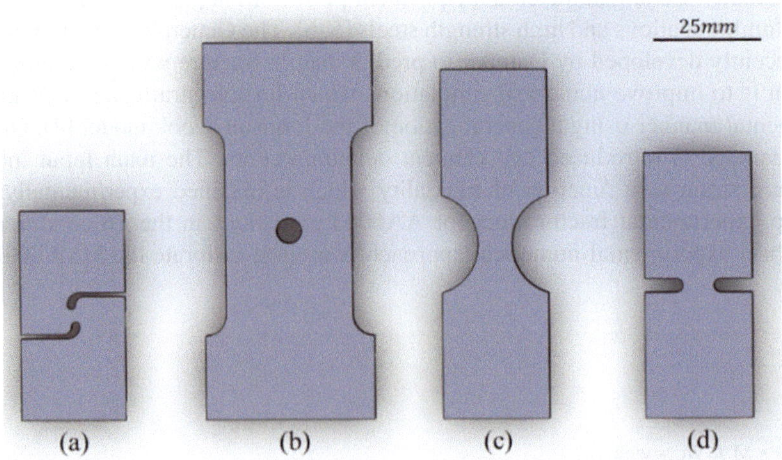

Fig. 35.1 Sample geometries (**a**) mini shear, (**b**) hole tension, (**c**) regular notch, (**d**) plane strain notch

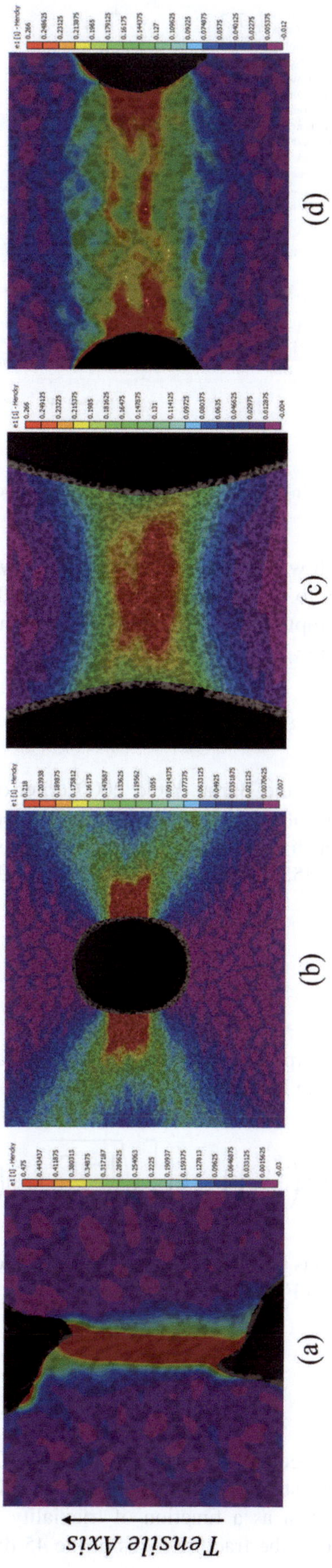

Fig. 35.2 Maximum principal strain contours at failure for **(a)** mini shear, **(b)** hole tension, **(c)** regular notch and **(d)** plane strain notch geometries

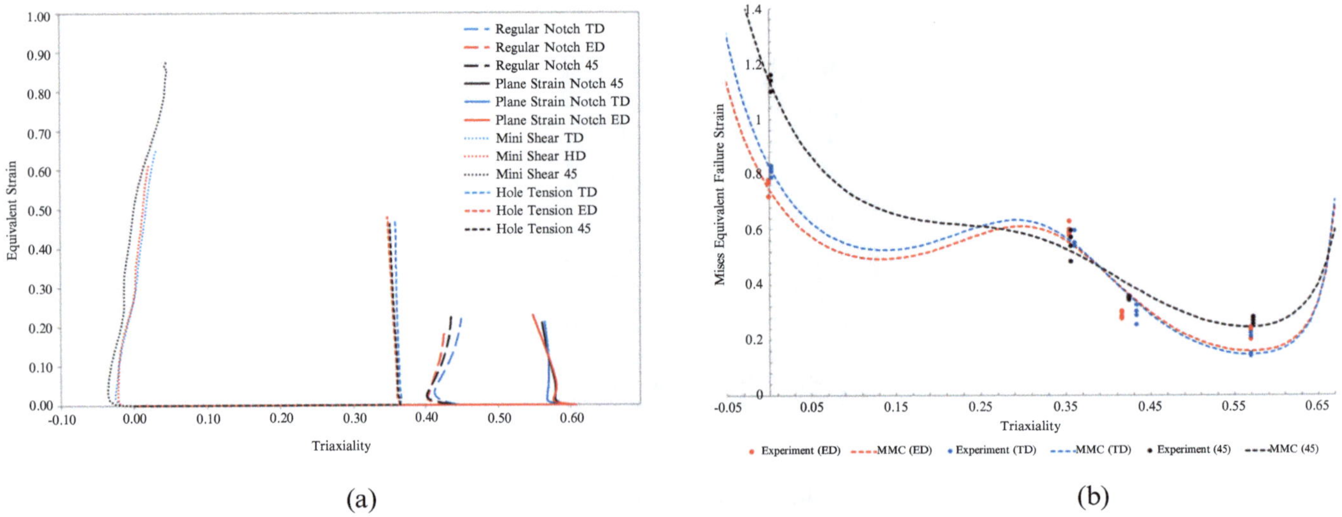

Fig. 35.3 (**a**) Equivalent strain vs. triaxiality for each loading case and (**b**) the MMC model fit to the hybrid experiment simulation data

to determine these parameters. Altair Hypermesh was used to mesh the parts shown in Fig. 35.1 with 2D elements. A mesh size of 0.3 mm was used in the critical regions to coincide with the VSGL. Fully integrated shell elements were used in the analysis which employ the plane stress assumption. Under this assumption a unique relation is established between the triaxiality (η) and the Lode parameter (ξ) given by:

$$\xi = -\frac{27}{2}\eta\left(\eta^2 - \frac{1}{3}\right)$$

The material model used in the simulation is the Barlat Yield 2000 criterion which has the capacity to capture the anisotropic nature of the extrusions. Calibration of the yield function is described in Zhumagulov et al. [9]. Simulations were carried out for each sample geometry in each of the three loading directions (ED, 45 and TD) and the triaxiality was exported from the simulations as shown in Fig. 35.3a. The average triaxiality value was calculated for each test condition according to:

$$\eta = \frac{1}{\bar{\varepsilon}_f}\int_0^{\bar{\varepsilon}_f} \eta\left(\bar{\varepsilon}\right) d\bar{\varepsilon}$$

Combination of the failure strains obtained experimentally and the numerical triaxiality values yields several critical points which can be used to calibrate the following MMC model:

$$\bar{\varepsilon}_f(\eta,\xi) = \left\{\frac{C_1}{C_2}\left[C_3 + \frac{\sqrt{3}}{2-\sqrt{3}}(1-C_3)\left(\sec\left(\frac{\xi\pi}{6}\right)\right) - 1\right]\left[\sqrt{\frac{1+C_4^2}{3}}\cos\left(\frac{\xi\pi}{6}\right) + C_4\left(\eta + \frac{1}{3}\sin\left(\frac{\xi\pi}{6}\right)\right)\right]\right\}^{\frac{-1}{C_5}}$$

where C_1 through C_5 are constants. The model has been fit to the hybrid experimental-numerical data points using the least squares methodology and the results are shown in Fig. 35.3b.

35.4 Conclusions

A hybrid experimental-numerical approach has been used to calibrate the MMC fracture criterion. The results reveal the anisotropic nature of fracture in the extrusions investigated. Although the ED and TD fracture loci are similar, the simulations suggest that the evolution of the equivalent strain as a function of triaxiality differs between these loading directions, particularly in the notch geometries. Furthermore, the fracture locus of the 45 direction is considerably different from the other directions, namely due to the large failure strain of the shear geometry. Nevertheless, the material reference system must be considered in order to accurately predict failure in the extrusions investigated. Future work will implement these fracture loci to evaluate the crashworthiness of this anisotropic extrusion.

References

1. Johnson, G.R., Cook, W.H.: Fracture characteristics of three metals subjected to various strains, strain rates, temperatures and pressures. Eng. Fract. Mech. **21**(1), 31–48 (1985)
2. Gao, X., Kim, J.: Modeling of ductile fracture: significance of void coalescence. Int. J. Solids Struct. **43**(20), 6277–6293 (2006)
3. Zhang, K., Bai, J., Francois, D.: Numerical analysis of the influence of the Lode parameter on void growth. Int. J. Solids Struct. **38**(32), 5847–5856 (2001)
4. Bai, Y., Wierzbicki, T.: A new model of metal plasticity and fracture with pressure and Lode dependence. Int. J. Plast. **24**(6), 1071–1096 (2008)
5. Beese, A.M., Luo, M., Li, Y., Bai, Y., Wierzbicki, T.: Partially coupled anisotropic fracture model for aluminum sheets. Eng. Fract. Mech. **77**(7), 1128–1152 (2010)
6. Luo, M., Wierzbicki, T.: Numerical failure analysis of a stretch-bending test on dual-phase steel sheets using a phenomenological fracture model. Int. J. Solids Struct. **47**(22), 3084–3102 (2010)
7. Basaran, M., Wölkerling, S.D., Feucht, M., Neukamm, F., Weichert, D.: An extension of the GISSMO damage model based on lode angle dependence. In: LS-DYNA Anwenderforum (2010)
8. Solutions, C.: VIC-3D User Manual. Correlated Solutions, Columbia, SC (2005)
9. Zhumagulov, A., Imbert, J., Kohar, C., Inal, K., Worswick, M., & Mishra, R.. Energy absorption characterization of multicellular AA6063-T6 extrusions. Paper presented at the ICILSM, 2016

Chapter 36
DIC Analysis for Crack Closure Investigations During Fatigue Crack Growth Following Overloads

G.L.G. Gonzáles, J.A.O. González, J.T.P. Castro, and J.L.F. Freire

Abstract The DIC technique was used to determine displacement and strain fields surrounding the tip of a fatigue crack in DC(T) (disk compact tension) specimen of normalized AISI 4340 steel, before and after an 120% overload was applied over otherwise fixed loading conditions in mode I $\Delta K = 31$ MP$\sqrt{m}$ and $R = K_{min}/K_{max} = 0.1$. The specimen has thickness of 4.6 mm, crack length to thickness ratio of 1.3 and width of 55.55 mm. From the DIC analysis, important parameters could be extracted, such as stress intensity factors, crack opening displacements, J-integrals, crack tip blunting, crack opening loads, and plastic zone sizes. These parameters were used to investigate the crack closure behavior before and after the application of the overload event, which produced a significant retardation in the subsequent crack growth rate. The role of the non-linearities observed in the experimental results during the crack closure phase, which are most important to correctly measure the above mentioned parameters, are analyzed and discussed.

Keywords DIC • Crack closure • Overload • Crack-tip fields • Fracture mechanics

36.1 Introduction and Experimental Methods

Since the pioneer investigations of Elber [1], many studies have been performed to investigate the crack closure phenomenon in order to model and hopefully to improve the understand of crack growth mechanics, in particular under variable amplitude fatigue loads [2, 3]. Since the Digital Image Correlation is a powerful tool to measure the full-field displacement and strain distribution surrounding the crack tip region with sufficient accuracy, it has potential to be very useful in crack closure measurements as well. However, as discussed following, such measurements are not as simple and straightforward as one could anticipate, due to the non-linearities introduced by the closure phase during the loading cycle.

In this work, DIC analysis was used to study the crack closure behavior during fatigue crack growth following overloads. For that purpose, a disk-shaped compact tension DC(T) specimen (see Fig. 36.1a) of AISI 4340 steel with 200 HV (Hardness Vickers) was used. In order to perform the DIC analysis, the sequences of images taken during the tests were exported to be analyzed using the commercial software VIC-3D developed by Correlation Solutions [4]. The loading history consisted of blocks of constant amplitude loading, including the application of two overload (OL) cycles (see Fig. 36.1b).

36.2 Results and Conclusion

The horizontal (u) and the vertical (v) displacement fields obtained from the DIC analysis, before (with A as reference image) and after (with B as reference image) the OL, were used to calculate the Mode I SIF through the J-integral approach. For that, a rectangular path enclosing the crack tip was selected (see Fig. 36.2a). Assuming linear elastic conditions prevailing along the integration path, Linear Elastic Fracture Mechanics (LEFM) concepts can be used in the formulation, and therefore the J-integral value can be related to the equivalent SIF.

From the SIFs values shown in Fig. 36.2b, it can be noted that a non-linear phenomenon related to a classical crack opening-closure behavior at the crack tip region occurs during the loading cycles before the OL, and that it affects the measured K_I at low loads. Moreover, after approximately 2 kN, the experimentally determined K_I results follow the same trend (slope) predicted by the ASTM equation, which assumes that the crack is fully opened during the entire loading cycle,

G.L.G. Gonzáles (✉) • J.A.O. González • J.T.P. Castro • J.L.F. Freire
Pontifical Catholic University of Rio de Janeiro, PUC-Rio, Rua Marquês de São Vicente 225, Gávea, Rio de Janeiro, RJ, Brazil
e-mail: gonzalesglg@aaa.puc-rio.br

M. Sutton, P.L. Reu (eds.), *International Digital Imaging Correlation Society*, Conference Proceedings of the Society for Experimental Mechanics Series, DOI 10.1007/978-3-319-51439-0_36

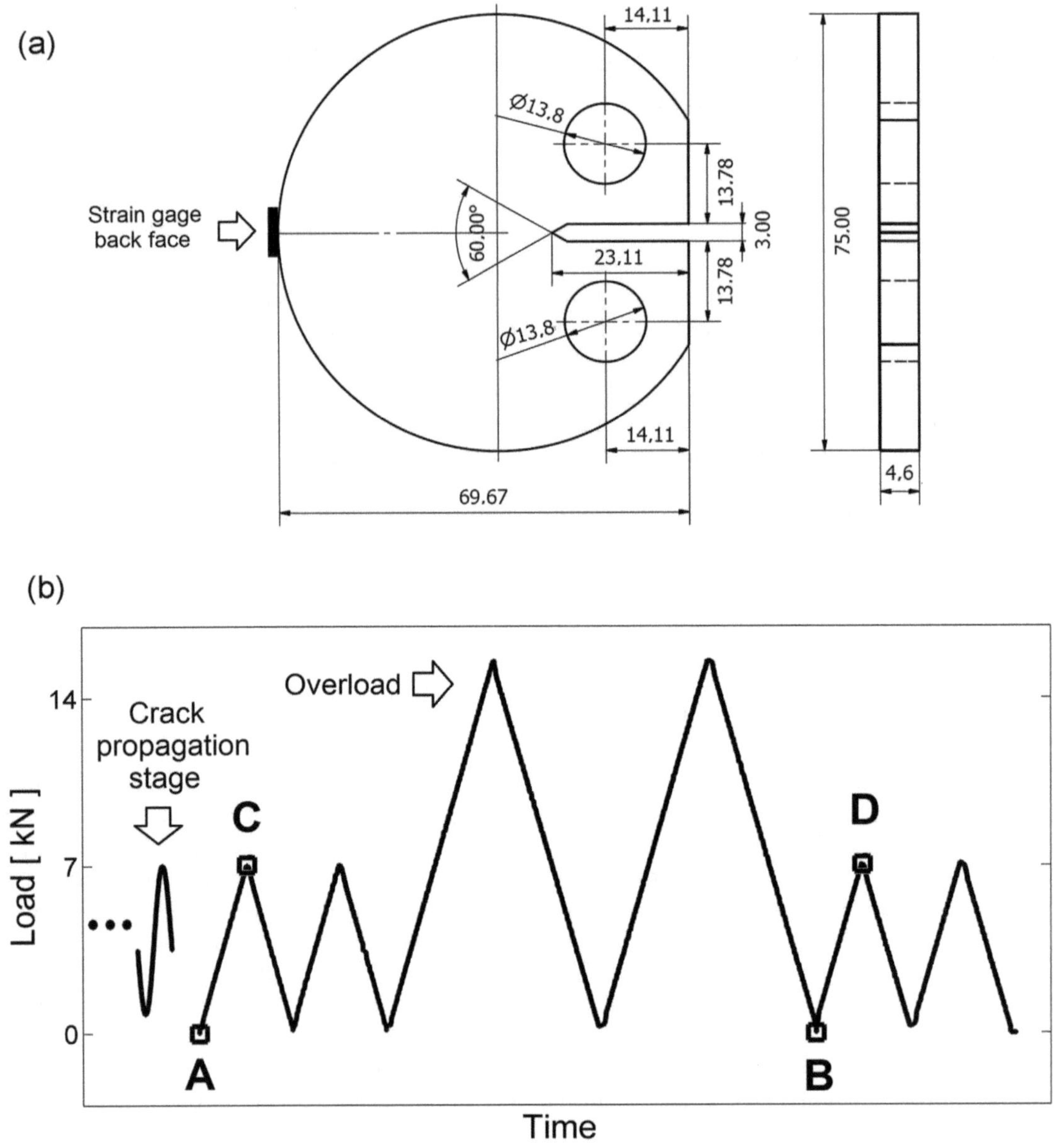

Fig. 36.1 (**a**) DC(T) specimen geometry (dimensions in mm) (**b**) Loading history applied to the specimen

neglecting non-linear behaviors observed in the crack tip region. These K_I results can and should be interpreted as a pseudo-K_I or a pseudo-SIF value calculated assuming a linear-elastic, thus closure-free, crack behavior. The analysis of the DIC data after the overload of 120% of the first loading block peak verified this claim. It can be notice that these overloads blunt the crack tip and almost remove the crack closure effects immediately after their application, so that the crack remains open even at the minimum load after it was returned to its previous value. Hence, in this case the K_I calculated from the measured data using LEFM assumptions can be directly compared with the ASTM equation.

In addition, the vertical (v) displacement field obtained from the DIC analysis was used to obtain COD measurements. For that, two symmetrical points located along the crack faces behind the crack tip are located at a distance 2.7 mm from the crack tip (see Fig. 36.3a). From Fig. 36.3b, it can be noted that the COD measurements before and immediately after the OL have the same behavior than the Mode I SIF presented in Fig. 36.2b. Moreover, Fig. 36.3b shows COD measurements for post overload cycles, OL+6 k and OL+1300 k. Due to crack propagation, for each increment of about 0.1 mm, the maximum peak load was updated by using the ASTM equation and the actual crack size to keep constant $\Delta K = 31$ MP$\sqrt{}$m. The experimental results (Fig. 36.3b) show that the crack opening load increased from 33% of the maximum load before the application of OL (2.3 kN/7 kN) to 86% (4.8 kN/5.6 kN) after the application of the OL+1300 k cycles. This observation is consistent with the retardation of the fatigue crack growth shown in Fig. 36.3c.

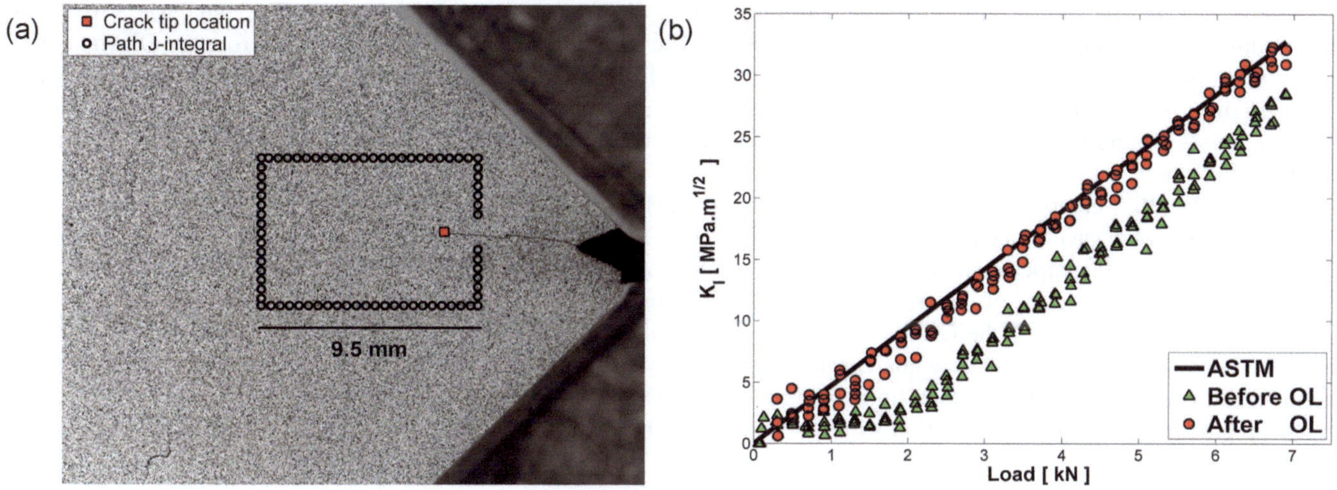

Fig. 36.2 (**a**) Path used in J-integral formulation (**b**) SIF evaluation with J-approach

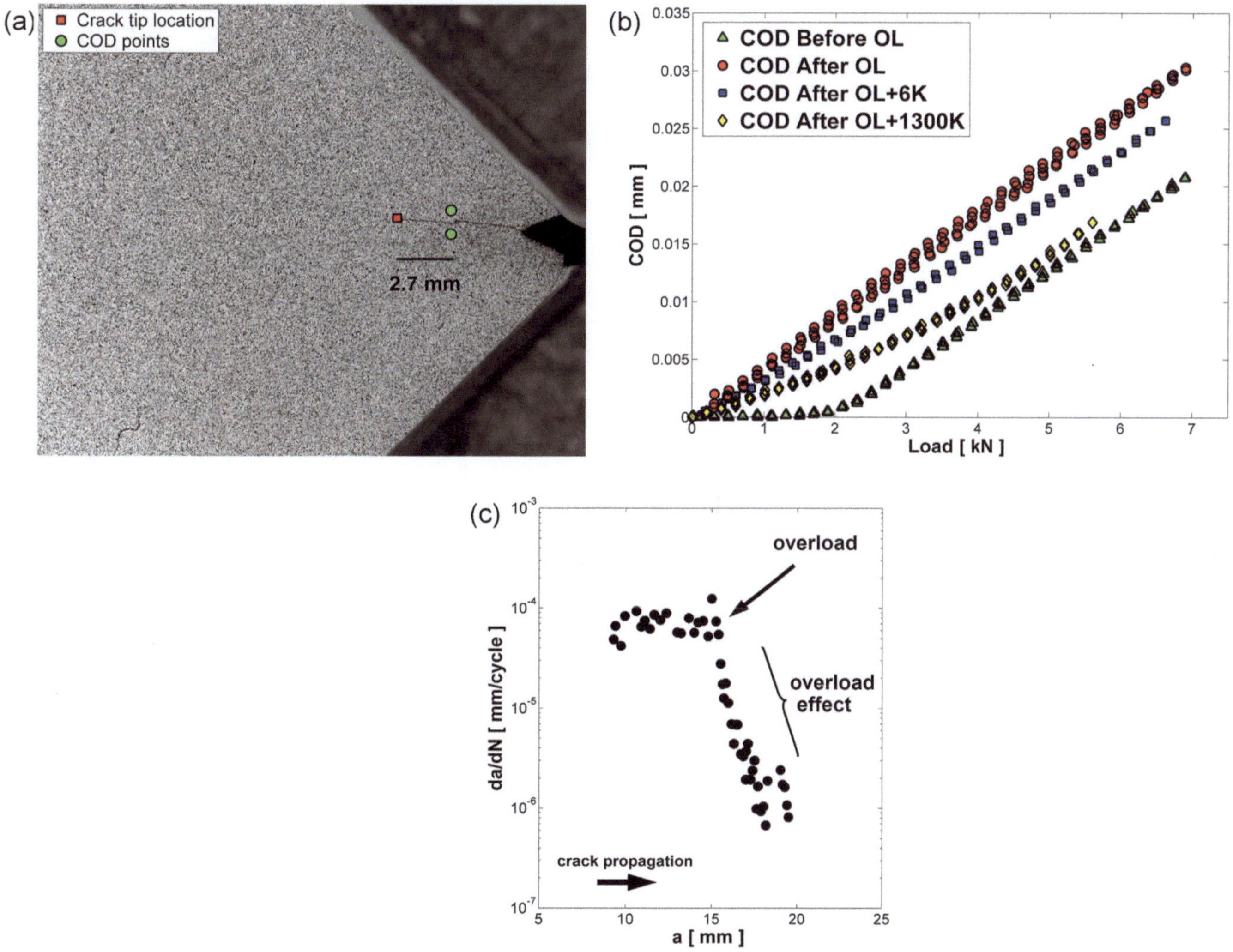

Fig. 36.3 (**a**) Example of symmetrical points selected for COD measurements (**b**) COD measurements following overloads (**c**) da/dN vs crack length

Figure 36.4 depicts the strain maps parallel to the loading direction (ε_y) obtained from the DIC analysis at the same maximum load 7 kN, before and immediately after the OL. The strain ranges around the crack tip before the OL (max. load at C and A as the reference image, see Fig. 36.1b) are shown in Fig. 36.4a. The residual strains induced by the OL at the crack tip (max. load at D and C as the reference image, see Fig. 36.1b) are shown in Fig. 36.4b. Moreover, it can be seen that the data points around the crack faces and the crack tip were excluded from this analysis to avoid their intrinsically higher noise level.

Finally, Fig. 36.5 compares the strain behavior of a point 0.5 mm ahead to the crack-tip with the strain information collected by a strain gage placed at the back face of the DC(T) specimen during the entire loading cycle shown in Fig. 36.1b. As expected, the experimental results have the same trend, including the non-linear behavior, a clear indication of their compatibility. The effect of the crack closure in strain distribution is observed in the first cycles before OL. Then, the crack tip blunting induced by the OL is observed in the loading part of the overload cycle. The second OL cycle has a hysteretic behavior and a significantly non-zero offset caused by the first OL cycle. Moreover, it can be noted that the initial nonlinearities were removed, the crack tip is being held open even at zero loading most probably as result of the induced crack tip blunting. The strain distribution without closure effects observed in the last loading cycle, similar to the first one, confirms this observation.

Acknowledgements This work was supported by CNPq-Conselho Nacional de Desenvolvimento Científico e Tecnológico (Brazil).

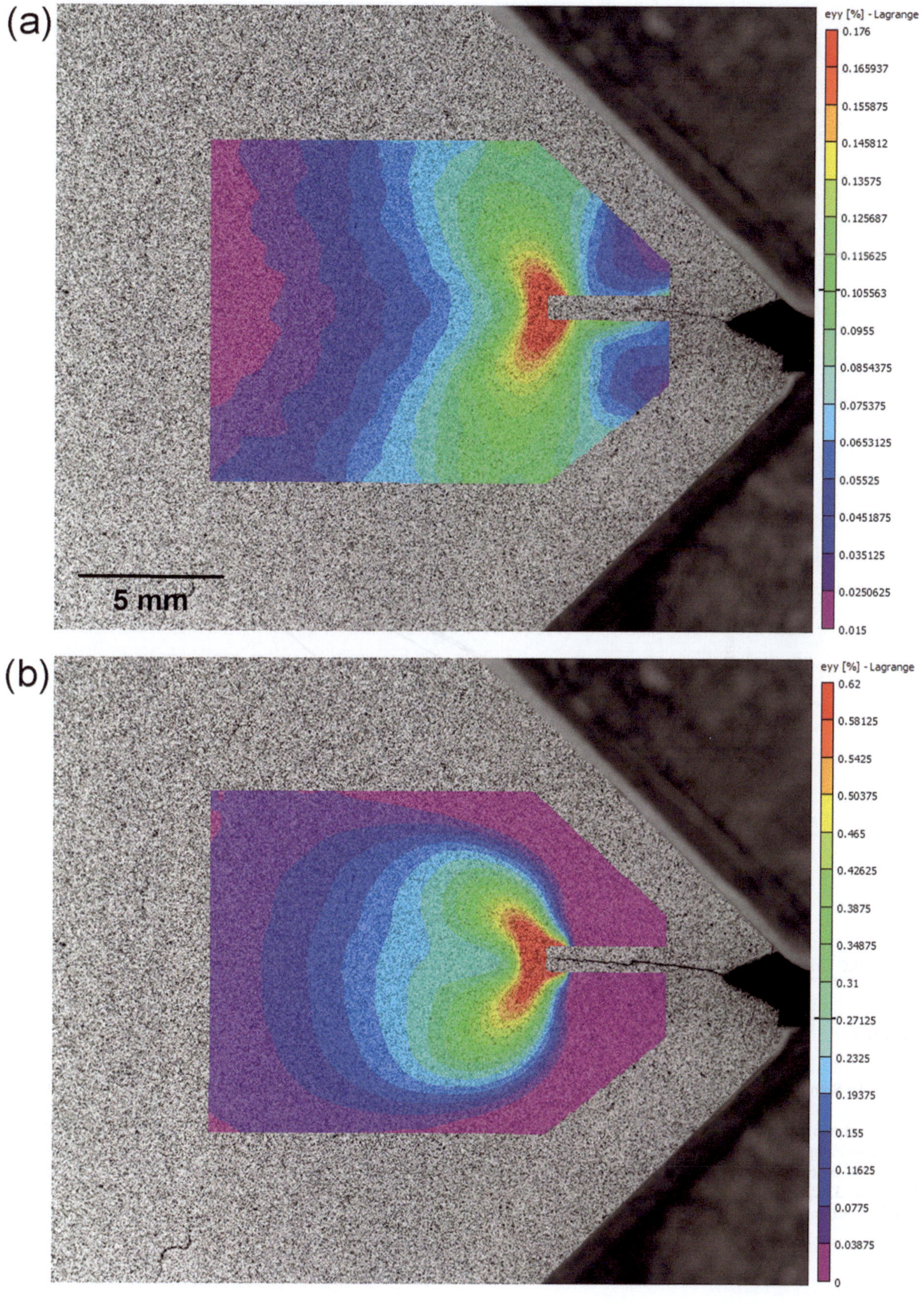

Fig. 36.4 Crack-tip strain field before (**a**) and after (**b**) overload

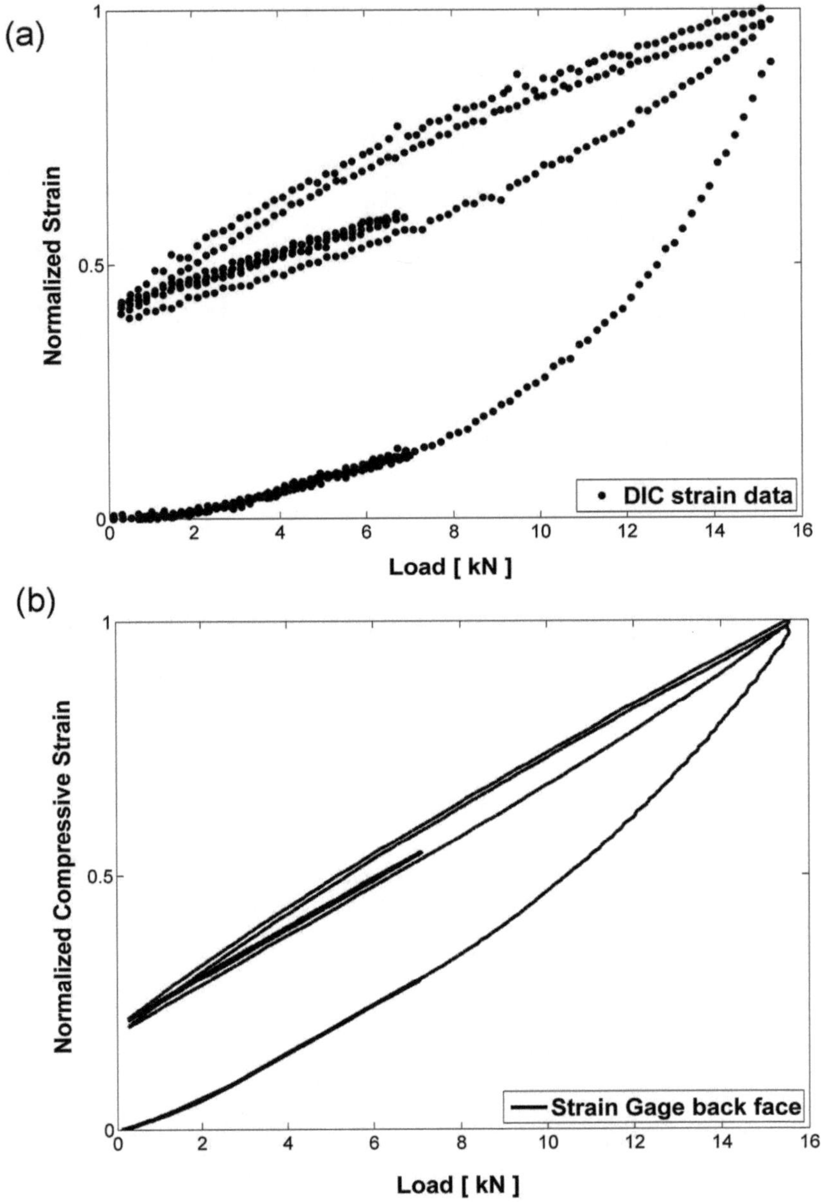

Fig. 36.5 (**a**) DIC strain data at 0.5 mm ahead the crack tip (**b**) Strain gage data at back face of the specimen.

References

1. Elber, W.: The Significance of Fatigue Crack Closure, Damage Tolerance in Aircraft Structures. ASTM International, West Conshohocken, PA (1971)
2. Skorupa, M.: Load interaction effects during fatigue crack growth under variable amplitude loading - a literature review. Part II: qualitative interpretation. Fatigue Fract. Eng. Mater. Struct. **22**, 905–926 (1999)
3. Lugo, M., Daniewicz, S.R., Newman, J.C.: A mechanics based study of crack closure measurement techniques under constant amplitude loading. Int. J. Fatigue. **33**(2), 186–193 (2011)
4. Correlated Solutions Inc., U. S. A. http://www.correlatedsolutions.com/

Chapter 37
On the Evaluation of Volume Deformation from Surface DIC Measurements

Marco Rossi, Marco Sasso, and Luca Cortese

Abstract Stereo DIC allows to measure with a good accuracy the displacement field and the shape changes of the surface of objects during deformation processes. Moreover, using multiple cameras makes possible to measure the deformation of the whole surface of the object in 360°. Such boundary information can be exploited to compute the internal displacement of a solid object using interpolation functions, especially if the deformation is large. This can be applied, for instance, to investigate the strain localization that occurs in metal specimen subjected to plastic deformation, and to study the necking initiation and the fracture development. In this paper, an interpolation method based on Bézier curves is developed and tested using simulated experiments on specimen with different shape. In particular, a flat and a round specimens are studied, looking at the deformation in the necking zone. The accuracy of the interpolation algorithm is evaluated at different level of deformation using both isotropic and anisotropic materials. Examples on real measurement are also discussed.

Keywords Digital image correlation • Stereo measurement • 3D reconstruction • Large deformation

37.1 Introduction

The use of stereo-DIC to measure the shape and the displacement field of three-dimensional surfaces is nowadays a well established experimental practice. Several commercial and academic software are available to perform this type of measurement with a high level of accuracy [1] (http://matchidmbc.be/). On the other hand, it is rather difficult to measure the volume displacement in solids, this can be achieved using X-ray tomography and digital volume correlation (DVC) [2–5], however, those techniques require the existence of a random internal pattern that has to be present already in the material or introduced artificially [6–8]. Another possibility is using the surface information obtained with stereo DIC around the surface of the specimen [9, 10] and interpolate the internal displacement using a sort of shape function [11]. In this work a study of the accuracy of this type of reconstruction is achieved using simulated experiments on a steel-like materials.

37.2 Reconstruction Algorithm

The reconstruction algorithm exploits 3D Bézier curves to interpolate the volume displacement and, in this form, it is applicable to simple geometries (flat or round specimens) of specimens that undergo large deformation, caused, for instance, by plasticity or hyperelastic behaviour. In case of flat specimens, it is necessary to know the displacement map along two faces while for round specimens the whole surface has to be measured [10]. For flat specimens, as first step, the surface is subdivided in a regular rectangular mesh. Then, the displacement measured by the stereo-DIC over the two faces is interpolated along the surface nodes of the rectangular mesh. In such way, each measurement point in one face corresponds to another one in the opposite face.

$$B(k) = (1-k)^2 N^L + 2(1-k)\,kP + k^2 N^U \tag{37.1}$$

M. Rossi (✉) • M. Sasso
Università Politecnica delle Marche, via Brecce Bianche, 60100, Ancona, Italy
e-mail: m.rossi@univpm.it; m.sasso@univpm.it

L. Cortese
“Sapienza” Università di Roma, Rome, Italy
e-mail: luca.cortese@uniroma1.it

M. Sutton, P.L. Reu (eds.), *International Digital Imaging Correlation Society*, Conference Proceedings of the Society for Experimental Mechanics Series, DOI 10.1007/978-3-319-51439-0_37

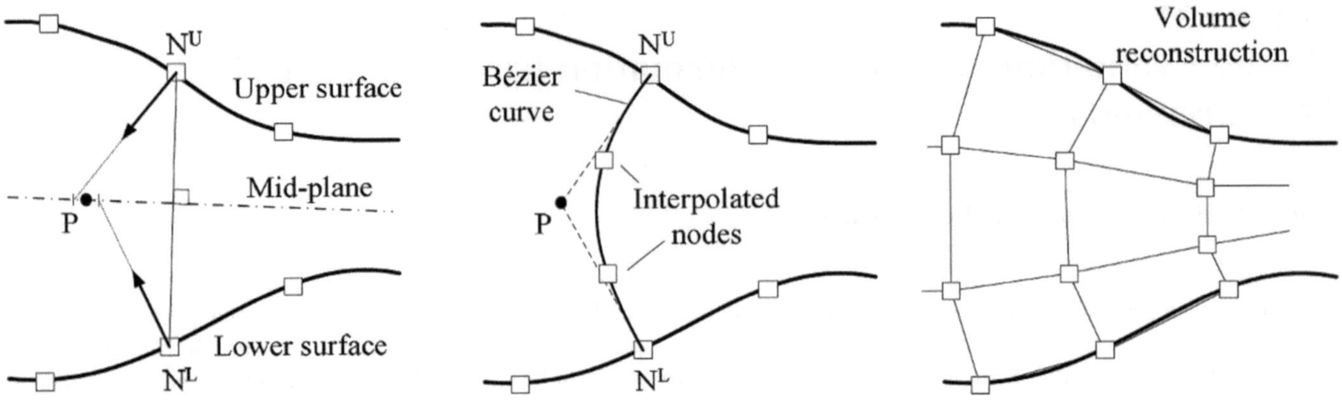

Fig. 37.1 Example of volume reconstruction

where N^L is the point in the lower surface, N^U is the node at the upper surface and P is a central node obtained as the closest point respect to the intersection of the normal of N^L and N^U in the mid-plane; k is a parameter that varies from 0 to 1 and can be used to divide the curve in different points. The derivation of the 3D strain map is obtained using the shape function of brick elements. A similar approach can be applied to round specimens, in this case the point P is chosen at the axis of the specimen, which can be obtained as an average axis from the stereo DIC measurement.

In Fig. 37.1, an example of the reconstruction process is illustrated. If noise is applied, obviously, it will impact also the displacement of the reconstructed internal nodes, thus this technique can be adopted only if the total amount of strain is sufficiently high.

37.3 Conclusion

The present work introduces a procedure to evaluate the volume displacement of solids from surface measurements. This method can be easily implemented on flat specimen and round specimen and allow to investigate to occurrence of instabilities process such as necking. In future, this approach has to be validated on actual experimental data.

References

1. Sutton, M.A., Orteu, J.J., Schreier, H.W.: Image Correlation for Shape, Motion and Deformation Measurements. Springer, New York (2009)
2. Bay, B.K., Smith, T.S., Fyhrie, D.P., Saad, M.: Digital volume correlation: Three-dimensional strain mapping using X-ray tomography. Exp. Mech. **39**, 217–226 (1999)
3. Smith, T.S., Bay, B.K., Rashid, M.M.: Digital volume correlation including rotational degrees of freedom during minimization. Exp. Mech. **42**, 272–278 (2002)
4. Verhulp, E., van Rietbergen, B., Huiskes, R.: A three-dimensional digital image correlation technique for strain measurements in microstructures. J. Biomech. **37**, 1313–1320 (2004)
5. Gates, M., Lambros, J., Heath, M.T.: Towards high performance digital volume correlation. Exp. Mech. **51**, 491–507 (2011)
6. Franck, C., Hong, S., Maskarinec, S., Tirrel, D., Ravichandran, G.: Three-dimensional full-field measurements of large deformations in soft materials using confocal microscopy and digital volume correlation. Exp. Mech. **43**, 207–218 (2007)
7. Germaneau, A., Doumalin, P., Dupré, J.: 3D strain field measurement by correlation of volume images using scattered light: recording of images and choice of marks. Strain. **43**, 207–218 (2007)
8. Roux, S., Hild, F., Viot, P., Bernard, D.: Three-dimensional image correlation from X-ray computed tomography of solid foam. Compos Part A Appl Sci Manuf. **39**, 1235–1265 (2008)
9. Rossi, M., Chiappini, G., Sasso, M.: Characterization of aluminum alloys using a 3D full field measurement. In: SEM Annual Concerence & Exposition on Experimental & Applied Mechanics, p. 2010 (2010)
10. Genovese, K., Cortese, L., Rossi, M., Amodio, D.: A 360-deg Digital Image Correlation system for materials testing. Opt. Lasers Eng. **82**, 127–134 (2016)
11. Rossi, M., Pierron, F.: Identification of the plastic behaviour in the post-necking regime using a three dimensional reconstruction technique. Key Eng. Mater. **504–506**, 703–708 (2012)

Chapter 38
DIC of Dual Thick-Wall Pressurized Pipe

Benjamin Cheung, Jeremy Goh, Megnath Ramesh, Jason Carey, and Simon Iremonger

Abstract The Multipurpose Composites Group (MCG), along with an industrial partner Sanjel Energy Services Inc. (Sanjel) is developing a coupled pressure-testing and digital image correlation (DIC) system for the testing and analysis of internally pressurized carbon steel pipes. In the context of this industry, 'failure' is considered to be the formation and propagation of a crack in a second cylindrical layer of cement external to the pressurized pipe. Simple elastic models predict failure in the cement layer prior to any yielding in the steel pipe. Certain mechanical additives to the cement are predicted to improve a number of mechanical properties, namely a decrease in tensile modulus, an increase in tensile strength, and an increase in fracture toughness. Initial testing has included Brazilian tensile tests with DIC strain measurement, along with Brazilian tensile tests of pre-cracked specimens to determine the critical stress intensity factor of the cement. The next phase of this project is to implement image capture for DIC strain measurement and visualization along with pressure testing, including both static and fatigue testing. Both the image capture and pressure control will be automated with custom-built software control, while 2D and 3D (via stereo image capture) DIC strain measurement will be performed using commercial software.

Keywords DIC • Modeling • Cement • Fracture toughness • Elastic modulus

38.1 Background

In hydrocarbon resource extraction activities it is critical that the hydrocarbon zone being extracted be isolated from the surrounding downhole environment, most importantly maintaining isolation from aquifers or other hydrocarbon containing formations. This can be performed by encasing steel casings in a sleeve of robust cements, isolating the formation environment from the fluid internal to the steel pipe. During resource extraction the region internal to the casing can undergo fluctuating pressures and temperatures. This cycling can result in increased potential of cracks in the cement, breaking the zonal isolation and resulting in both operational and environmental risks for the activity. In this project, the University of Alberta Multipurpose Composites Group (MCG), along with Sanjel Energy Services Inc. (Sanjel) has developed an experiment that will replicate these industrial scenarios in order to investigate the failure modes of the cement layer.

A concentric cylinder model was used to predict the failure of the outer cement layer. For an internally pressurized cylinder, the expected failure mode is in tension due to radial expansion of the cylinder. Both hoop and radial stress are greatest on the innermost surface of an internally pressurized cylinder, so it was critical to determine the effective pressure on the inner surface of the cement layer to predict cement failure. A 2D dual wall cylinder model was used to predict the failure of the cement, as shown in Fig. 38.1. For this experiment the outer pressure p_o has been assumed to be at ambient pressure to simplify the experimental setup. A plane strain assumption was used in the initial modeling, simply to determine the necessary operating conditions of the experiment design. The inner cylinder is assumed to be composed of carbon steel while the outer cylinder is a specific cement blend. The tensile failure stress of the cement is much lower than that of the steel, thus the failure of the cement can be safely tested without concern of failing the inner steel cylinder.

B. Cheung (✉) • J. Goh • M. Ramesh • J. Carey
Department of Mechanical Engineering, University of Alberta, 10-203 Donadeo Innovation Centre for Engineering, 9211-116 St NW, Edmonton, AB, Canada T6G 1H9
e-mail: bcheung2@ualberta.ca

S. Iremonger
Sanjel Energy Services Inc, 1400, 700–9th Ave SW, Calgary, AB, Canada T2P 3V4

M. Sutton, P.L. Reu (eds.), *International Digital Imaging Correlation Society*, Conference Proceedings of the Society for Experimental Mechanics Series, DOI 10.1007/978-3-319-51439-0_38

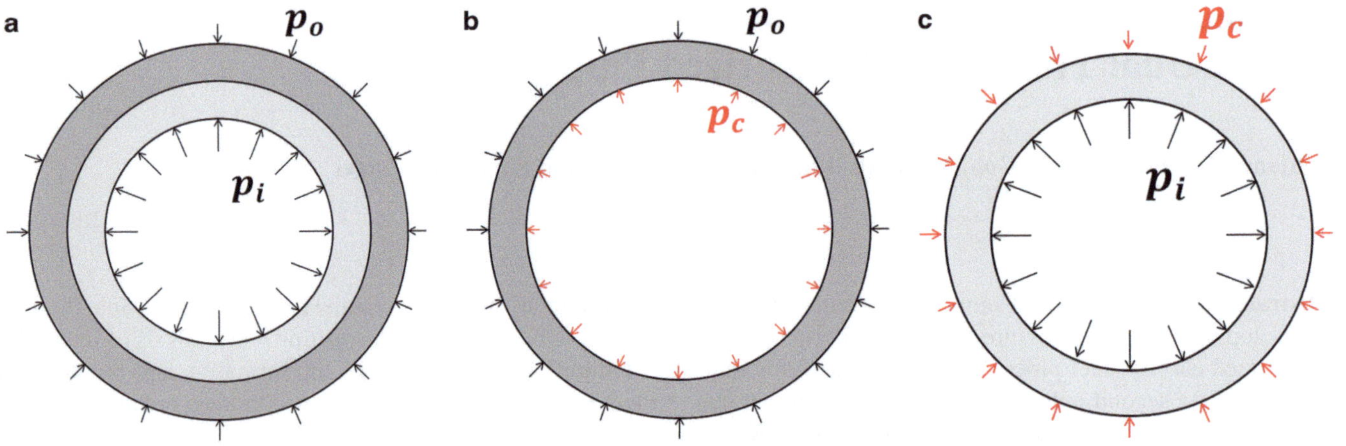

Fig. 38.1 Free body diagram of composite cylinder analysis, including (**a**) the composite cylinder assembly; (**b**) the outer cylinder (cement); and (**c**) the inner cylinder (steel), where p_i, p_o, and p_c are the inner, outer, and contact pressures, respectively

Fig. 38.2 Image of pre-cracked Brazilian test specimen

One of the major objectives of this project was to test the performance of cement with and without fiber additives. These are purely 'mechanical additives' as they do not change the chemical structure or composition of the cement; rather, the final cement structure simply becomes a heterogeneous mix of the cement and the fibers. The impact of the fiber addition is modeled through the rule-of-mixtures approach as well as with fracture mechanics theory. With the rule of mixtures approach, there is simply not enough fiber (0.5% by weight of blend) to significantly impact the elastic modulus. In considering fracture mechanics, two extremes are considered based on the quality of the bond between the fibers and the cement. If the fiber had poor bonding qualities, the addition of fibers would increase the initial crack size within fibrous samples and lower the critical failure stress. However, if the fiber-cement interface resulted in good bonding, then the fibers would increase fracture toughness by limiting the propagation and growth of cracks. Significant effort has been done by Sanjel previously to optimize the fiber selected and the fiber content in a fiber blend [1].

To test these theories, the fracture toughness of the cements had to be determined. As previously mentioned, the failure mode of an internally pressurized cylinder is expected to be tensile fracture in the radial direction (hoop stress) and thus the Mode I stress intensity factor (K_I) was required to predict the critical failure stress. Compressive Brazilian disk tests with prescribed initial cracks (as shown in Fig. 38.2) were used to determine the K_{IC} (critical stress intensity factor for pure Mode I fracture) of the cement with and without fiber additives, using a linear regression of the normalized fracture load and a function of the normalized crack length [2].

Digital image correlation (DIC) has and will continue to be used throughout to perform all the strain measurements of the tested samples. Brazilian disk tests with DIC are used to determine the tensile modulus of the cured cement blend with and without fiber additives. As well, DIC of the crack tip region is possible if further analysis into the fracture mechanics behavior of the cement is required. Finally, the ultimate goal of the project is to determine the behavior of the outer cement layer while the system is under internal pressurization. 3D DIC will be used to evaluate the stress-strain behavior of the cement under these conditions.

38.2 Method

Sample preparation and testing has been performed at MCG's laboratories. To ensure consistency across samples, the industry partner Sanjel supplied the MCG with a pre-mixed dry cement blend. Specific blends have already been tailored to meet the Sanjel's viscosity, density and thermal diffusivity requirements and are prepared at the Sanjel's laboratory and shipped to the MCG in bulk quantity. Sample preparation at MCG's laboratories involves only addition of specific fiber additives and water following industry guidelines [3]. For pressure testing, samples are prepared by curing a sleeve of cement around a 2 in. schedule 40 pipe. A tested specimen is shown in Fig. 38.3. Wall thicknesses of the pipe and cement sleeve are designed to reflect the ratios seen in industry applications. Two main factors that can vary between samples: cure time and the inclusion of fiber additives. All other sample preparation factors (cement blend, curing pressure, etc.) are held to limit variation between experiments. Curing time must also be considered in sample preparation; this can affect mechanical properties on a yearly scale, though this is impractical for testing purposes. Most samples are targeted with a curing time between 2 and 10 weeks. Fiber addition is binary ('without' and 'with' fiber samples) to limit variation.

The experimental setup is divided into two main aspects: the mechanical design and the data acquisition. For the mechanical side, a high pressure syringe pump (Teledyne ISCO 500D, Lincoln, NE, USA) was used to provide pressure to the prepared samples. Samples were held in place by a set of hydrostatic test plugs (Curtiss-Wright Griptight High Pressure Test Plugs, Hatfield, PA, USA) and torqued to maintain seal at the connections. For data acquisition, dedicated software was written using the App Designer toolbox in Matlab (Matlab R2016a, The MathWorks Inc., Natick, MA, USA) to interface with the pump to control pressure testing profiles, retrieve pressure data, and trigger image capture. The software has been tested so far with up to two cameras in synchronization, with both GigE and CameraLink protocols. Post-processing and DIC of the images is performed using commercial DIC software (DaVis 8.2.0, LaVision Inc., Göttingen, Germany).

38.3 Results and Discussion

To date, most of the project efforts have been focused on the design and assembly of the experimental setup. However, with the Brazilian disk testing, a few conclusions can already be drawn as to the performance of the fibrous cement versus the cement without fiber additives from a few exploratory tests done at the outset of the project. The failure strain of the fibrous samples was on average higher than that of the non-fibrous samples. Some testing has shown a decrease in the effective tensile Young's modulus of the fibrous samples; however has not yet been confirmed to be statistically significant and the mechanism of how this is occurring must be determined. As mentioned previously, the rule of mixtures approach cannot

Fig. 38.3 Image of cracked cement-on-pipe sample

fully describe this behavior given the content of fibers. As well, the ultimate failure stress of the fibrous samples was also higher, which is an indication that there is good bonding between fibers and cement, as stated in [1], and that the fibers are indeed increasing cement fracture toughness. The evaluation of the fracture toughness is still ongoing for both the fibrous and non-fibrous samples.

With the completion of the experimental setup, 3D DIC of the casing samples can now be performed. One challenge that must be addressed prior to regular testing is contrast patterning of the samples. For the Brazilian disk samples, it was found that the natural grain and texture of the cement samples provided sufficient contrast to calculate the strain, but this will need to be evaluated for the cylindrical sample. A concern in testing is the dehydration of the cement sample, therefore if a patterning is required then any painting and speckling must be performed in a relatively short period of time. Further planned improvements for the experimental setup include refining the sample preparation to limit the formation of defects which induce unwanted stress concentration locations in the samples.

References

1. Iremonger, S.S., Bolt, M., Lawrence, S.: Enhanced thermal well integrity through the use of a new cement tensile strength-enhancing fiber. In: SPE Canada Heavy Oil Technical Conference, Calgary, AB, Canada (2015)
2. Atahan, H.N., Tasdemir, M.A., Tasdemir, C., Ozyurt, N., Akyuz, S.: Mode I and mixed mode fracture studies in brittle materials using the Brazilian disc specimen. Mater. Struct. **38**, 305–312 (2005)
3. "Specification for Cements and Materials for Well Cementing" ANSI/API Specification 10A, effective June 1, 2011

Chapter 39
Digital Image Correlation for Large Strain

J.-F. Witz, P. Lecomte-Grosbras, A. Morch, C. Martel, F. Lesaffre, and M. Brieu

Abstract Due to their high deformability, the elastomers are widely used in various fields of applications. Their hyper-elastic behavior requires means of characterization adapted to contactless, high strain as well as heterogeneous strain field. Digital Image Correlation (DIC) have been used to characterize such material. In case of large strain it might be necessary to re-actualize the reference image to compute the displacement field on every images of the test. Different strategy to apply DIC in large strain are presented. The direct addition of the successive calculated displacement leads to inconsistent results, it is required to make a composition of the successive calculated displacements. Some artefacts due to the multiple interpolation are highlighted and may leads to filter the results. In case of high strain gradients, a method with very small element size and median filter applied directly in correlation process has been used highlighting the ability of large strain DIC in heterogeneous media.

Keywords Rubber-like material • Large strain • Digital image correlation • Heterogeneous strain

39.1 Introduction

The elastomers are widely used in various fields of applications (transportation, leisure, biomedical, etc.) due to their high deformation capability. Their hyper-elastic behavior requires means of characterization adapted for identification of their behavior in a consistent strain level range related with their real using conditions. It is thus necessary to propose experimental setup allowing to apply large and multiaxial strain and contactless measurements techniques to measure large strain and high gradients on soft and highly deformable materials. A review of different measurement techniques applicable in high strain can be found in [1]. The Digital Image Correlation (DIC) [2] is a contactless measurement technics allowing to measure heterogeneous fields by comparison of images at different strain level. In case of large strain it is usually required to perform incremental measurement re-actualizing the reference image that might induce high uncertainties accumulation leading to inconsistent results. The aim of this study is to analyze the different strategy to apply DIC under large strain as well as the uncertainties and induced artefacts. Two different types of samples will be studied: homogenous rubber sample in equi-biaxial loading under high strain and heterogeneous material, silicon sample including rigid inclusions, with high local strain gradients.

J.-F. Witz (✉)
LML/Lam3, CNRS, Cité scientifique, 59651 Villeneuve d'Ascq, France
e-mail: jean-francois.witz@ec-lille.fr

P. Lecomte-Grosbras • A. Morch • F. Lesaffre • M. Brieu
LML/Lam3, CentraleLille, Cité scientifique, 59651 Villeneuve d'Ascq, France
e-mail: pauline.lecomte@ec-lille.fr; annie.morch@ec-lille.fr; francois.lesaffre@ec-lille.fr; mathias.brieu@ec-lille.fr

C. Martel
LML/Lam3, Université Lille 1, Cité scientifique, 59651 Villeneuve d'Ascq, France
e-mail: corentin.martel@ec-lille.fr

M. Sutton, P.L. Reu (eds.), *International Digital Imaging Correlation Society*, Conference Proceedings of the Society for Experimental Mechanics Series, DOI 10.1007/978-3-319-51439-0_39

Fig. 39.1 (**a**) Biaxial machine (**b**) biaxial sample for large strain measurement (**c**) Heterogeneous sample for high gradients analysis

39.2 Material and Methods

A biaxial tensile machine has been developed at Laboratoire de Mécanique de Lille-LaM^3 (LML/LaM^3). It is composed of a frame, four motors controlled independently, allowing to apply complex and non-proportional loading path. The clamping is provided by self-tightening grips in order to avoid sample sliding induced by the thickness reduction due to the large strain. For the measurement and control, the machine is equipped of four load cells allowing to measure until 2 kN as well as two cameras on both opposite faces for strain measurement by video-extensometry and DIC (Fig. 39.1a). A DIC software, Yadics, developed at LML/Lam^3 has been used [3].

Two types of samples have been tested to study Digital Correlation efficiency in case of large homogeneous strain or high strain gradients. The first one is a homogeneous biaxial sample composed of carbon reinforced rubber as shown on Fig. 39.1b. The sample has been covered with black and white ink in order to provide adequate texture made of random gray level distribution in order to perform DIC calculation. The second type of sample is a heterogeneo us sample composed of silicone polymer and rigid wood inclusions (Fig. 39.1c). These inclusions will induce strain heterogeneities and high strain gradients. These samples have also been covered with black and white ink to perform DIC analysis.

The Digital Image Correlation Method consists in finding the spatial relation between pixels of the initial image captured at time t_0 and the one in the deformed image taken at time t. The method is based on optical flow assumption: $f(X) = g(X) + U(X)$ with X the position of the pixels, f and g respectively the gray level function of the initial and deformed image and U the unknown displacement. To solve this equation a metric is required to compare the images (SSD (Sum of Square Difference), NCC (Normalized Cross Correlation), etc.) and minimizing a cost function with an optimization algorithm. It is also required to define the form of the displacement field (Rigid Body, Homogeneous strain, Polynomial, Bspline, Finite Element Basis, etc.). In this study the SSD metric has been used and the displacement has been decomposed over a Finite Element Basis [4].

In the context of large strain, it might be impossible to directly compare initial and deformed image. Reactualizing the reference image might be required to compute the displacement field on all the image sequence. In such a case, the total displacement field has to be reconstructed from all the intermediate calculated displacements but simple addition of all successive displacements may lead to inconsistent results as shown on followings equations. Indeed the displacement cannot be added directly and has to be defined in the same frame of reference to apply the composition as shown in the following equations:

$$T(t_0, t_1) = X_1(X_0) = X_0(X_0) + U_{0\leftrightarrow 1}(X_0) \tag{39.1}$$

$$T(t_0, t_i) = X_i(X_0) = X_0(X_0) + U_{0\leftrightarrow i}(X_0) \tag{39.2}$$

$$T(t_0, t_i) = X_i(X_0) = X_0(X_0) + U_{0\leftrightarrow i}(X_0) \tag{39.3}$$

$$T(t_i, t_{i+1}) = X_{i+1}(X_i) = X_{i+1}(X_i) + U_{i\leftrightarrow i+1}(X_i) \tag{39.4}$$

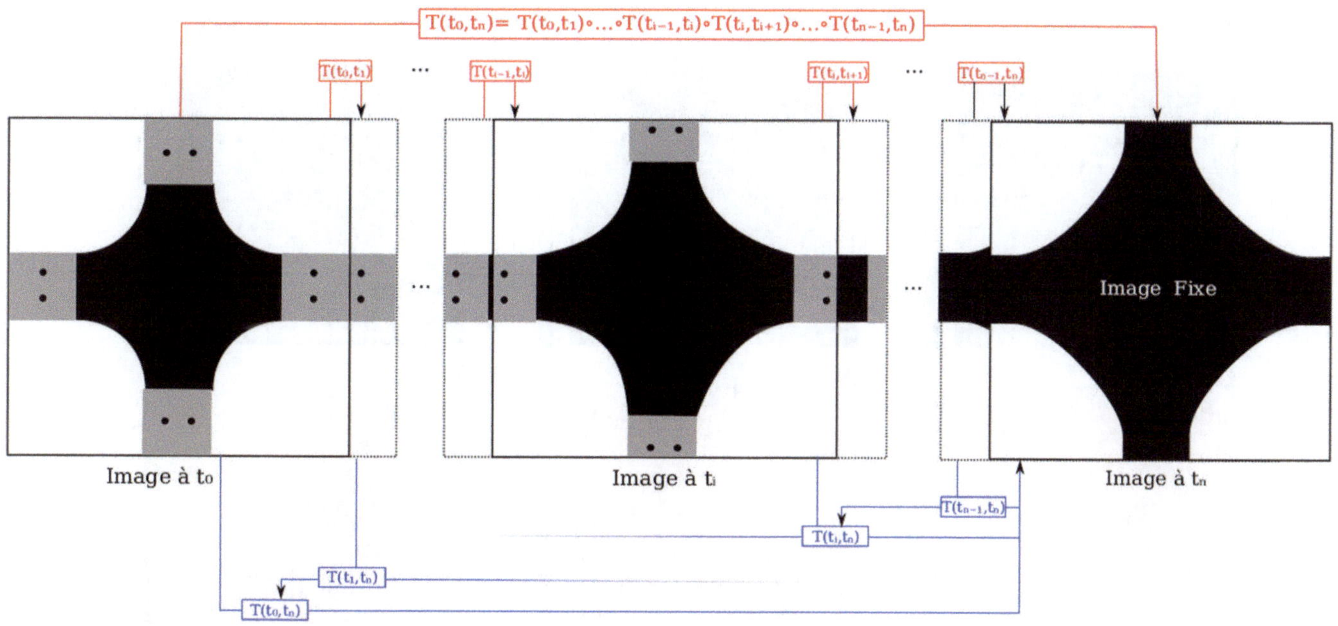

Fig. 39.2 Composition (in *red*) and reactualization (in *blue*) methods in case of large strain

$$U_{0\leftrightarrow i+1}(X_0) = U_{0\leftrightarrow i}(X_0) + U_{i\leftrightarrow i+1}(X_0) \neq U_{0\leftrightarrow i}(X_0) + U_{i\leftrightarrow i+1}(X_1) \quad (39.5)$$

$$U_{0\leftrightarrow i+1}(X_0) = U_{0\leftrightarrow i}(X_0) + U_{i\leftrightarrow i+1}(X_0(X_0) + U_{0\leftrightarrow i}(X_0)) \quad (39.6)$$

This composition is computationally costly and may become impossible if there is too many images to be analyzed. The change of frame of reference leads to perform this composition of the displacement field that requires an interpolation process. The choice of the type of interpolation induces uncertainties and errors on the displacement fields. To improve the convergence of the calculation and reduce the number of reactualization of the reference image and number of composition it is possible to initialize the calculation with the results obtained on the previous image of the sequence. The Fig. 39.2 summarizes the different methods.

39.3 Results

An equi-biaxial tensile test has been performed on rubber sample. The Fig. 39.3a, b shows the reference and moving images. The sequence of images has been analyzed reversely: the last image of the sequence where the deformation is the highest has been chosen as the reference one in order to avoid perturbations due to the lack of information on edges that move out of the observation field of the camera during the test. A sequence of 1520 images has been captured. The displacement field has been decomposed over a finite element basis of 32 × 32 pixels (the image resolution being 2048 × 2048 pixels) and reactualization with composition of the transformations has been performed every 20 images. To validate computation, the displacement field has been applied to the moving image to correspond to the reference one as shown on Fig. 39.3c. The Fig. 39.3d–f show the strain fields calculated under large transformations. One can clearly observe the artefacts due to the accumulation of successive interpolations on the strain fields in the case of addition method.

A sample constituted of silicone matrix and a rigid inclusions has been tested too in order to study a case with non-homogenous strain field and localized large strain. Increasing displacements have been applied in axial and transversal directions in order to apply multi-axial loading as shown on Fig. 39.4a. As previously, the displacement field has been decomposed over a finite element basis. To highlight the strain gradients, the element size has been chosen very small (4 × 4 pixels) in order to have a good spatial resolution. A small median filter with a size of 3 pixels has been applied on the displacement field during the optimization process [5]. The Fig. 39.4b presents the Von Mises equivalent strain obtained on such sample.

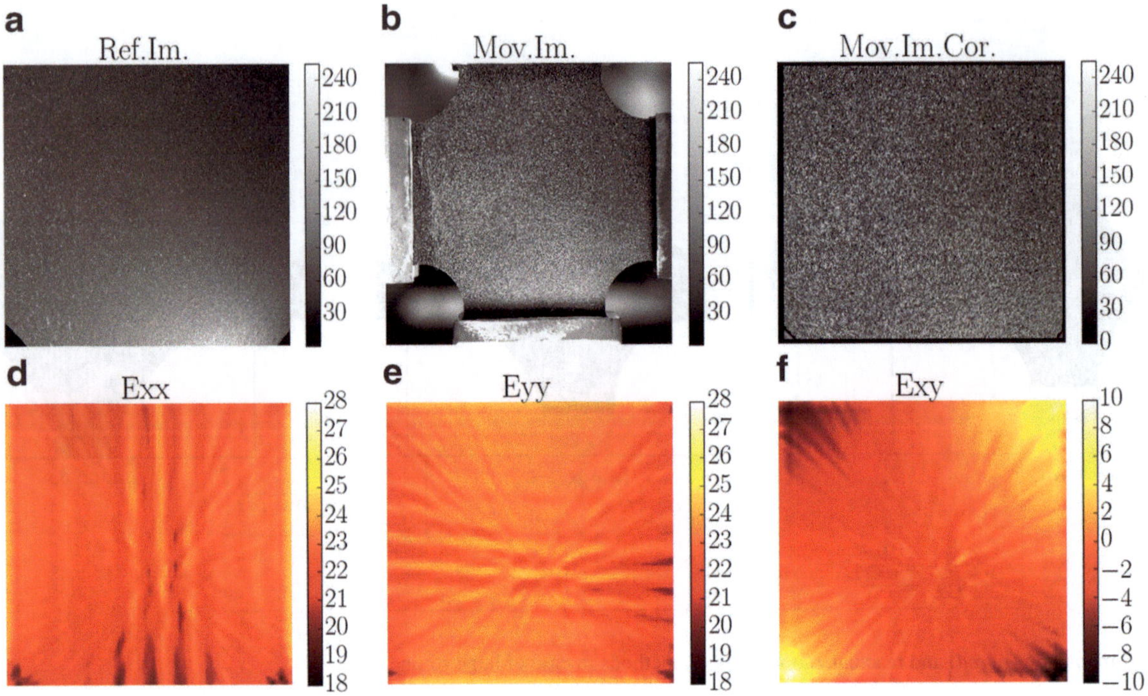

Fig. 39.3 (**a**) Reference image (last picture of the sequence), (**b**) moving image, (**c**) Moving image corrected by the calculated displacement, (**d**) longitudinal strain ε_{xx} (%), (**e**) transversal strain ε_{yy} (%), (**f**) shear strain ε_{xy} (%)

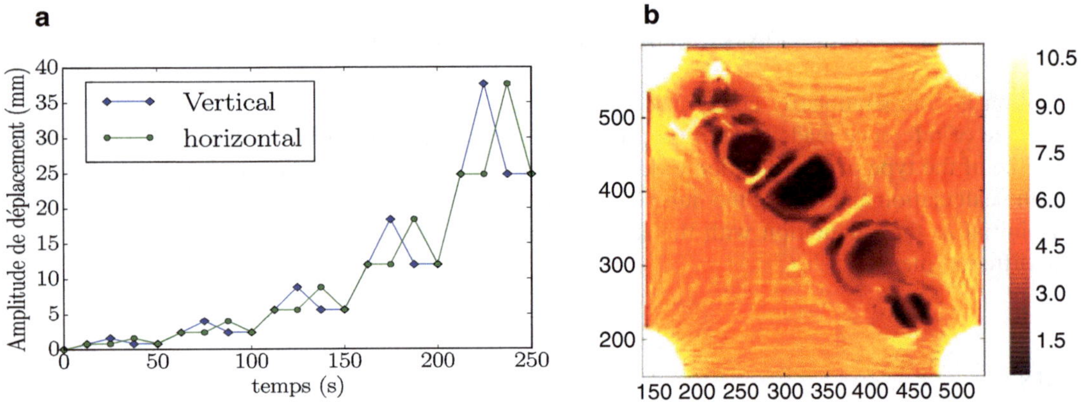

Fig. 39.4 (**a**) Loading path, (**b**) Von Mises strain (%)

One can see the strong localization of strains. This figure highlights the ability of large strain DIC on heterogeneous media.

39.4 Conclusion

Different samples have been tested in bi-axial tensile loading. The first type of sample is an homogeneous rubber sample. The influence of the different strategy for digital image correlation, requiring reactualization of the reference image in case of large strain, has been study. It highlights that the direct addition of the displacement field leads to inconsistent results. Composition with transportation of the displacement field in the right reference frame has to be done but leads to artefacts due to accumulation of successive interpolations. In case of high strain gradients it is necessary to have a good spatial resolution.

The DIC calculation have been performed in case of high strain gradients with very small elements size (4 × 4) and median filter applied on the displacement field during the optimization process highlighting the ability of DIC in the context of large strain in heterogeneous media.

References

1. Le Cam, J.-B.: Strain. **48**, 174 (2012)
2. Sutton, M.A., et al.: Image Vis. Comput. **1**(3), 133–139 (1983)
3. Yadics software. www.yadics.univ-lille1/wordpress/
4. Besnard, G., et al.: Exp. Mech. **46**, 789 (2006)
5. Dahdah, N., et al.: Strain. **52**, 324–335 (2016). doi:10.1111/str.12193

Chapter 40
Shock Response of Composite Materials Subjected to Aggressive Marine Environments

C. Javier, J. LeBlanc, and A. Shukla

Abstract The marine community has an interest in using composite materials for the construction of vehicles and other marine structures. The use of composite materials is appealing due to their abilities in keeping naval vehicles in stealth. These structures are subjected to aggressive marine environments during their service life, such as high salinity water and ultraviolet radiation that can degrade their performance over time. These effects are of particular concern when composite vehicles are subjected to shock and blast loading. Therefore, there is a need to investigate how composite materials that have been exposed to marine environments respond to shock events, and how these responses differ from an equivalently loaded virgin structure. An experimental investigation was conducted to understand the mechanical response of carbon/epoxy composites subjected to long-term exposure of aggressive marine environments. In order to simulate marine environments, an elevated temperature, salt water solution was created. The material was submerged in a 3.5%, 65°C salt solution for 0, 30 and 60 days. The temperature was maintained at 65°C to accelerate diffusion of water into the epoxy, thus accelerating the weathering process. Accelerated weathering simulates years of actual service life in short time. The specimens were then subjected to a controlled and concentrated air blast loading using a shock tube. High speed photography coupled with 3D DIC is utilized to provide full field displacements, strain, and velocities of shock tube specimens during in air shock loading. The specimens for all of the weathering cases recovered elastically after the shock event. The large difference in displacements between the non-weathered and weathered specimens is due to the degradation of the material from water absorption. This confirms that submergence in extreme marine environments degrades the material and diminishes the material properties.

Keywords Composite material • Shock loading • Marine environment • Water diffusion • Accelerated weathering

40.1 Introduction

The use of composite materials in the marine community is appealing due to their strength to weight to strength ratio, increase fuel economy, and because of their abilities to aid in keeping naval vehicles in stealth. Due to these advantages, unmanned vehicles are being constructed from composite materials. These structures are subjected to aggressive marine environments during their service life, including high salinity water and/or salt spray and ultraviolet radiation that can significantly degrade their performance over time. These effects are of particular concern when composite vehicles are deployed in a military setting where they may be further subjected to shock and/or blast loading. Therefore, there is a need to investigate how composite materials that have been exposed to marine environments for extended periods of time respond to shock events, and how these responses differ from an equivalently loaded virgin structure.

40.2 Experimental Setup

An experimental investigation was conducted to understand the mechanical response of carbon fiber reinforced composites subjected to long-term exposure of aggressive marine environments. Carbon/epoxy was chosen due to its significance in

C. Javier (✉) • A. Shukla
Dynamic Photo-Mechanics Laboratory, Department of Mechanical, Industrial and Systems Engineering University of Rhode Island, Kingston, RI 02881, USA
e-mail: carlos_javier@my.uri.edu

J. LeBlanc
Naval Undersea Warfare Center, Newport, RI, USA

M. Sutton, P.L. Reu (eds.), *International Digital Imaging Correlation Society*, Conference Proceedings of the Society for Experimental Mechanics Series, DOI 10.1007/978-3-319-51439-0_40

recent naval applications and its relevance in future designs. The composite's fiber construction is a 2 × 2 twill weave cured in an epoxy resin, with a total of four plies. An image of the composite material and the fiber construction is show in Fig. 40.1a. In order to simulate marine environments, an elevated temperature, salt water solution was created. The material was submerged in a 3.5% salinity water solution at 65°C for 0, 30 and 60 days. The temperature was maintained at 65°C to accelerate water diffusion into the composite matrix, thus accelerating the weathering process. The composite specimens were clamped on two ends and subjected to a controlled and concentrated air blast loading using a shock tube. The shock tube is composed of three separate sections: driver section, driven section, and reduced diameter muzzle. A Mylar diaphragm separates the driver and driven sections, while the driver section is pressurized using helium gas. Under critical pressure, the diaphragm bursts releasing a high pressure wave. The high pressure travels down the length of the driven section and develops into a shock wave front. Typical velocities of the gas range from Mach 1 to Mach 2. The shockwave then leaves the muzzle, impacts the specimen and the pressure from the impact is reflected back into the muzzle. The reflected pressure is the pressure that loads the specimen. High speed photography coupled with 3D DIC was utilized to provide full field displacements, strain, velocities, and strain rates of shock tube specimens during in air shock loading. Two cameras are used to track the out of plane displacement of the composite panels during a shock loading. The cameras were placed at an angle, perpendicular to specimen and provide stereo vision to determine changes in 3 dimensional space. The third camera was placed on the side of the specimen to validate the DIC measurements, as shown in Fig. 40.1b. Images of the event from all cameras were taken at 50,000 fps.

40.3 Results

The weathered (30 and 60 days) and non-weathered (0 day) specimens displayed dramatically different behavior after being subjected to a dynamic shock loading, as seen in Fig. 40.2. On average, non-weathered specimens displayed a maximum out of plane mid-point deflection of 5.57 mm. The 30 day weathered specimens had a maximum out of plane mid-point deflection of 6.89 mm on average. The 60 day weathered specimens had a maximum out of plane mid-point deflection of 6.96 mm. The strains in the strains in the longitudinal direction for the 0, 30, and 60 day cases are 6.24, 6.55 and 6.94% respectively. The velocities in which all cases deflect outward are 14.27 m/s, 17.02 m/s, and 17.00 m/s for the 0, 30, and 60 day cases. None of the weathering cases fractured. The specimens for all of the weathering cases recovered elastically, and went through beam oscillations due to the elastic strain energy stored from the shock event. The difference in out of plane deflection between the 0 day weathering case and the 30 day weathering case is 1.32 mm. Similarly, the difference in longitudinal strains and velocity of the specimens as it deflects between the non-weathered case and the two weathering cases is relatively large. This confirms that submergence in extreme marine environments degrades the material and diminish its material properties. However, the difference between the 30 and 60 days case is 0.07 mm of out of plane deflection. This small difference is due to the fact that the epoxy matrix had reached a point of saturation in which water absorption was at a minimum. A series of experiments were conducted with higher pressure in order to achieve complete fracture of all cases. The failure strains for all weathered cases (0, 30, 60 days) were in the range of 1.6–1.8% indicating that the composite fibers were unaffected by the weathering. The matrix of the material is the main factor in degradation of composites, due to water absorption.

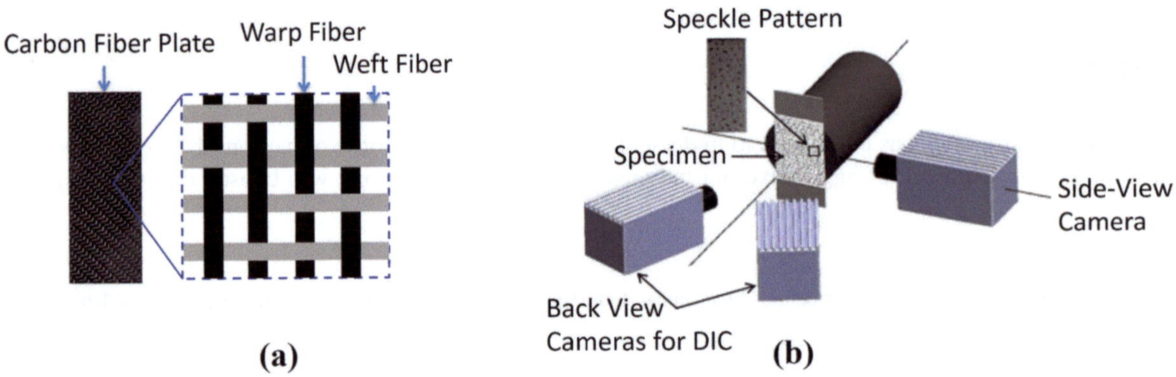

Fig. 40.1 (**a**) Zoomed representation of the 2 × 2 twill weave. (**b**) High speed camera setup for DIC

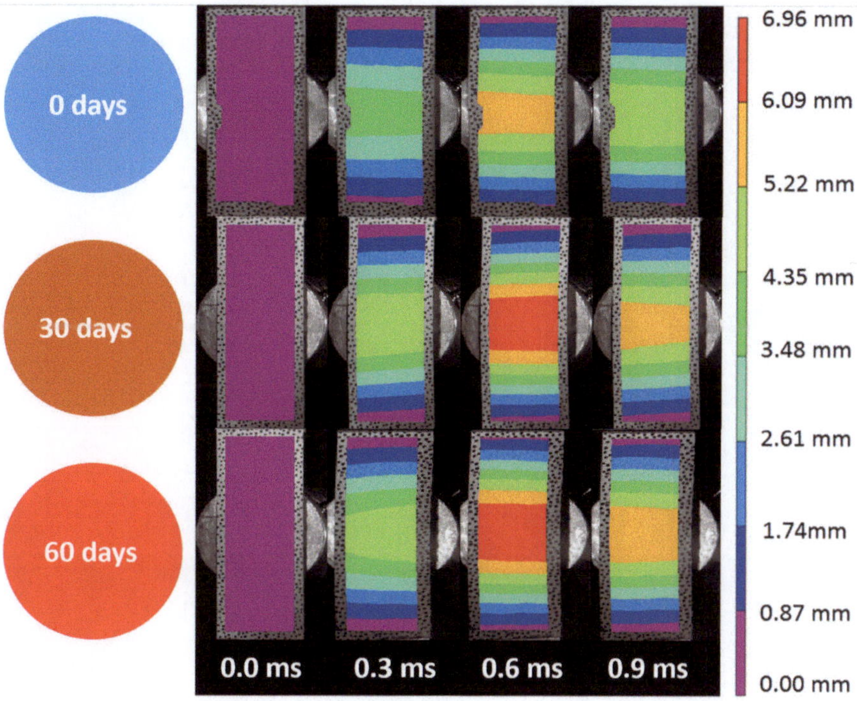

Fig. 40.2 Full field out of plane deflection throughout the shock loading event

Acknowledgements The authors acknowledge the financial support provided by NEEC (NUWC and NSWC) under contract number N00174-16-C-0012 and the Naval Undersea Warfare Center, Newport RI.

Chapter 41
Investigations on Cyclic Flexural Behavior of Fiber Reinforced Cementitious Composites Using Digital Image Correlation and Acoustic Emissions

Muhammad M. Sherif, Osman E. Ozbulut, and Jonathon Tanks

Abstract Superelastic shape memory alloys (SE SMAs) are smart materials that recover 6–8% of inelastic strains upon unloading and exhibit good energy dissipation. In this study, the mechanical behavior of cementitious composites, reinforced with steel and SE SMA fibers, under flexure was examined. Fiber reinforced concrete, with a total fiber volume ratio of 0.6%, was prepared and a third point bending flexural test was conducted using an incrementally increasing displacement loading protocol. DIC was used to measure the strain and displacement contours and detect the crack width propagation. The results of the cyclic flexural testing were analyzed to assess the re-centering and crack recovery capabilities. Acoustic emissions (AE) were monitored using AE transducers to predict the failure modes (fiber pullout/matrix cracking). The AE were analyzed using average frequency, cumulative energy, duration and hits. The correlation between the DIC and AE was investigated. The results suggest that the crack data can be correlated with AE.

Keywords Shape memory alloys • Fiber reinforced cementitious composites • Digital image correlation • Acoustic emissions

41.1 Introduction

Cementitious composites are among the most used constructive material in the design of civil structures due to their relatively low cost and high compressive strength. However, due to their low tensile strength, steel reinforcement is used at the tension fibers. Nevertheless, due to the cracking of concrete members under service loads, steel reinforcement corrodes causing deteriorations. A number of researchers have investigated the use of fiber as reinforcements in cementitious composites. Fiber reinforced cementitious composites display an increase in the mechanical properties such as the tensile strength and ductility as well as corrosion resistance in the case of using PVA fibers. Such techniques resolve crack propagations and corrosion problems, but lack the capabilities of re-centering and/or crack closure for cementitious composites.

Smart materials such as shape memory alloys (SMAs) have a unique property of remembering their original shapes. Superelastic SMAs can fully recover to their original shapes upon unloading after being stressed to reach 6–8% strains. Figure 41.1a illustrates a typical stress-strain response of superelastic SMAs. Furthermore, SMAs possess good energy dissipation capacity, high corrosion resistance as well as long fatigue life. Researchers in the civil engineering community have implemented superelastic SMAs in various fields ranging from damping devices to bridge restrainers [1]. In the past decade, several studies investigated the use of SMAs fibers or reinforcement in concrete members to enhance the performance of structures within the aspects of crack-propagation, re-centering and energy dissipation [2].

Digital Image Correlation (DIC) is an optical surface measurement system that monitors the full-field of displacement and strain evolution. Special preparations such as applying a speckle pattern to specimens monitored by the DIC system is required to be capable of capturing strains. DIC tracks and correlates the speck patterns within a subset space in a series of sequential images captured during the experimental procedure. A single optical device can acquire the two-dimensional (2D) deformations and strains when placed perpendicular to the speckle pattern. Moreover, crack width, length and propagation rate can be analyzed accurately as they are represented by discontinuities in the displacement field.

M.M. Sherif (✉) • O.E. Ozbulut
Department of Civil and Environmental Engineering, University of Virginia, Charlottesville, VA 22901, USA
e-mail: mms4bx@virginia.edu

J. Tanks
Department of Civil and Environmental Engineering, Tokyo Institute of Technology, Tokyo 152-8550, Japan

M. Sutton, P.L. Reu (eds.), *International Digital Imaging Correlation Society*, Conference Proceedings of the Society for Experimental Mechanics Series, DOI 10.1007/978-3-319-51439-0_41

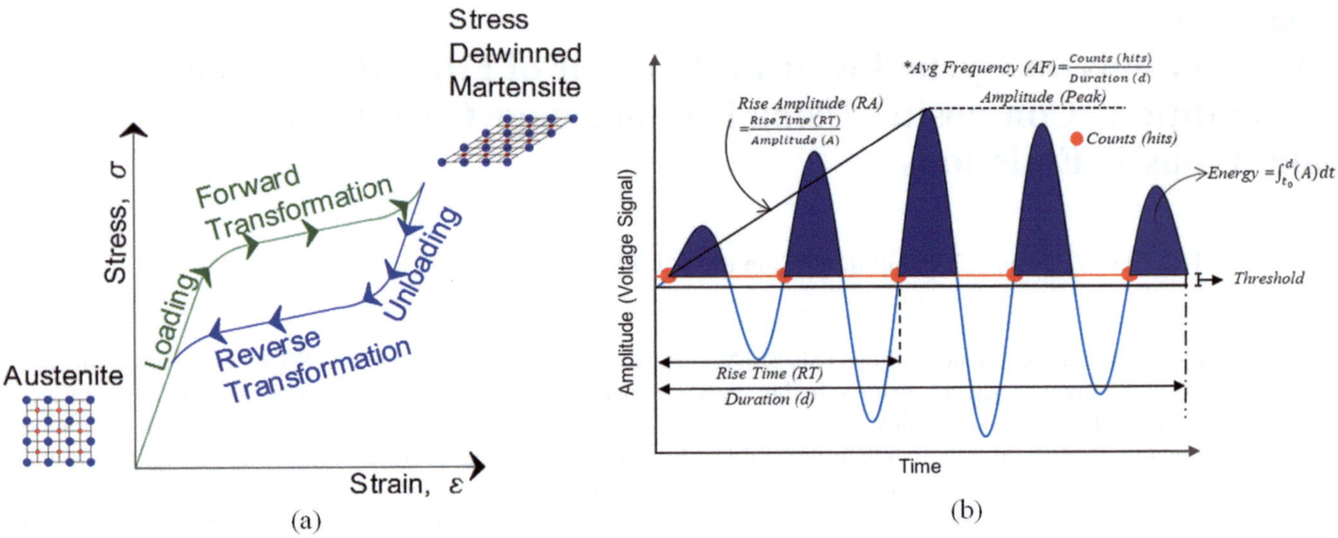

Fig. 41.1 (**a**) Typical stress-strain response of superelastic SMA. (**b**) Typical acoustic emission signal and AE parameters

Deteriorations within a material releases energy in terms of transient elastic waves known as acoustic emissions (AE). A typical AE signal with its properties is illustrated in Fig. 41.1b. Acoustic waves can be detected by attaching AE sensors (usually made of piezoelectric material) to the specimen. AE transducers record the acoustic responses as an electric waveform after an energy release event in the system. The analysis of the acoustic emissions can provide information about the event weather it is due to a matrix cracking, reinforcement rupture or pull-out, and determine the rate of crack propagation. The scale of crack depends on the energy emitted at the source and can be identified by the amplitude of AE. Micro-cracking emits waves with low energy and low amplitudes while macro-cracks emit waves with high amounts of energy and large amplitudes. The failure mechanisms of cementitious materials can be categorized under tensile or shear failure modes. Tensile mode occurs due to a motion perpendicular to the sides of the crack face. The shear mode of failure occurs due to a motion parallel, but opposite in direction, at the crack face.

This study investigates the correlation between DIC and AE results for randomly distributed SMA fibers in mortar specimens. The fiber reinforced mortar is subjected to cyclic flexural loads by using the center point bending test. The performance of the SMA fiber reinforced mortar is analyzed based on the mid-span displacements, residual strength and crack width propagation. To further investigate the re-centering and crack recovery, the acoustic emissions of the samples are analyzed with regards to cumulative energy and moving averages of hits and average frequency.

41.2 Methodology

Specimen preparations and mortar mixture design were based on the ASTM C348-02 standards [3]. A total of five different mixtures with various SMA fiber volume ratios were prepared. Table 41.1 presents the mixture design for the SMA fiber reinforced mortar. ASTM Type I/II Portland cement and natural sand fine aggregates were used for all specimens. SMA fibers are strands with a 0.350 mm diameter. The ratios of fiber volumes that were investigated were 0.3, 0.5 and 1.0% and the fiber length was 30 mm. For the 0.3% fiber volume ratio, an additional mixture was prepared with 20 mm SMA fibers. A plain mortar specimen was prepared to serve as control specimens. Three specimens with dimensions 40 × 40 × 160 mm were casted for each mixture. The samples were moisture cured at a 95% relative humidity and a temperature of 23 ± 1.5 °C for 28 days. The mortar specimens were tested using three-point loading flexural test. The loading protocol was set in displacement-control. The protocol consisted of 13 cycles with an increasing displacement increment of 0.1 mm to a total displacement of 1.3 mm and with a displacement rate of 1.0 mm/min. Figure 41.2a displays the displacement protocol for the experimental test. A two-dimensional system was used to monitor the full field of strain and displacement. An acoustic transducer was attached to one side of the specimen to record the acoustic emissions. Figure 41.2b illustrates an image taken by the optical camera for the specimen and test set-up.

Table 41.1 Specimen mix design.

Mixture id	M-control	M-SMA-1	M-SMA-2	M-SMA-3	M-SMA-4
Sand/cement	2.75	2.75	2.75	2.75	2.75
Water/cement	0.48	0.48	0.48	0.48	0.48
SMA fiber volume ratio (%)	–	0.30	0.30	0.50	1.00
SMA fiber length (mm)	–	20	30	30	30
No. of specimens	3	3	3	3	3

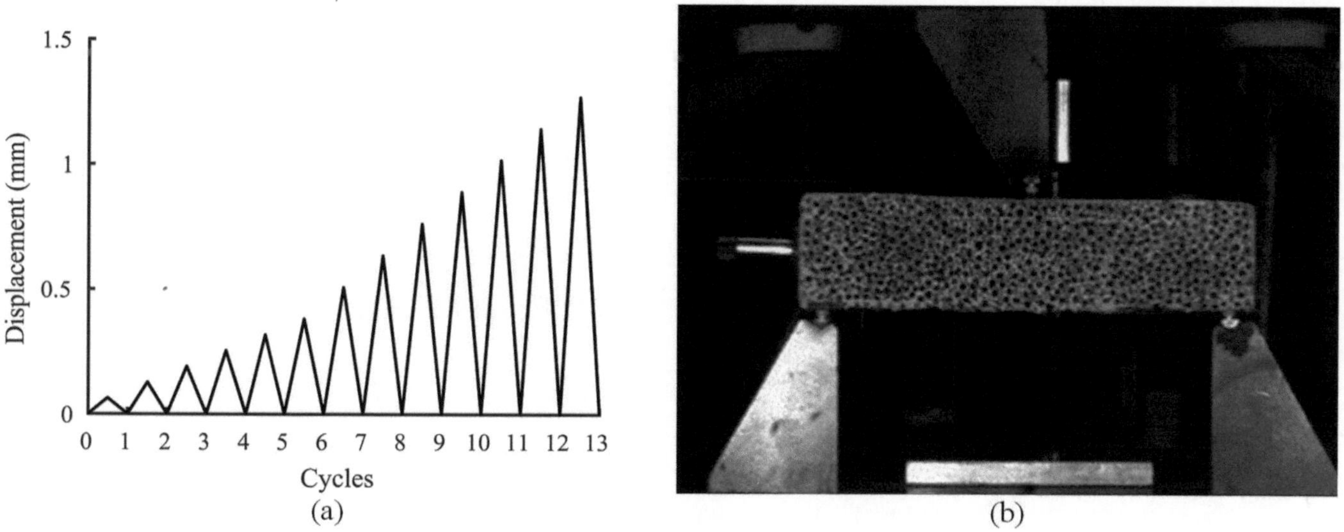

Fig. 41.2 (**a**) Cyclic flexural displacement protocol. (**b**) Test setup captured by an optical camera

41.3 Conclusion

This research investigates the use of digital image correlation as well as acoustic emissions to better understand the behavior of SMA fiber reinforced mortar beams. The performance of the fiber reinforced mortar specimens was analyzed using mid-point deflection, as well as crack width propagation and full field strain. The acoustic emissions were analyzed based on the average frequency, cumulative energy, duration and the moving averages of hits and average frequency.

References

1. Ozbulut, O.E., Hurlebaus, S., DesRoches, R.: Seismic response control using shape memory alloys: a review. J. Intell. Mater. Syst. Struct. **22**(14), 1531–1549 (2011)
2. Ozbulut, O.E., Hurlebaus, S.: Neuro-fuzzy modeling of temperature- and strain-rate-dependent behavior of NiTi shape memory alloys for seismic applications. J. Intell. Mater. Syst. Struct. **21**(8), 837–849 (2010)
3. ASTM C348-14: Standard test method for flexural strength of hydraulic-cement mortars. ASTM International, West Conshohocken, PA (2014)

Chapter 42
Method for Determining Temperature Dependence of Material Properties and Failures on the Meso-Scale

Charles M. Spellman, Vishaal B. Verma, and G. Alex Arzoumanidis

Abstract Composite layup delamination and damage is caused by a number of phenomena, including manufacturing quality, environmental conditions, and loading history. Simulating the mechanical properties of these materials requires both advanced modeling and validation procedures; meso-scale materials testing equipment is fundamental to both qualitative and quantitative model generation and confirmation. Studying the small-scale effects with DIC and environmental control has thus far been quite difficult. Psylotech has recently developed an under-microscope compatible temperature chamber and bath module specifically for its μTS platform, a micro-scale universal load frame. The turnkey system enables efficient testing of composites (including separating the constituent polymer matrix material and reinforcement fibers) biological samples, shape-memory alloys, and other small-scale specimens exhibiting time-temperature dependence. Extreme dynamic sensor resolution enables studying wide changes in modulus and an advanced controller allows for complex waveforms with tight time-syncing to the image capture system. An example application including optical DIC analysis of composite layup failure will be presented.

Keywords Meso-scale • DIC • Microscopy • Temperature • Characterization

42.1 Introduction

Meso-scale testing with DIC at elevated temperatures enables experimental validation of small-scale theory. In metals, one can study inter-granular creep mechanisms in situ. In polymer composites, creep at the matrix/fiber interface reveals microstructural contribution to the macro material response and can be used to validate multi-scale FEA simulation codes. Testing small scale samples has additional advantages. Less energy is required to raise specimen temperature. Moreover, small samples makes it easier to have a uniform temperature distribution. The proposed chamber leverages convection to effectively control specimen temperature. This work focuses on a relatively limited range of −100 to 200°C, but the fundamental concept can be applied to much higher temperatures.

This work provides an early example of testing a chopped fiber composite at high and low temperature. The following describes the system and tests that were done to both qualify the design specifications and show a proof of concept for use in the field. Load frame data was recorded in addition to DIC imagery.

42.2 Mechanical Apparatus

The Psylotech μTS Temperature Chamber (TC) (Fig. 42.1) includes a temperature control module, supply and return plumbing system, and an environmental chamber with temperature range of −100 to 200°C. The control module includes a blower, internal plumbing, a thermal mass capacitor, and a cryogenic solenoid valve. The blower provides a pressure drop across the module inlet and outlet, providing a steady stream of air to the chamber in a closed loop. The cryogenic solenoid valve includes an inlet for an external 22 PSI liquid nitrogen (LN2) connection. The valve allows for a simple on/off control of LN2 into internal plumbing and air stream of the control module. In order to smooth out the control of cold air generation during temperatures below 0°C, a thermal mass capacitor was utilized. It was constructed of a heavy stainless steel tube inline with the module's internal air stream. The mass of the capacitor prevents large inflections or spikes in cold input into

C.M. Spellman (✉) • V.B. Verma • G. Alex Arzoumanidis
Psylotech, Inc., 1616 Payne Street, Evanston, IL, 60201, USA
e-mail: cspell@psylotech.com

M. Sutton, P.L. Reu (eds.), *International Digital Imaging Correlation Society*, Conference Proceedings of the Society for Experimental Mechanics Series, DOI 10.1007/978-3-319-51439-0_42

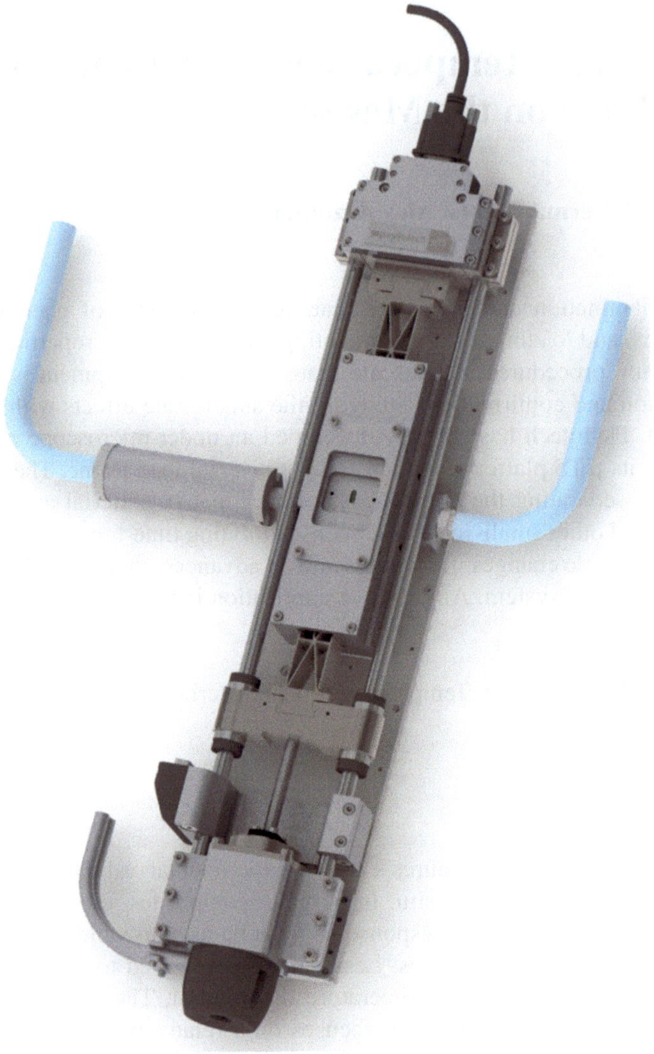

Fig. 42.1 Psylotech μTS load frame with temperature chamber

the air stream by exchanging energy between its mass and the air. Two type-K thermocouple devices are used to measure both chamber air temperature and specimen local temperature.

The chamber, separated mechanically from the control module, includes an enclosure which completely surrounds the specimen and grips during testing. It includes a primary cover with integral self heating to prevent frost formation and a variety of secondary cover solutions which are designed to mount to the primary cover. Various secondary cover attachments are available including a glass slide for large scale image capture and a recessed cover for microscope objective clearance. The chamber includes one central inlet at its base and two outlets located at opposing ends. This configuration provides a consistent convective air stream throughout the chamber from center to opposing ends regardless of grip location or specimen length. The chamber is heated via a heating element mounted directly at its inlet and in line with the incoming air stream, providing increased heating efficiency. It also improves safety and convenience as the user does not have to worry about the plumbing system or control module reaching dangerously high temperatures.

Connecting the control module to the chamber involves the use of insulated silicone tubes; the flexibility of the tubing minimizes vibration transfer as opposed to a rigid conduit or braided metal hose option. The tubing is easy to route and relatively inexpensive, making many positional configurations possible.

The TC is designed to be used with a variety of grip types including: dogbone wrap-around, clamped tension, compression, and platform (Fig. 42.2). Grip extensions allow offset mounting to reach into the chamber and allow use of full actuator stroke. The extensions are thin walled to minimize thermal conductivity to sensors and are manufactured with wire EDM to maintain the same high precision and accuracy of the grips.

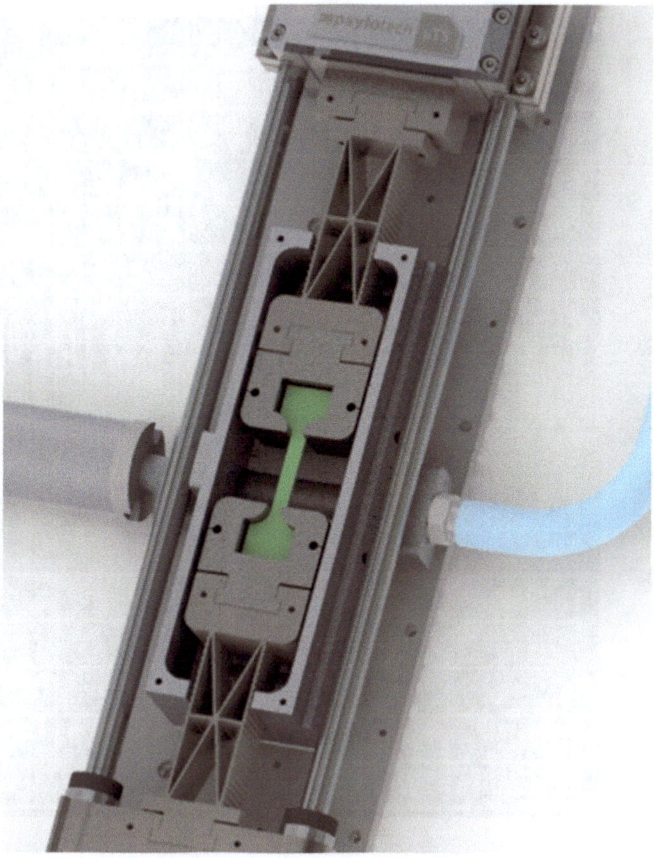

Fig. 42.2 Cutaway of temperature chamber mounted to the μTS

42.3 Software Interface

Psylotech's LabVIEW based load frame control software readily accepts modification due to its open-source status. A specific software module within Psylotest was written to facilitate temperature control and display information to the user. Scheduled PID settings are available to account for control schemes at varied temperatures. Further, the user can specify control based on local specimen temperature measurement or the air temperature of the surrounding chamber space. There are limits for specimen melting temperature, air-specimen differential temperature (to reduce ill-effects of initial gradients), and parameters for measuring stability.

The loading profile can be run as normal through Psylotest, with the addition of an extra temperature setpoint parameter and a button to enable the temperature controller. Using the aforementioned parameters as a stability checking regime, the system will go to each segment and wait until the new temperature setpoint has settled to the user's specification. The algorithm ensures minimal temperature gradients throughout the specimen while also maintaining an acceptable steady state temperature error.

42.4 Experimental Tests and Analysis

Once the system was properly tuned to hold different temperatures, it was found to meet its stability criterion of ±0.5 C within 1 standard deviation (Fig. 42.3). Additionally, sensor noise, sensor drift, and time to temperature were found to be within target constraints.

GPO3 is a fiberglass reinforced polyester composite commonly used as an electrical insulator in industrial applications. In addition to strength and electrical isolation, high temperature resistance is a key parameter. Starting with raw material stock #8549K79 from McMaster-Carr, a 12″ × 12″ × 1/32″ sheet was cut into strips roughly 1/2″ wide. These strips were then

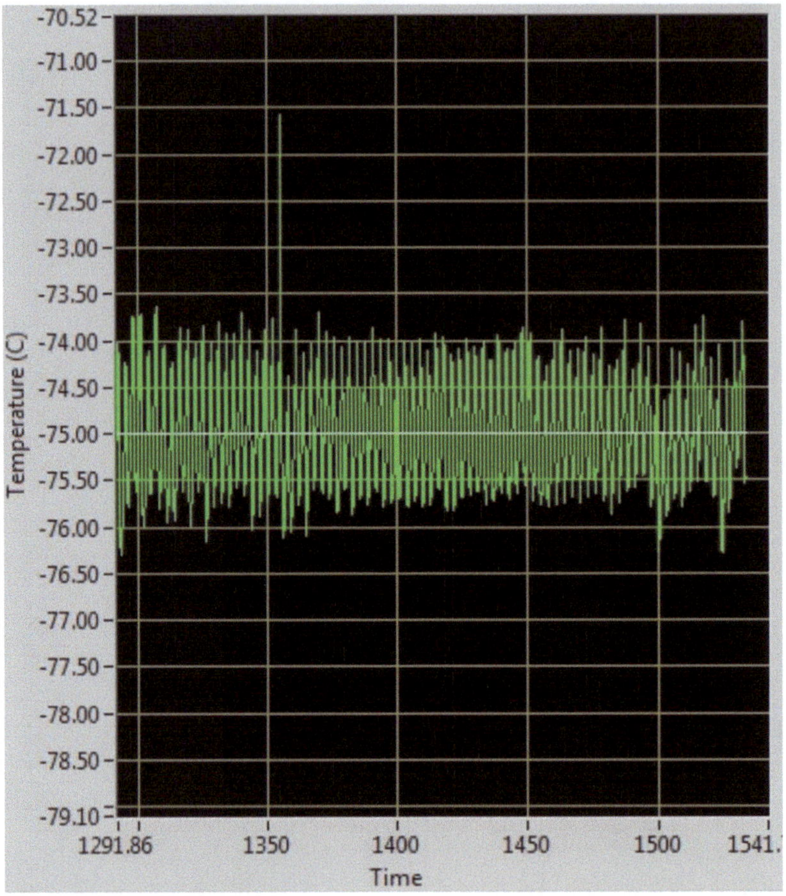

Fig. 42.3 High temperature stability

cut to length to fit the chamber with Psylotech's clamp grips. Specimens were sanded with 280 grit sandpaper in a random fashion, and a felt tip marker was used to apply a speckle pattern that provided sufficient spatial resolution to identify fiber and matrix sections separately.

With the specimen in place, the load frame was jogged until the specimen pre-load limit of 5 N was reached. The top cover of the chamber was attached and the temperature control routine was started. A 0.1% strain ramp in encoder control was run up and down at a 1E-4/s strain rate at temperatures from 25 to 175°C. Data was recorded at 20 Hz on the load frame, and a camera image was taken for every 4th load frame point (5 Hz). A digital low pass filter of 100 Hz was applied to the recorded load frame data to minimize noise.

The Psylotest program outputs sensor data as tab separated values. The raw image output of the 2048 × 2048 pixel camera was analyzed using Correlated Solutions' Vic-2D 6 software (Fig. 42.4). With a post-processing program written in LabVIEW, triggered image captures were easily synced to load frame data (Fig. 42.5).

42.5 Conclusion

It is clear that an under-microscope load frame with high resolution sensors, intuitive software control, high speed, and now coupled with temperature control allows unprecedented test capability. Fibers were visible and high shear concentrations in the 2D strain field were a leading indicator of failure points as expected. By comparing the data of the 2D strain field to simulation results, it is possible to fine tune models and develop new techniques. Furthermore, the ability to test at various temperatures allows unique tests that evaluate the temperature dependence of these phenomena.

The Temperature Chamber for Psylotech's μTS platform is priced competitively and available for order starting Fall 2016. Visit www.psylotech.com, or call 847-328-7100 for more information.

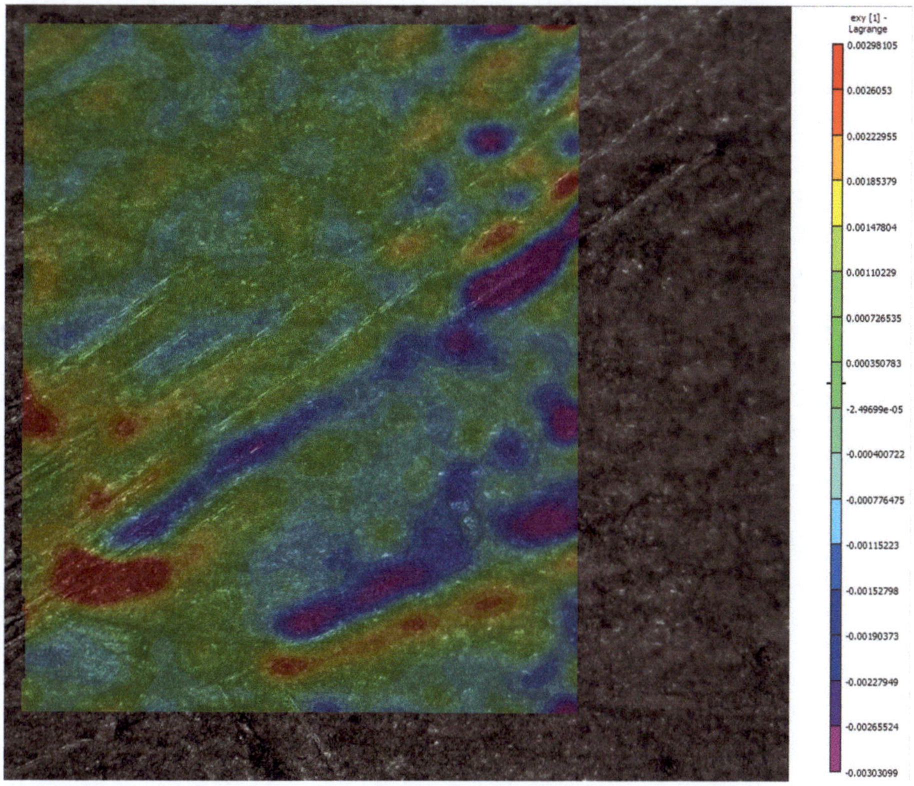

Fig. 42.4 Shear strain field at 175°C

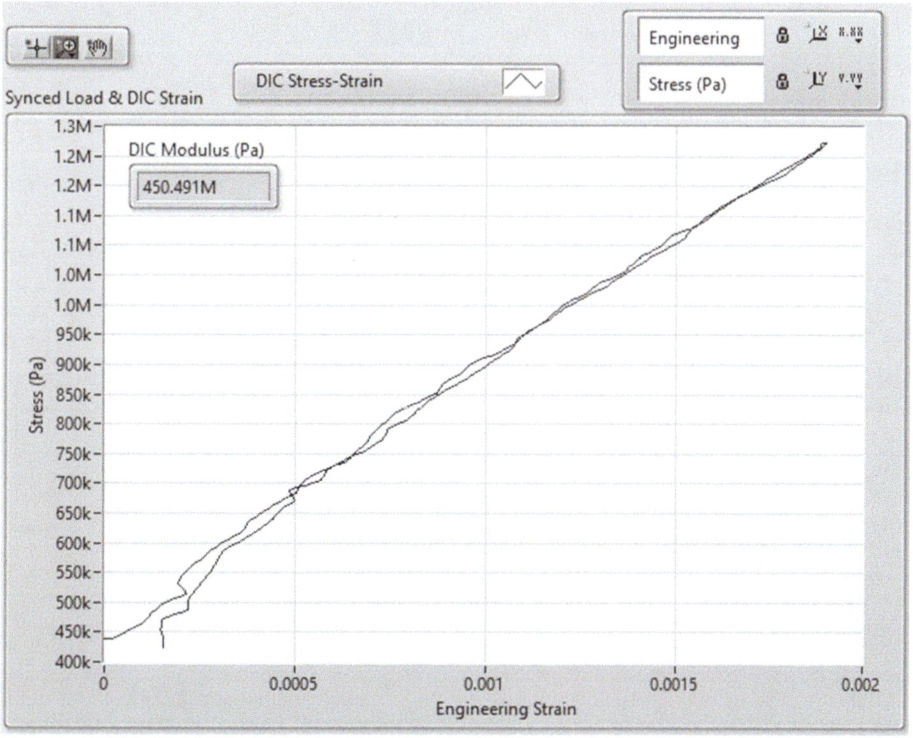

Fig. 42.5 DIC strain and load frame stress plotted together

Acknowledgments The authors would like to thank Psylotech technical staff member Zachary Zanzinger for help throughout the prototyping phase. Additionally, thanks are due to the undergraduate team of Northwestern University's Segal Design Institute for initial concept screening: Travis Cottle, Megan Renner, Daniel Russman and Yang Yu.

Chapter 43
Shock-Structure Interaction Using Background Oriented Schlieren and Digital Image Correlation

S. Kishore, M. Pinto, and A. Shukla

Abstract Flight-critical structures can be subjected to blast or shock loadings, e.g., during conflict situations or during supersonic flight operations. It is therefore necessary that structural behavior of the materials used be well understood in such extreme environments. The Background Oriented Schlieren (BOS) technique is applied to study the reflection of shockwave over an inclined aluminum plate in tandem with the 3D DIC technique to study the full field deformation of the back face of the aluminum plates. The mathematical formulation of the BOS technique is used to observe pressure and shock wave fronts. Back face DIC of the aluminum plates can be correlated to the BOS data at the same time and hence providing a more comprehensive way to study fluid structure interaction during a shock wave reflection.

Keywords Shock reflection • BOS technique • DIC • Mach stem • Shock tube

43.1 Introduction

Flight-critical structures can be subjected to blast or shock loadings, e.g., during conflict situations or during supersonic flight operations. It is therefore necessary that structural behavior of the materials used be well understood in such extreme environments. Using Digital Image Correlation (DIC) to obtain full field deformation information has been a useful tool in the study of the response of aerospace materials to shock loading. However, in the case of oblique shock-structure interaction various shock reflection phenomenon may occur depending on the Mach number of the incident shock and the structural orientation with respect to the shock. Hence, the knowledge of shock reflection, flow, and pressure fields will help in the better understanding of the structural response of aerospace materials under these conditions. This paper describes the implementation of a technique called Background Oriented Schlieren (BOS) that can produce quantitative visualization of shock-structure interaction. Correlating the combination of the full-field deformation information and knowledge of the shock-structure interaction phenomena, it becomes possible to have a deeper understanding of the behavior of the plate under oblique shock loading.

43.2 Experimental Setup

An experimental setup was developed using a high intensity light sources, translucent speckle pattern, and high-speed cameras to obtain information about the shock and flow using this technique. Figure 43.1 gives a schematic representation of the BOS experimental setup. In addition, three dimensional, full-field structural deformation information of a speckled aluminum plate was obtained using high speed photography and DIC. A shock tube apparatus was used to load an aluminum plate of 50.8 mm by 152.4 mm dimensions, clamped at the two ends and inclined at an angle of 60° with respect to horizontal plane and was subjected with a planar shockwave of 0.8 MPa peak incident pressure. Full field deformation of the back face was obtained using 3D DIC data captured two using high speed cameras looking a speckle pattern painted on the back face of the specimen. Another speckle pattern was illuminated using light passing through a suitable focal length lens in order to make the incident light parallel. The speckle pattern was observed using a high speed camera as a shock passed in front of

S. Kishore (✉) • M. Pinto • A. Shukla
Dynamic Photo-Mechanics Laboratory, Department of Mechanical Industrial and Systems Engineering, University of Rhode Island, Kingston, RI 02881, USA
e-mail: shyamal_kishore@uri.edu

M. Sutton, P.L. Reu (eds.), *International Digital Imaging Correlation Society*, Conference Proceedings of the Society for Experimental Mechanics Series, DOI 10.1007/978-3-319-51439-0_43

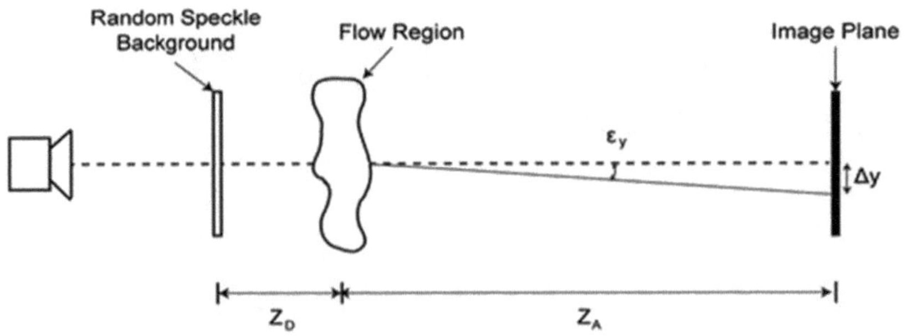

Fig. 43.1 Schematic of background oriented Schlieren method

it. Deflections of the light from refractive index changes in the air due to the passing shock enable the visualization of the shock propagation and reflection over the inclined aluminum plate.

43.3 Mathematical Formulation of Bos Technique

The deflection of the light beam as it passes through the flow region, ϵ_y may be expressed as, [1]

$$\epsilon_y = \frac{\Delta y\left(Z_D + Z_A\right)}{Z_D f} \tag{43.1}$$

$$\epsilon_y = \frac{1}{n}\int_{z_D}^{Z_D+Z_A} \frac{\partial n}{\partial y} dZ \tag{43.2}$$

where, $\boldsymbol{\Delta y}$ is the pseudo-displacements measured on the background plane, $\boldsymbol{Z_D}$ is the distance from the background to the boundary of the flow region and $\boldsymbol{f}$ is the focal length of the lens. This deflection may also be expressed in terms of the change in refractive index of the flow region using Snell's Law.

n_0 is the refractive index of the unperturbed fluid and $\frac{\partial n}{\partial y}$, the change in refractive index in the y-direction. After introduction of the Gladstone-Dale relation, which links the refractive index of a fluid to its density, the above equations can be recast into the Poisson equation. In this way, the gradient of the fluid density may be related to the pseudo-displacements measured on the background plane. This relation suggests that the pseudo strains are also relatable to pressure in the flow [1].

$$\frac{\partial^2 \rho\left(x, y\right)}{\partial x^2} + \frac{\partial^2 \rho\left(x, y\right)}{\partial y^2} = k\left(\frac{\partial u\,(x, y)}{\partial x} + \frac{\partial v\,(x, y)}{\partial y}\right); \text{Where, } k = \frac{n_0\left(Z_D + Z_A\right)}{\kappa Z_A Z_D f} \tag{43.3}$$

43.4 Results

The BOS technique helps us visualize the shock front, its reflection off the plate and formation of Mach Stems and their growth. In addition to the visualization of the fluid structure interaction, the 3D DIC data obtained from the back face of the plate can help us correlate the plate response to shock to the fluid structure interaction and this combined approach can serve as a tool for a more comprehensive study. In Fig. 43.2, the contours of sum of pseudo strains in x and y directions have been shown from the BOS data after 60 μs from the time the shock first hits the plate. These contours are representative of the densities and hence the pressures. A shock front can be easily observed by looking for regions of high negative sum of strains. A Mach stem can be observed attached to the plate, along with incident and the reflected wave can also be observed. In Fig. 43.3, the out of plane displacement contours can be seen from the 3D DIC data of the back-face at the same time instant.

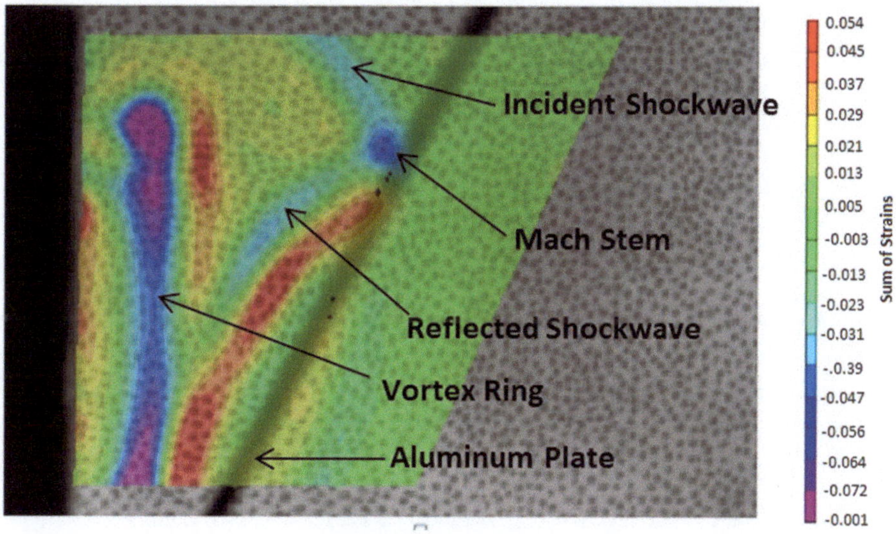

Fig. 43.2 Visualization of shock wave reflection on an aluminum plate with its features using sum of pseudo strains in x and y directions

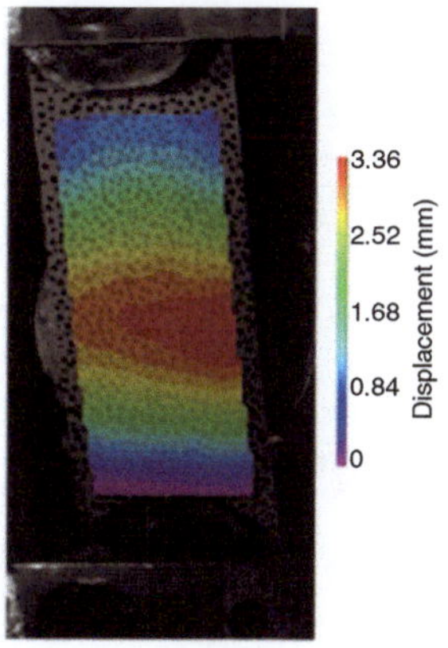

Fig. 43.3 Back face DIC with out of plane displacement contours

Acknowledgements The authors acknowledge the financial support of *Air Force Office of Scientific Research (AFOSR)* for the financial support provided under the Grant No. FA9550-13-1-0037.

Reference

1. Meier, G.E.A.: New optical tools for fluid mechanics. In: Proceeding of 8th International Symposium Flow Visualization, Sorrento, Italy, 1–4 September 1998, ISBN 0 9533991 0 9 (CD-ROM), Paper 226 (1999)

Chapter 44
Identification of Heterogeneous Elastoplastic Materials by Constitutive Equation Gap Method

T. Madani, Y. Monerie, S. Pagano, C. Pelissou, and B. Wattrisse

Abstract A strategy for identifying the shape and the parameters of elastoplastic laws in heterogeneous material is proposed here. It allows the local characterization of both mechanical properties and stress fields. This method relies on the constitutive equation gap principles and on the measurement of overall loads and strain fields. These fields are obtained by the numerical differentiation of displacement fields measured by Digital Image Correlation. The identification method is based on the iterative minimization of an energy norm involving the secant elastoplastic tensor.

We propose here to characterize the robustness of the method and its stability (to noise, initial parameters and mesh) on numerical data sets obtained on numerically-deformed images. Finally, we illustrate the ability of the method to identify the local material properties of a heterogeneous material on real experimental data.

Keywords Digital image correlation • Identification • Constitutive equation gap • Elastoplasticity

44.1 Introduction

Optical measurement techniques applied in the field of Experimental Mechanics have improved significantly in the last decades. Thanks to the large amount of information given by full-field measurement techniques, it is now possible to tackle complex identification problems such as strongly localized phenomena (Lüders bands, crack propagation ...). Mechanical properties are generally identified from overall loading data and kinematic fields (obtained by Digital Image Correlation, interferometric techniques, grid methods, etc.). The Finite-Element Updating Method is probably the most widespread identification method, as it allows identifying very different physical properties. Specific methods, adapted to the identification of mechanical behaviors were also proposed. A general overview over existing identification techniques can be found in [1]. The Constitutive Equation Gap Method (CEGM) method was chosen because of its adaptability to a wide range of behaviors but also for its ability to identify heterogeneous behaviors.

The method presented here is an extension of the one proposed in [2]. It relies on the analytic expression of the local elastoplastic secant stiffness tensor $\underline{\underline{B^s}}$. For a linear kinematic model, the tensor $\underline{\underline{B^s}}$ can be expressed directly as a function of the material properties (e.g. for J2-plasticity: yield stress σ_0 and hardening coefficient h) and of the loading (through the plastic increment $\Delta\gamma$). After briefly presenting the method, we illustrate its performances on numerical and experimental data.

T. Madani • Y. Monerie • S. Pagano • B. Wattrisse
Université de Montpellier, CNRS, Montpellier, France

Laboratoire de Micromécanique et Intégrité des Structures, MIST, IRSN-CNRS-Université de Montpellier, France
e-mail: tarik.madani@umontpellier.fr; yann.monerie@umontpellier.fr; stephane.pagano@umontpellier.fr; bertrand.wattrisse@umontpellier.fr

C. Pelissou
Institut de Radioprotection et de Sûreté Nucléaire, PSN/SEMIA, Saint-Paul-Lez-Durance, France

Laboratoire de Micromécanique et Intégrité des Structures, MIST, IRSN-CNRS-Université de Montpellier, France
e-mail: celine.pelissou@irsn.fr

M. Sutton, P.L. Reu (eds.), *International Digital Imaging Correlation Society*, Conference Proceedings of the Society for Experimental Mechanics Series, DOI 10.1007/978-3-319-51439-0_44

44.2 Identification Methodology

We propose a method to identify the parameters of an elastoplastic constitutive law in a 2D (plane stress) framework for a kinematic multi-linear hardening associated with a Von Mises criterion. The CEGM is based on the minimization of a functional expressing the gap in the constitutive equation. In its simplest form (small strain hypothesis, equilibrium, and linear elastic constitutive behavior) the cost-function reads:

$$F_{CEGM}\left(\overrightarrow{u^c}, \underline{\underline{B}}^s(p)\right) = \frac{1}{2}\int_{\Omega}\left[\underline{\varepsilon}\left(\overrightarrow{u^c}\right) - \underline{\varepsilon}\left(\overrightarrow{u^m}\right)\right] : \underline{\underline{B}}^s(p) : \left[\underline{\varepsilon}\left(\overrightarrow{u^c}\right) - \underline{\varepsilon}\left(\overrightarrow{u^m}\right)\right] d\Omega \tag{44.1}$$

where vector p regroups the material parameters, $\overrightarrow{u^m}$ is a measured displacement field, $\overrightarrow{u^c}$ a displacement field compatible with the local and global equilibrium of the studied domain Ω. For an elastoplastic material and at a given time step n, the secant stiffness tensor $\underline{\underline{B}}_n^s$ reads [3]:

$$\underline{\underline{B}}_n^s(p) = \left[\underline{\underline{B}}^{e-1} + \frac{\Delta\gamma(n)}{1 + \frac{2}{3}h\Delta\gamma(n)}\underline{\underline{P}}\right]^{-1} \tag{44.2}$$

where $\underline{\underline{B}}^e$ is the elastic tensor (depending, for a cubic material, on the three elastic constants: e.g. Young modulus E, shear modulus G, and Poisson ratio ν), $\Delta\gamma(n)$ is the plastic multiplier increment at time step n and $\underline{\underline{P}}$ is a constant mapping matrix.

The secant tensor depends on five material constants $p = \{E, G, \nu, \sigma_0, h\}$ (three for elasticity and two for plasticity). The identification procedure consists in minimizing functional F_{CEGM} over all admissible displacements $\overrightarrow{u^c}$ (such as the derived stress field satisfies the equilibrium equations, the constitutive equations and the global equilibrium conditions) and all admissible material parameters (such as $\underline{\underline{B}}_n^s$ is positive-definite). Thanks to the properties of convexity of functional F_{CEGM}, the minimization can be performed in two consecutive steps: first with respect to its first argument (to determine a displacement field associated with a statically-admissible stress fields) and then with respect to the secant stiffness tensor (to identify the material parameters). In order to increase the robustness of the procedure, the elastic parameters are determined separately from the plastic ones on purely elastic load steps. The procedure is controlled through an optimization algorithm. The iterative procedure is started with an initial set of parameter chosen arbitrarily. The procedure is stopped using a convergence criterion on the norm of the secant tensor, and the optimal material parameters are obtained.

44.3 Identification of Heterogeneous Mechanical Properties

The method is firstly assessed on numerical data obtained on a finite-element simulation of a heterogeneous elastoplastic polycrystal with four phases. The geometry and the material parameters used for the direct computation are given in Fig. 44.1a. Supposing that the phase distribution is known and introducing the displacement field and the overall loads in the identification procedure, the material parameters distribution (see Fig. 44.1b) and the stress field (see Fig. 44.1c) are identified. In this example (isotropic J2-plasticity with linear kinematic hardening), the material parameters of each phase are: $E^R = 210$ GPa, $\upsilon^R = 0.3, \sigma_0^R = 300$ MPa, $h^R = 2$ GPa, $E^O = 180$ GPa, $\upsilon^O = 0.25, \sigma_0^O = 300$ MPa, $h^O = 2.5$ GPa, $E^G = 100$ GPa, $\upsilon^G = 0.15, \sigma_0^G = 300$ MPa, $h^G = 1$ GPa, $E^Y = 150$ GPa, $\upsilon^Y = 0.2, \sigma_0^Y = 300$ MPa, $h^Y = 1.5$ GPa. Figure 44.2 shows some identified quantities (hardening modulus h and axial stress ${\sigma_{yy}}^c$). The identification, performed here without superimposed noise, is quite consistent with the imposed fields.

The method has been applied on experimental data obtained on a heterogeneous material with a known heterogeneity: a Dual Phase steel (DP600) tensile specimen machined in a trapezoidal sample having undergone a known pre-hardening. The method correctly identifies the heterogeneous behavior of the material as can be seen in Fig. 44.2 showing the "theoretical" yield stress associated with the known pre-hardening (Fig. 44.2a) and the identified one (Fig. 44.2b).

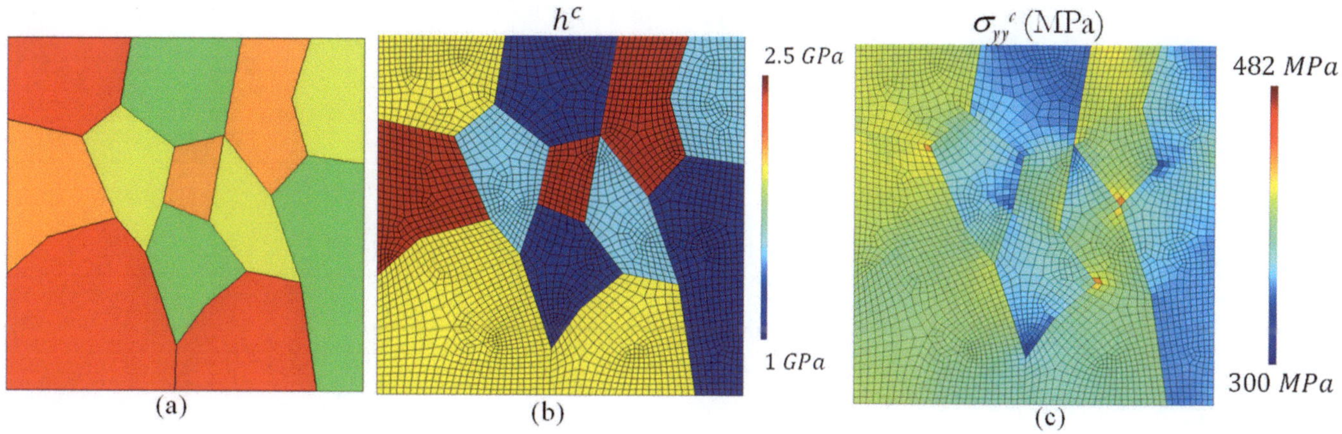

Fig. 44.1 (**a**) Geometry of the polycrystalline specimen composed of four phases (R, O, Y, and G) and 12 grains. (**b**) identified hardening modulus and (**c**) identified axial stress field component

Fig. 44.2 Identified fields of yield stress σ_0: (**a**) theoretical values and (**b**) identified values

44.4 Conclusions

We illustrate the use of the CEGM to identify the parameters of an elastoplastic behavior from full-field measurements. Here we identify mechanical stresses and a distribution of elastic and plastic coefficients. The originality of this work resides in its ability to tackle heterogeneous stress fields associated with either heterogeneous materials or complex structures for elastic or plastic materials. The method is developed for multi-linear hardening behaviors. It requires only an estimation of the elastic properties. The choice of the initial guess was shown not to affect the identified values. The proposed procedure was checked on the basis of numerically-obtained displacement fields given by a FE simulation. Results of the study show the ability of the method to deal with strongly heterogeneous situations. Gaussian white noise was also superimposed to the numerical data in order to assess the robustness of the method with respect to noise. This method was successfully applied on an experimental situation associated with a heterogeneous material.

References

1. Avril, S., Bonnet, M., Bretelle, A.S., Grediac, M., Hild, F., Ienny, P., Latourte, F., Lemosse, D., Pagano, S., Pagnacco, E.: Overview of identification methods of mechanical parameters based on full-field measurements. Exp. Mech. **48**, 381–402 (2008)
2. Latourte, F., Chrysochoos, A., Pagano, S., Wattrisse, B.: Elastoplastic behavior identification for heterogeneous loadings and materials. Exp. Mech. **48**, 435–449 (2008)
3. Simo, J.C., Hughes, T.J.R.: Computational Inelasticity. Springer, New York (1998)

Chapter 45
Metrological Analysis of the DIC Ultimate Error Regime

M. Bornert, P. Doumalin, J.-C. Dupré, C. Poilâne, L. Robert, E. Toussaint, and B. Wattrisse

Abstract In DIC, the "ultimate error regime" corresponds to situations for which the shape function used to describe the material transformation perfectly matches the actual one. We propose to confront results obtained from numerically-shifted images with the predictions of theoretical models developed in the literature to describe bias and random error evolutions with respect to the imposed displacement. Results show the overall good predictions of these models but small deviations arise, mainly around integer values of imposed displacements for noisy images. These deviations are interpreted as the unrepresentativeness of the underlying hypotheses of the theoretical models in these particular cases.

Keywords Digital image correlation • Metrology • Ultimate error

45.1 Introduction

The characterization of metrological performances of DIC techniques has been widely studied from experimental and theoretical point of view. Due to the complexity of the problem, these approaches generally deal with a limited number of parameters of the overall DIC problem. The first theoretical approaches proposed a noise propagation analysis in DIC procedure [1]. They were later complemented by adding the effect of image interpolation [2]. The collaborative work carried out by the workgroup "Metrology" of CNRS research network 2519 aims at contributing to a systematic approach to this question by dissociating the elements of the measurement chain. In [3], the effect of the image transformation was investigated. Two error regimes can be encountered: (1) the ultimate error regime, when the shape function is rich enough to represent the actual the displacement field, and (2) the shape function mismatch regime, when the shape function is too poor to reproduce the real displacement field on the chosen subset.

We propose to compare the predictions of theoretical models used to describe the random and bias errors to results obtained on translated computer-generated images. In these situations, the image transformation is perfectly known and the displacement measurements correspond to the ultimate error regime (no shape-function bias).

M. Bornert
Laboratoire Navier (UMR 8205), CNRS, ENPC, IFSTTAR, Université Paris-Est, Marne-la-Vallée, France

P. Doumalin • J.-C. Dupré
Institut P', UPR 3346 CNRS, Université de Poitiers, SP2MI, Futuroscope Chasseneuil, France

C. Poilâne
Normandie Univ, ENSICAEN, UNICAEN, CEA, CNRS, CIMAP, 14000 Caen, France

L. Robert
Université de Toulouse, Mines Albi, INSA, UPS, ISAE, ICA (Institut Clément Ader), Albi, France

E. Toussaint
Institut Pascal, UMR 6602, Université Blaise Pascal – IFMA, Aubière, France

B. Wattrisse (✉)
LMGC, Univ. Montpellier, CNRS, Montpellier, France
e-mail: bertrand.wattrisse@umontpellier.fr

M. Sutton, P.L. Reu (eds.), *International Digital Imaging Correlation Society*, Conference Proceedings of the Society for Experimental Mechanics Series, DOI 10.1007/978-3-319-51439-0_45

45.2 Theoretical Models for Random and Systematic Errors

The first theoretical analysis was proposed by Roux and Hild [1]. It consists in a statistical description of the sensitivity of displacement measurements with respect to superimposed image noise. This approach requires major hypotheses: perfect reconstruction of the grey level images (*I* being the initial image and *T* the final one), optical flow conservation and statistical model of the image noise (a zero-mean with a σ_n standard deviation uniform white noise). Authors suggested an image noise propagation in an analytical description of the correlation criterion (namely, the classical "Sum of Squared Difference" SSD criterion). In this case the bias error is null. Using similar hypotheses, Wang et al. [2] consider an imperfect reconstruction based on some interpolation of grey levels. The expression of random (σ_u) and systematic ($\overline{\Delta u}$) errors is then:

$$\begin{cases} \sigma_u \propto \dfrac{\sigma_n}{d\sqrt{\overline{\nabla I^2}}} \\ \overline{\Delta u} = -\dfrac{\overline{h(\tau)\nabla I}}{\overline{\nabla I^2}} + f_i(\tau)\dfrac{\sigma_n^2}{\overline{\nabla I^2}} \end{cases} \tag{45.1}$$

where d stands for the subset size, ∇I the grey level gradient. In this model, the random error σ_u is independent of the actual displacement. It has the same expression than the one obtained in [1]. The first part of the expression (denoted E_{interp}) of the bias error is noise-independent and shows a dependence with the fractional part τ of the imposed displacement through the local grey level residual $h(\tau)$ at the correlation optimum. The second part (denoted E_{noise}) shows a quadratic dependence with noise. Function f_i depends on the chosen grey level interpolation: it has been shown to be linear in case of a linear interpolation and quintic for a cubic one. Note that in this expression $\overline{A}$ stands for the average of A over the given subset for a set of drawing. It should be emphasized that this way of defining the bias error differs from the classical approaches where it is averaged over all the subset realizations, with various grey level and random noise distributions.

45.3 Experimental Results

Figure 45.1 represents the evolution of both random and bias errors determined from computer-generated images obtained using the TexGen software for different levels of superimposed noise. Contrary to the predictions of (45.1), the random error is not constant with respect to the imposed displacement for a given noise level. But the value of the plateau linearly increases with the imposed noised in accordance with (45.1), and the slope of this evolution is consistent with the expression given in (45.1) for the tested image and the chosen subset size. Figure 45.1b illustrates the evolution of the bias error with the imposed displacement τ for a linear interpolation scheme. Classical S-shaped curves are obtained, with an inversion of concavity when increasing noise level. Whatever the noise level, the systematic error is null for integer imposed displacements.

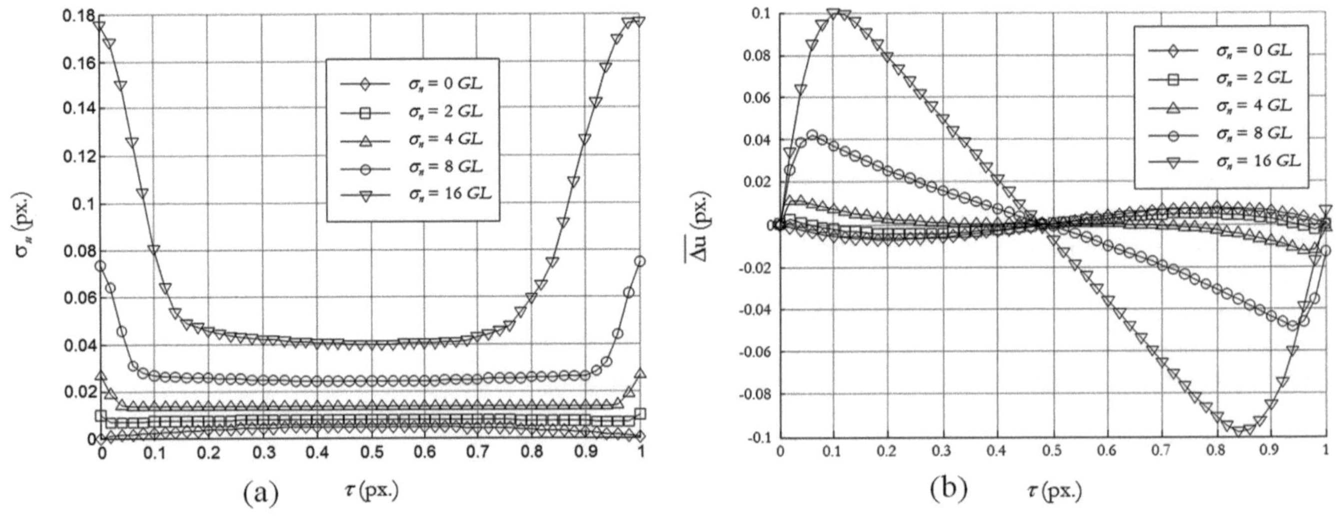

Fig. 45.1 (**a**) Evolution of the random error σ_u and (**b**) evolution of the bias error $\overline{\Delta u}$ as a function of imposed displacement τ for different noise levels σ_n for $d = 16$ px

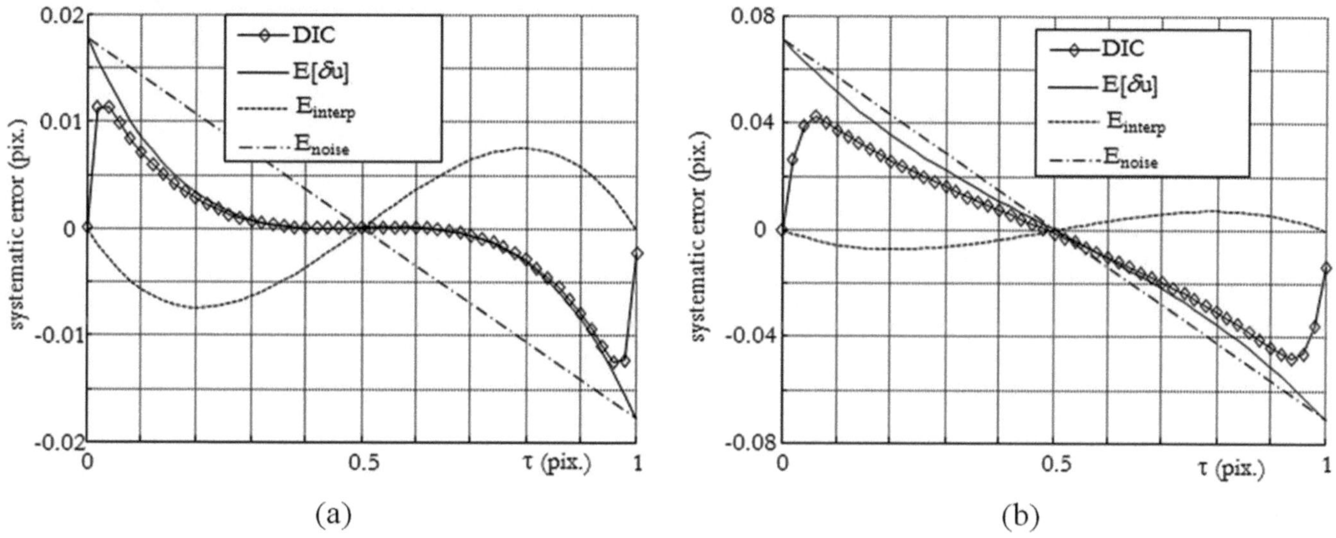

Fig. 45.2 Comparison of the experimental (DIC) and the theoretical (E[δu]) bias errors as a function of imposed displacement τ for $\sigma_n = 4GL$ (**a**) and $\sigma_n = 8GL$ (**b**)

Figure 45.2 compares the experimental bias error and the theoretical expression given in ((45.1)). When considering un-noised images, the results are perfectly consistent. When introducing image noise, some discrepancies appear: the experimental bias error is null for a null displacement while it is not null in the analytic expression. The difference is higher for higher noise levels.

45.4 Conclusions

The theoretical model given in ((45.1)) gives very satisfactory results for the description of the random and bias errors as a function of image characteristics (gray level gradients, noise), DIC parameters (subset size, interpolation scheme) and imposed displacement as long as the imposed displacement is not "too close" to integer values for noisy images. The discrepancy is related to the hypotheses made in the expectancy computations which are obtained on a fixed subset for several realizations of image noise with the model, while they are determined on various subsets for one realization of image noise for the experimental data. In practice, this latter method is more representative of displacement fields computed on a pair of images. Theoretical models should be complemented to account for these experimental observations.

Acknowledgements On behalf of the Workgroup "Metrology" of the French CNRS research network 2519 "Mesures de Champs et Identification en Mécanique des Solides/Full field measurements and identification in solid mechanics". This work is dedicated to our colleague Laurent Robert who passed away in 2015.

References

1. Roux, S., Hild, F.: Stress intensity factor measurements from digital image correlation: post-processing and integrated approaches. Int. J. Fract. **140**, 141–157 (2006)
2. Wang, Y.Q., Sutton, M.A., Bruck, H.A., Schreier, H.W.: Quantitative error assessment in pattern matching: effects of intensity pattern noise, interpolation, strain and image contrast on motion measurements. Strain. **45**, 160–178 (2009)
3. Bornert, M., Brémand, F., Doumalin, P., Dupré, J.-C., Fazzini, M., Grédiac, M., Hild, F., Mistou, S., Molimard, J., Orteu, J.-J., Robert, L., Surrel, Y., Vacher, P., Wattrisse, B.: Assessment of digital image correlation measurement errors: methodology and results. Exp. Mech. **49**, 353–370 (2009)

Chapter 46
Inverse Identification of the High Strain Rate Properties of PMMA

Frances Davis, Clive Sivour, and Fabrice Pierron

Abstract Due to its low density and transparency, PMMA is frequently used as a glass substitute in aircraft and automotive components. Polymer windows are expected to resist impact loads and can be subjected to high strain rate loading ($>1000\,s^{-1}$). While the quasi-static behavior of PMMA is well characterized, significantly fewer studies have looked into its high strain rate response. PMMA exhibits rate-dependent elastic and failure behavior, with both the modulus and the yield stress becoming sensitive to strain rate above a threshold of approximately $100\,s^{-1}$. A new experimental technique for quantifying the response of glassy polymers at high strain rates using the virtual fields method is proposed. Samples were subjected to inertial impact tests and images of the sample deformation during the impact was captured using an ultra-high speed camera, recording at 2 Mfps. The full field displacement, strain, and acceleration were input to the virtual fields method to perform an inverse identification of the material parameters.

46.1 Introduction

The goal of this study is to propose a new experimental technique for quantifying the viscoplastic behavior of glassy polymers at high strain rates. In this paper, the results of a preliminary test campaign to validate this novel approach at a single strain rate are presented. First, the protocol for performing inertial impact tests, which relies upon ultra-high-speed imaging, is described. Next, an inverse technique used to identify parameters from full field imaging data called the virtual fields method is introduced. Finally, the results of the test campaign and the parameter identification are discussed.

46.2 Experimental Protocol

The gas gun facilities in the Solid Mechanics and Materials Engineering Group at Oxford were used to perform the impact tests. The specimen is L=50 mm long by H=31 mm tall with a thickness of 4 mm. A grid with a pitch of 1 mm was applied to the specimen using transfer paper. Figure 46.1 shows a schematic of the experiment. An ultra-high-speed camera (HPV-X, Shimadzu) was used to image the sample during the impact test. A total of 128 frames were recorded at a rate of 2 Mfps. Using the grid method [1], the collected images were analyzed to produce time-resolved displacement maps. To minimize the influence of measurement noise, the displacement field at each pixel was filtered in time. The displacement was then numerically differentiated to produce maps the acceleration and strain.

46.3 Virtual Fields Method

The virtual fields method (VFM) is an inverse technique that utilizes a weak form of the equilibrium equations to identify material properties. For the specific case of plane stress, the principle of virtual work can be written as:

F. Davis (✉) • F. Pierron
Faculty of Engineering and the Environment, University of Southampton, Highfield SO17 1BJ, UK
e-mail: frances.davis@soton.ac.uk

C. Sivour
Department of Engineering, Science University of Oxford, Oxford OX1 3PJ, UK

M. Sutton, P.L. Reu (eds.), *International Digital Imaging Correlation Society*, Conference Proceedings of the Society for Experimental Mechanics Series, DOI 10.1007/978-3-319-51439-0_46

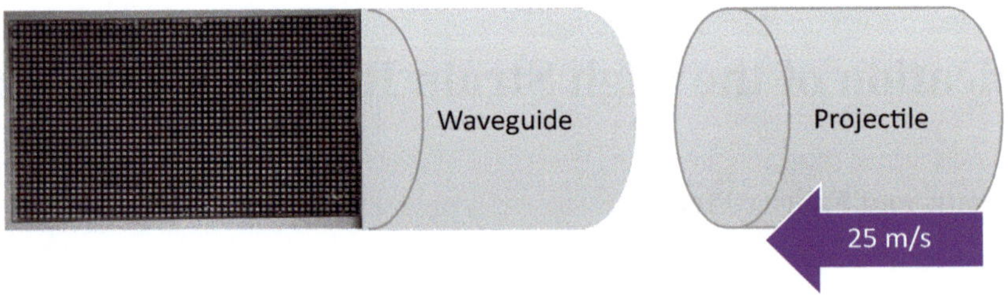

Fig. 46.1 Schematic of inertial impact test showing an image of actual sample

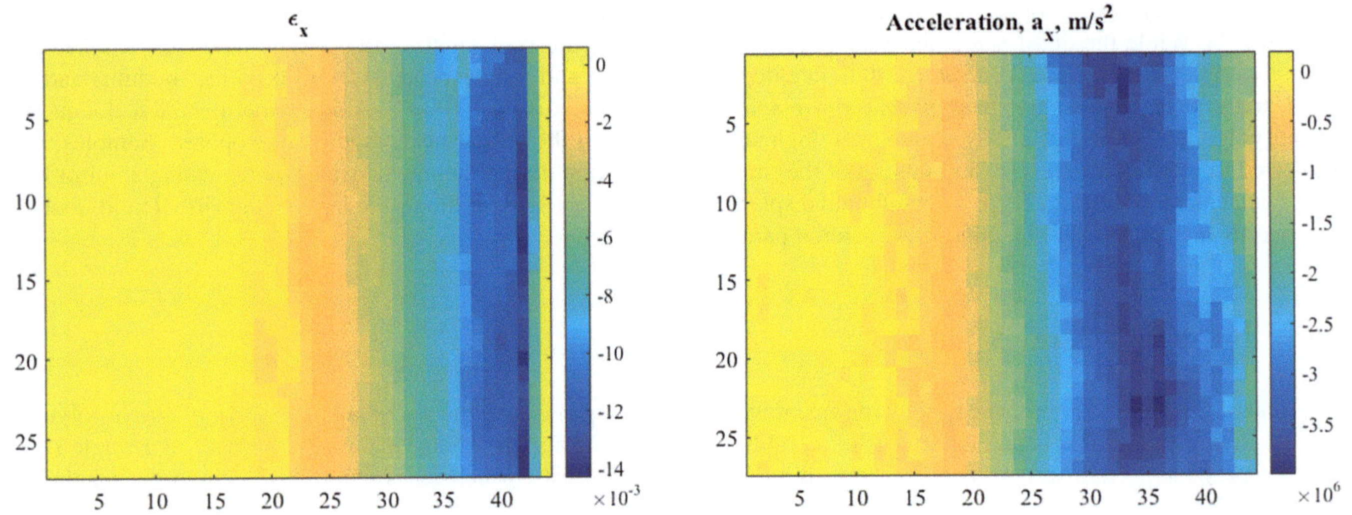

Fig. 46.2 Maps the strain and acceleration in the impact direction 8 μs after impact

$$-\int_S \boldsymbol{\sigma}(x,y,t) \cdot \boldsymbol{\varepsilon}^*(x,y,t)\, dS + \int_{\partial S} \boldsymbol{T} \cdot \boldsymbol{u}^*(x,y,t)\, dL = \int_S \rho\, \boldsymbol{a} \cdot \boldsymbol{u}^* dS \qquad (46.1)$$

where $\boldsymbol{\sigma}(x,y,t) = [\sigma_{11}, \sigma_{22}, \sigma_{12}]$ is the stress tensor, $\boldsymbol{u}^*$ and $\boldsymbol{\varepsilon}^*$ are the virtual displacement and strain, respectively, $\boldsymbol{T}$ is the traction vector, $\boldsymbol{a}$ is the acceleration vector, ρ is the density of the material, S is the surface of the solid, and ∂S is the edge of solid where tractions are applied. For high strain rate tests, the VFM treats the acceleration as a volume distributed load cell eliminating the need to take force measurements. The VFM has been well developed for linear elastic problems and the material parameters can be identified by defining two independent virtual fields and solving a set of linear equations [2].

46.4 Results and Discussion

Three PMMA specimens were impacted at speed of 30 m/s. The maximum strain rate attained in the tests was approximately $1100\,\text{s}^{-1}$. The strain and acceleration in the impact direction 8 μs after impact are shown Fig. 46.2. The blue band seen on the left hand side the acceleration map is the stress wave which is traveling from right to left in the sample. The strain maps clearly shows compressive strain building at the impact edge of the sample.

Using the strain and acceleration data as inputs to the VFM (Eq. (46.1)), the Young's modulus and Poisson's ratio for each sample were determined. The values identified are listed in Table 46.1. The identified values correspond well the values reported in literature for PMMA where E and ν were reported to be 3–6 GPa and 0.33, respectively. Additional tests are planned increasing the speed of the projectile to produce plastic deformation so that the rate dependent yield stress of PMMA can be studied.

Acknowledgements Dr. Frances Davis and Prof. Fabrice Pierron acknowledge support from EPSRC through grant EP/L026910/1. Prof. Fabrice Pierron also expresses gratitude to the Wolfson Foundation for support through a Royal Society Wolfson Research Merit Award.

Table 46.1 Identified elastic properties of PMMA

Specimen number	E [GPa]	ν
103	5.20	0.38
104	5.18	0.35
108	5.20	0.36

References

1. Badulescu, C., Grédiac, M., Mathias, J.D.: Investigation of the grid method for accurate in-plane strain measurement. Meas. Sci. Technol. **20**(9), 095102 (2009)
2. Pierron, F., Grédiac, M.: The Virtual Fields Method: Extracting Constitutive Mechanical Parameters from Full-Field Deformation Measurements. Springer, New York (2012)

Chapter 47
Quantification of the Compressibility of Elastomers Using DIC

Frances Davis, Jason L'Hommel, Jean-Benoît Le Cam, and Fabrice Pierron

Abstract Both filled and unfilled elastomers are generally modeled as incompressible solids. However studies on the compressibility of elastomers have indicated that volume expansion is observed at large stretches coinciding with the onset of cavitation. Varying methods such as dilatometry and hydrostatic weighting have been used to calculate volume changes that elastomers undergo during stretching. However, these techniques cannot map volume variations for a heterogeneous state of stress, motivating the present work. In this study, carbon black filled elastomer samples were subjected to a uniaxial stretch and the deformation was recorded using two back-to-back stereo-correlation systems. The back-to-back stereo-correlation systems allow the in-plane strains on the front and back face of the specimen to be calculated along with the normal strain through the thickness. Using the assumption of plane stress, the volume variation of the elastomer as a function of the applied longitudinal strain was determined.

Keywords High strain rate • Inertial impact • PMMA • Grid method • Virtual fields method

47.1 Introduction

Elastomers are soft rubbers that can undergo very large reversible deformations (up to 300%). They are routinely used for wide variety of engineering applications due to their damping qualities and high elongation at failure. The mechanical response of rubbers is generally modeled by assuming that they are incompressible solids. Early studies examining the volume variation in rubbers used dilatometry and observed volume changes when large stretches were applied [1]. Recently, digital image correlation has been used to examine the volume change in natural rubbers [2]. A uniaxial stretch was applied to the rubber and a single camera was used to record the deformation. The stretch through the thickness was calculated assuming that the rubber was transversely isotropic. While this approach can be used to study uniaxial tension it would not be appropriate to examine heterogeneous states of stress. In this paper, two back-to-back stereo-correlation systems were used to examine the volume variation in a filled elastomer.

47.2 Experimental Protocol

As proof of concept, a carbon black filled elastomer was subjected to a uniaxial stretch and the volume variation was measured using two back-to-back stereo-correlation systems. The uniaxial tensile test was performed on an Instron 5569 equipped with a 2 kN load cell. Samples were cut from rubber sheets with a gauge length and width of 50 mm. The initial thickness was 2 mm. A white speckle pattern was applied to both sides of the elastomer using transfer paper before installing the sample in the grips (Fig. 47.1). The imaging setup consisted of four cameras (Manta, Allied Vision, 2452 × 2056 px) with 50 mm AF Nikkor lenses. The cameras were positioned as shown in Fig. 47.1a. Nila LED lights were used to illuminate the field of view. To eliminate specular reflections, polarized light filters were used on each of the lights and lenses. The samples were loaded at a rate of 10 mm/min up to a displacement of 175 mm. Images of the elastomer were collected at a rate of 0.3 Hz.

F. Davis (✉) • J. L'Hommel • F. Pierron
Faculty of Engineering and the Environment, University of Southampton, Highfield SO17 1BJ, UK
e-mail: frances.davis@soton.ac.uk

J.-B. Le Cam
Université de Rennes, Institut de Physique de Rennes, 35042 Rennes, France

M. Sutton, P.L. Reu (eds.), *International Digital Imaging Correlation Society*, Conference Proceedings of the Society for Experimental Mechanics Series, DOI 10.1007/978-3-319-51439-0_47

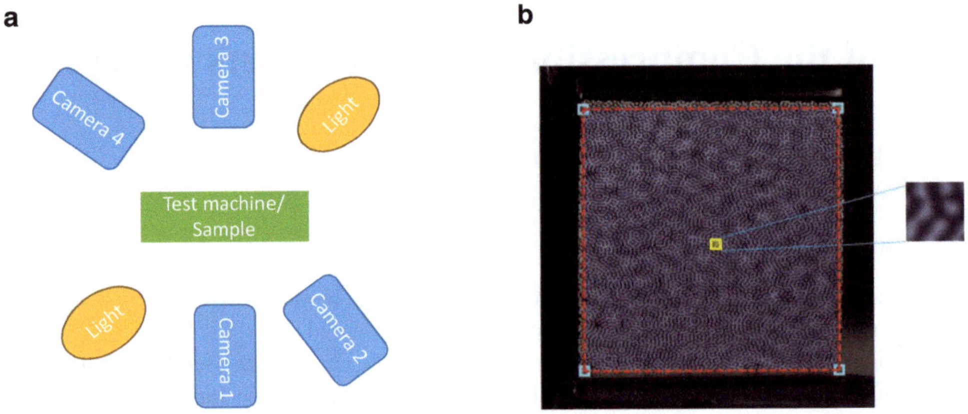

Fig. 47.1 Schematic of the back-to-back camera test configuration showing (**a**) the position of the cameras and lights relative to the test machine and (**b**) a sample image of the elastomer with the speckle pattern applied. The *small block* shows a representative subset

The commercial code MatchID (v. 2.1) was used to capture images, calibrate the stereo systems, and analyze the images. To calculate the change in volume the displacement data from the front and back camera systems must be obtained in a single coordinate system. To achieve this first, the front and back stereo systems were calibrated individually. Then a secondary calibration was performed between cameras 1 and 3. A black speckle pattern was transferred onto a thin sheet of PMMA and several images of speckle pattern were captured using both camera systems. A custom code was built to determine the coordinate transformation between cameras 1 and 3. The images collected during the tension test from the front and back stereo systems were analyzed individually both using a subset size of 33 × 33 px and a step size of 15 px. Due to the large deformation that occurred during the test the reference image was updated after every image. The resulting displacements were exported from MatchID in the coordinate systems of cameras 1 and 3 for the front and back camera systems, respectively. The coordinate transformation obtained from the secondary calibration was then used to bring all of the displacement data into a single coordinate system (Camera 1).

Using the transformed displacement data, the deformation gradient, $\mathbf{F} = \mathbf{1} + \nabla \mathbf{u}$, was calculated assuming a state of plane stress. The volume variation is then given by

$$\frac{dv}{dV_0} = \det \mathbf{F}\,. \tag{47.1}$$

In addition, the Green-Lagrange strain was calculated, $\mathbf{E} = 1/2\,(\mathbf{F}^{\mathrm{T}}\mathbf{F} - \mathbf{1})$.

47.3 Results and Discussion

The sample was first stretched to a displacement of 200 mm and then unloaded and allowed to recover. This initial stretch was applied to avoid observing the Mullins effect during the test which has been shown to have no impact on the volume variation [2]. The carbon black filled elastomer was then subjected to the experimental protocol described. The strain maps collected during the test revealed slight misalignment in the grips. This led to a small twist being applied to the sample in addition to the uniaxial stretch. The carbon filled elastomer was found to be incompressible at low strains (<15%). As the strain increased small increases volume (<5%) were observed. These findings agree well with the observed behavior reported for the uniaxial stretch of elastomers [2, 3]. In the future, the goal is to use this back-to-back camera technique to study more complex stress states and determine if the incompressibility assumption is valid. In addition, since this technique can produce maps the volume variation it will be possible to study the volume variation locally.

Acknowledgements Dr. Frances Davis and Prof. Fabrice Pierron acknowledge support from EPSRC through grant EP/L026910/1. Prof. Fabrice Pierron also expresses gratitude to the Wolfson Foundation for support through a Royal Society Wolfson Research Merit Award.

References

1. Cam, J.-B.L.: A review of volume changes in rubber: the effect of stretching. Rubber Chem. Technol. **83**(3), 247–269 (2010)
2. Le Cam, J.B., Toussaint, E.: Volume variation in stretched natural rubber: competition between cavitation and stress-induced crystallization. Macromolecules **41**(20), 7579–7583 (2008)
3. De Crevoisier, J., et al.: Volume changes in a filled elastomer studied via digital image correlation. Polym. Test. **31**(5), 663–670 (2012)

Chapter 48
Inverse Identification of the Elasto-Plastic Response of Metals at High Strain Rates

Sarah Dreuilhe, Frances Davis, Clive Sivour, and Fabrice Pierron

Abstract The study of the mechanical behavior of materials at high strain rates is of growing interest in many engineering applications such as automotive crash-tests and high speed machining. Current high strain rate techniques such as the split-Hopkinson bar fail to provide reliable experimental data for material model identification, particularly at low strains. In the paper, inertial impact tests, ultra high speed imaging, and the virtual fields method are used to identify the elasto-plastic properties of aluminum.

Keywords Virtual Fields Method • High strain rate testing • Impact • High speed imaging • Grid method

48.1 Introduction

The behavior and failure mechanisms of materials often change at high strain rates when compared with their quasi-static response. In automotive manufacturing the use of aluminum alloys has doubled in the last decade as the introduction of aluminum frames can reduce the body weight. During crash events these aluminum alloys can reach strain rates of $10^4\ s^{-1}$ [1] and understanding their dynamic response is necessary to design safe vehicles. In this paper, inertial impact tests coupled with ultra-high speed imaging and the virtual fields method were used to identify the high strain rate elasto-plastic properties of Al6082-T6.

48.2 Materials and Methods

A series of inertial impact tests were performed using the gas gun facilities at the Solid Mechanics and Material Engineering Group at Oxford. The gas gun was used to fire cylindrical projectiles at a nominal speed of 50 m/s. Each aluminum specimen was $L = 48$ mm by $w = 31$ mm with a thickness of 4 mm (Fig. 48.1). A grid with a pitch of 0.6 mm was adhered to the surface of the sample using an epoxy glue. An ultra-high speed camera (HPV-X, Shimadzu) was used to capture images of the sample during the impact test. Light gates inside the gas gun barrel were used to trigger the flashes and a custom make trigger between the waveguide and the projectile was used to trigger the camera. A total of 128 images were recorded using the HPV-X at a rate of 5 Mfps.

The grid method [2] was used to obtain time-resolved maps of displacement from the collected images. The resulting displacement was smoothed temporally using a 3rd order low pass Butterworth filter with a normalized cut-off frequency of 0.09. The smoothed displacement was double differentiated to obtain maps of the acceleration. Before calculating the strain, the displacement was spatially smoothed using a robust filter. The calculated strains and acceleration fields were used as inputs to the virtual fields method.

S. Dreuilhe • F. Davis • F. Pierron (✉)
Faculty of Engineering and the Environment, University of Southampton, Highfield SO17 1BJ, UK
e-mail: F.Pierron@soton.ac.uk

C. Sivour
Department of Engineering Science, University of Oxford, Oxford OX1 3PJ, UK

M. Sutton, P.L. Reu (eds.), *International Digital Imaging Correlation Society*, Conference Proceedings of the Society for Experimental Mechanics Series, DOI 10.1007/978-3-319-51439-0_48

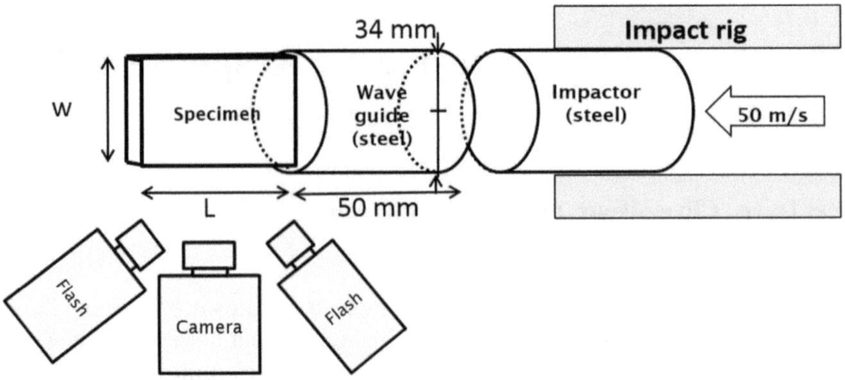

Fig. 48.1 Schematic of the impact rig

48.3 Virtual Fields Method

The principle of virtual work is derived from an integral form of equilibrium equations [3]. Ignoring body forces the following equation is obtained:

$$\overbrace{\int_{S_f} \boldsymbol{T} \cdot \boldsymbol{u}^* \,\mathrm{d}S}^{W^*_{ext}} - \overbrace{\int_V \boldsymbol{\sigma} : \boldsymbol{\epsilon}^* \,\mathrm{d}V}^{W^*_{int}} = \overbrace{\int_V \rho\, \boldsymbol{a} \cdot \boldsymbol{u}^* \,\mathrm{d}V}^{W^*_{acc}} \tag{48.1}$$

where $\boldsymbol{T}$ is the traction vector, $\boldsymbol{\sigma}$ is the stress tensor, $\boldsymbol{a}$ is the acceleration vector, ρ is the density of the material, and $\boldsymbol{u}^*$ and $\boldsymbol{\epsilon}^*$ are the virtual displacement vector and associated virtual strain tensor, respectively. Assuming a state of plane stress and imposing a virtual displacement of $u_1^* = 1$ and $u_2^* = 0$, the first integral in Eq. (48.1) will equal the average stress in the x_1 direction. This set of virtual fields also cancels out the contribution of internal virtual work, W_{int}, leaving only the inertial virtual work, W_{acc}. Replacing the integrals with discrete sums the following equation was found:

$$\overline{\sigma_1(x_1,t)}^{x_2} = \rho\, x_1\, \overline{a_1(x_1,t)}^{S}. \tag{48.2}$$

In Eq. (48.2), the overbars indicate averages over the line or area specified in the superscript. By varying the value of x_1, the average stress for any transverse section can be identified from the acceleration. To identify the elasto-plastic material parameters, a set of non-linear equations is minimized. The cost function to be minimized is sum of the squared difference between the average stress in each section calculated from the acceleration (Eq. (48.2)) and the average stress in each section calculated using the constitutive model. The identification is iterative; the cost function is evaluated for several different combinations of the material parameters (through the constitutive model) searching for the set of parameters that minimize the cost function. In this study, a linear elastic model with linear hardening was used to describe the elasto-plastic response requiring two parameters to be identified, the yield stress, σ_0, and hardening modulus, H.

48.4 Results and Discussion

A total of four aluminum specimens were successfully tested. The values of the yield stress and hardening modulus identified can be found in Table 48.1. Previous studies in the literature show that Al6082-T6 shows limited strain rate sensitivity. At quasi-static rates the yield stress and hardening modulus in tension were found to be 280 and 900 MPa, respectively. The values identified in Table 48.1 correspond well with the expected yield stress. However the values of the hardening modulus were not reliably identified with the values varying greatly. Examining the cost functions plots it was revealed that there is limited sensitivity to hardening modulus making it difficult to accurately identify.

In the future, optimized virtual fields which minimize the influence of experimental noise will be explored to determine if the hardening modulus can be more accurately identified.

Table 48.1 Identified plastic properties of Al 6082-T6

Specimen number	AL7	AL14	AL23	AL26
σ_0 [MPa]	261	262	255	256
H [MPa]	640	557	883	1028

Acknowledgements This material is based on research sponsored by the Air Force Research Laboratory, under agreement number FA8655-13-1-3041. The U.S. Government is authorized to reproduce and distribute reprints for Governmental purposes notwithstanding any copyright notation thereon. The views and conclusions contained herein are those of the authors and should not be interpreted as necessarily representing the official policies or endorsements, either expressed or implied, of the Air Force Research Laboratory or the U.S. Government.

References

1. Moćko, W., et al.: Compressive viscoplastic response of 6082-T6 and 7075-T6 aluminium alloys under wide range of strain rate at room temperature: experiments and modelling. Strain **48**(6), 498–509 (2012)
2. Badulescu, C., Grédiac, M., Mathias, J.D.: Investigation of the grid method for accurate in-plane strain measurement. Meas. Sci. Technol. **20**(9), 095102 (2009)
3. Pierron, F., Grédiac, M.: The Virtual Fields Method: Extracting Constitutive Mechanical Parameters from Full-Field Deformation Measurements. Springer, New York (2012)

Chapter 49
Viscoelastic Properties Identification Through Innovative Image-Based DMTA Strategy

Rian Seghir and Fabrice Pierron

Abstract We present here an innovative identificatfion strategy, using high power ultrasonic loadings together with both InfraRed Thermography and Ultra-High-Speed Imaging, able to fully-characterize the viscoelastic behaviour of polymer materials from a single sample. The main originality lies in the fact that contrary to conventional DMT Analysis, no frequency or temperature sweep is required since the experiment is designed to simultaneously produce both a heterogeneous strain-rate state and an heterogeneous temperature state allowing a local and multi-parametric identification.

Keywords Ultra-High Speed • Visco-elasticity • PMMA • Infrared-thermography • Identification

The identification of the high strain-rate properties of materials is a very important topic for many engineering applications like crash, blast, forming among others. It is also a very challenging experimental task mainly because of the difficulty in measuring impact loads accurately in regimes where inertia effects are significant (transient and elastic regimes) as well as in presence of heterogeneous deformation states. Especially, the most popular technique used to obtain material parameters at high strain-rates, i.e. the Split Hopkinson Bar, doesn't overcome such limits. A recent alternative is to take advantage of the acceleration information using it as an image-embedded load cell [3]. Even more recently, a very interesting article was published [4] where a sawbone specimen was submitted to longitudinal vibrational excitation in the ultrasonic range, and its response measured by DIC and IR thermography. The objective of the present work is to show that both approaches can be combined to extract stiffness and damping properties from these images. Such technique goes further than standard DMT Analysis, in the sense that deformations, stresses and temperatures are extracted locally to constitute an impressive experimental database ready for identification.

The concept of this work consists in producing within a visco-elastic sample, both a heterogeneous deformation-rate state and a heterogeneous temperature state. This is achieved loading the sample cyclically at high frequency using reasonably high loading amplitude. In these conditions, and if the sample is designed to be resonant (1st longitudinal mode) at the loading frequency, it undergoes a harmonic deformation state where both the strain and the strain-rate are null at sample extremities (free and fixed edges—see Fig. 49.1) and are maximum within its centre. In parallel, the sample is self-heated cycle after cycle due to the visco-elastic dissipation. This dissipation contributes to gradually heat-up the centre of the sample while the edges remain almost at room temperature. In other words, the test produces room temperature quasi static uniaxial tests close to sample edges and high deformation rate uniaxial tests at varying temperature within the centre. Points located between these points provide a range of experimental conditions. The challenge is to take profit of all this heterogeneity. Assuming that the displacement field, within the sample, can be the described as the product of a spatial and temporal sine function, a first estimation of the reachable strain-rate range can been obtained:

$$\dot{\varepsilon}_y^{max} = \left(\frac{\rho}{E}\right)^{\frac{1}{2}} U_0 \omega_f^2 \tag{49.1}$$

where U_0 is the loading amplitude, $\omega_f^2 = 2f$ is the angular frequency, f is the loading frequency, and E and ρ are the stiffness and the density of the material respectively. Therefore, the strain-rate roughly varies, along the sample length, from 0 to a level function of: (1) the square root of some material properties, (2) the loading amplitude and (3) the square of the loading

R. Seghir (✉) • F. Pierron
Faculty of Engineering and the Environment, University of Southampton, Highfield Campus, SO17 1BJ Southampton, UK
e-mail: r.seghir@soton.ac.uk

M. Sutton, P.L. Reu (eds.), *International Digital Imaging Correlation Society*, Conference Proceedings of the Society for Experimental Mechanics Series, DOI 10.1007/978-3-319-51439-0_49

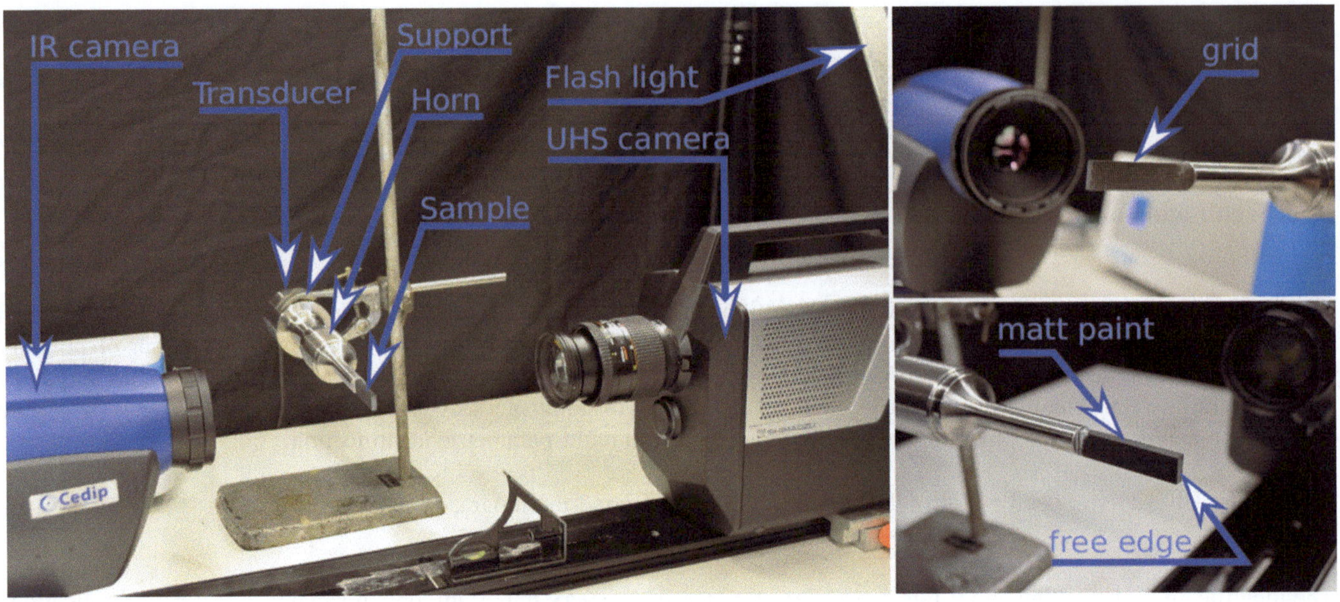

Fig. 49.1 Image-Based DMTA experimental setup

frequency. Then, to perform a local identification, the key is to reconstruct the local stress-strain curves from the dynamic equilibrium. One can easily demonstrate from the equilibrium of a free-standing homogeneous flat sample loaded along its longitudinal axis that:

$$\overline{\sigma_{yy}}^{x}(y,t) = \rho y \overline{a_y}(y,t) \tag{49.2}$$

where $\overline{\sigma_{yy}}^{x}$ is the averaged longitudinal Cauchy-stress over the section y and $\overline{a}_y$ is the averaged acceleration between the free edge and the section y.

The experimental setup consists in four main parts as shown in Fig. 49.1. (1) A high power Ultrasonic transducer which allow cyclically deforming the sample at 20 kHz—(2) An InfraRed camera which allows reconstructing surface temperature fields—and (3) an Ultra-High speed camera which allows capturing images of the deforming sample to obtain space and time resolved displacement and acceleration (after differentiation) fields. Finally, a regular and adhesive grid is used to extracts the in-plane displacement fields from the so-called phase-shifting algorithm [1]. The test consists basically in a set of successive short ultrasonic runs (1 s). Within each run, a 256 μs (at 500 kfps) UHS image sequence of the deforming sample is grabbed. In parallel, the IR data are continuously recorded until the sample melt after a certain number of run. For this experiment, the tested sample is a flat and transparent PMMA (Poly-Methyl-MethAcrylate) sample simply glued on the tip of the horn (see Fig. 49.1) and covered with a black matt paint on one face, to achieve efficient IR measurements, and with a regular grid, on the other face.

Figure 49.2 presents some stress-strain curves obtained during the 5th ultrasonic run. At that stage the temperature reaches 80°C within the sample centre. One can clearly see a variation of the stiffness as well as a variation of the damping along the sample length. Such variation is the result of two main phenomena. A 1st order one, the material temperature sensitivity, which leads to a drop of the stiffness and to the appearance of hysteresis loops when one approaches the central section. A 2nd order one—a material strain-rate sensitivity, which should mitigate the stiffness drop by making the material stiffer within the central region [2]. To go further and make a clear partition between both phenomena, it is then necessary to build a 3D representation of the material behaviour such as $E = f(\dot{\varepsilon},T)$. The Fig. 49.3 focuses on the material stiffness. Figure 49.3a evidences and allows quantifying the drop of the material stiffness as the sample temperature increases but also evidences a slight rising of the stiffness when the strain-rate grows. In the present case, the stiffness varies from less than 4 GPa to more than 5.5 GPa. Figure 49.3b presents some isothermal curves extracting from Fig. 49.3a. The stiffness variation allows observing two main things: (1) whatever the temperature, the material hardens as the strain-rate increases. Values and trends are totally in line with results obtained from more standard procedure [2]. (2) Higher the temperature, lower is the strain-rate sensitivity. Considering the time-temperature superposition principle, such a figure allow predicting that the variation of the material stiffness tends to saturate at quasi-static strain-rate (≡high temperature) while it rises at high rates.

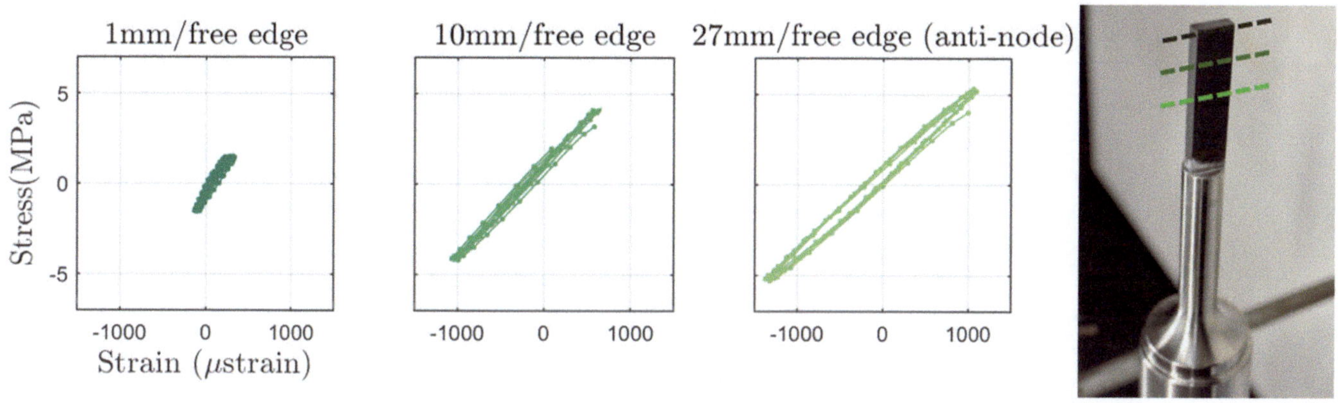

Fig. 49.2 Stress-strain curves for different sections of the sample obtained after 5 ultrasonic runs

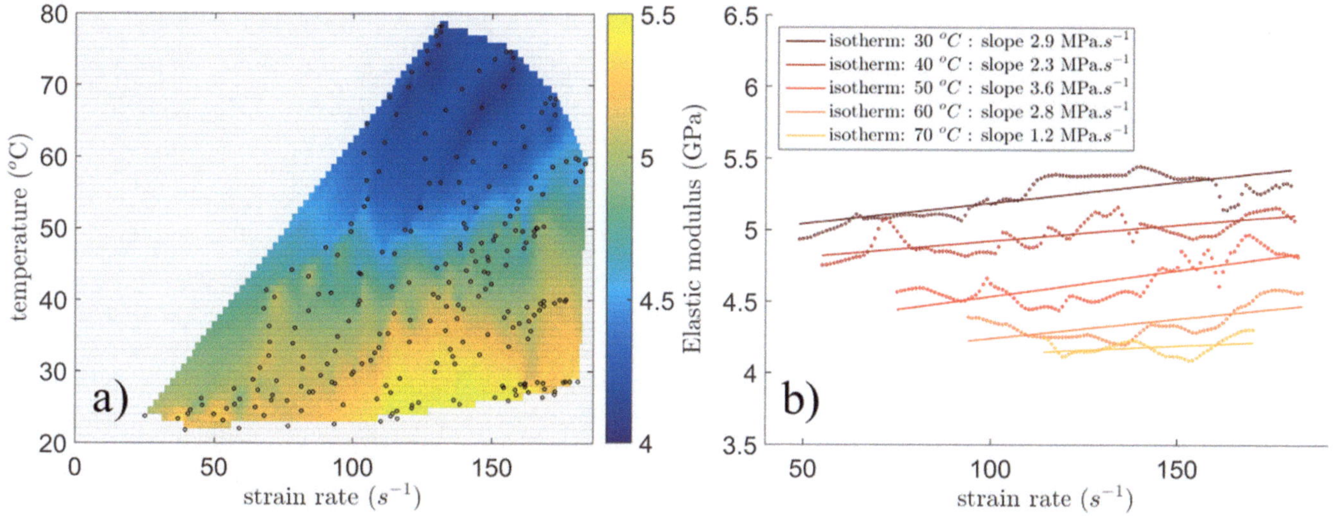

Fig. 49.3 Identified PMMA properties from 7 ultrasonic runs on a single sample. (**a**) Stiffness variation as a function of the $\dot{\varepsilon}$ and T and (**b**) isothermal cuts

The present work demonstrates the feasibility of a multi-parametric identification on a single sample and falls within an effort to invent new high-strain test methodologies based on full field imaging and inverse identification, to both overcome the limits of standard experimental strategies and take advantage of the deformation heterogeneities to achieve a full-characterization of a material from a "one-shot" test.

References

1. Grediac, M., Sur, F., Blaysat, B.: The grid method for in-plane displacement and strain measurement: a review and analysis. Strain. **52**(3), 205–243 (2016)
2. Li, Z., Lambros, J.: Strain rate effects on the thermomechanical behavior of polymers. Int. J. Solids Struct. **38**(20), 3549–3562 (2001)
3. Pierron, F., Zhu, H., Siviour, C.: Beyond hopkinson's bar. Philos. Trans. R. Soc. A Math. Phys. Eng. Sci. **372**(2023), 20130195 (2014)
4. Wang, D., Lucas, M., Tanner, K.: Characterising the strain and temperature fields in a surrogate bone material subject to power ultrasonic excitation. Strain. **49**(5), 409–419 (2013)

References

Chapter 50
Experimental Study of Measurement Errors in 3D-DIC Due to Out-of-Plane Specimen Rotation

Farzana Yasmeen, Sreehari Rajan, Michael A Sutton, and Hubert W. Schreier

Abstract In practice, out-of-plane motions usually are not avoidable during experiments. Since 2D–DIC measurements are vulnerable to parasitic deformations due to out of plane specimen motions, three-dimensional digital image correlation (3D-DIC) oftentimes is employed. The 3D-DIC method is known to be capable of accurate deformation measurements for specimens subjected to general three-dimensional motions, including out of plane rotations and displacements. As a result, there has been limited study of the deformation measurements obtained when using 3D-DIC to measure the displacement and strain fields for a specimen subjected only to out-of-plane rotation. This paper presents experimental results regarding the effect of out-of-plane rotation on strain measurement using 3D-DIC. Specifically, full-field deformation results are obtained during rigid body, out-of-plane rotation in the range $-40° \leq \theta \leq 40°$ using a two-camera stereovision system. Results indicate that (a) the measured normal strain in the foreshortened direction increases in a non-linear manner with rotation angle, (b) the normal strain along the direction of the rotation axis is essentially zero for all rotation angles and (c) the in-plane shear strain is small but increases linearly with rotation angle. Results also indicate that the magnitude of the errors in the strain are a strong function of how the calibration process is performed, with measurement errors exceeding ± 1400 $\mu\varepsilon$ for what would normally be considered "small angle" calibration processes.

Keywords Stereo digital image correlation • Rigid body out of plane rotation • Stereo calibration • Strain errors

50.1 Introduction

Digital image correlation (DIC) is an optical non-contact deformation measurement technique capable of measuring full field deformation on a specimen's surface [1]. DIC was first originated and devised for planar surfaces undergoing in-plane deformations in the early 1980s [2] and extended to general deformations of curved or planar surfaces in the early 1990s [3–5]. Since that time, the DIC technique has grown into one of the most popular measurement method worldwide. In principle, the method known as 2D-DIC gives accurate results when a nominally planar specimen is subjected to nominally in-plane mechanical loading. In such cases, the planar surfaces experience in plane deformation. If the single viewing camera is placed perpendicular to the specimen, then the measured deformations would reflect the true deformations under ideal conditions. However, as shown previously [6], any out of plane displacement or out of plane rotation of the specimen will induce in-plane displacement gradients that deleteriously affect the measured strains. In this work, the authors have focused on the effects of out-of-plane rotation on 3D-DIC deformation measurements using experimental result. This paper presents details regarding out-of-plane rotation experiments that were performed with a two-camera stereovision system (3D-DIC or StereoDIC) to investigate the effect of out of plane rotation on 3D-DIC measurements.

F. Yasmeen (✉) • S. Rajan • M.A. Sutton
Department of Mechanical Engineering, University of South Carolina, Columbia, SC 29208, USA
e-mail: fyasmeen@email.sc.edu

H.W. Schreier
Correlated Solutions Incorporated, 121 Dutchman Blvd, Irmo, SC 29063, USA

M. Sutton, P.L. Reu (eds.), *International Digital Imaging Correlation Society*, Conference Proceedings of the Society for Experimental Mechanics Series, DOI 10.1007/978-3-319-51439-0_50

50.2 Experiment

Figure 50.1 shows the experimental setup. The two LED light sources illuminated the specimen at the angles shown to minimize the potential for imaging of reflections. The two cameras were mounted firmly to the optical bench to eliminate unwanted camera motions during the experiment. The distance from the camera lenses to the specimen was approximately 0.3048 m. The resulting speckle pattern and grey level histogram for speckle image are shown in Fig. 50.2. Table 50.1 summarizes the components in the stereovision optical system. Table 50.2 shows the parameters used to analyze the images and acquire the displacements and strains.

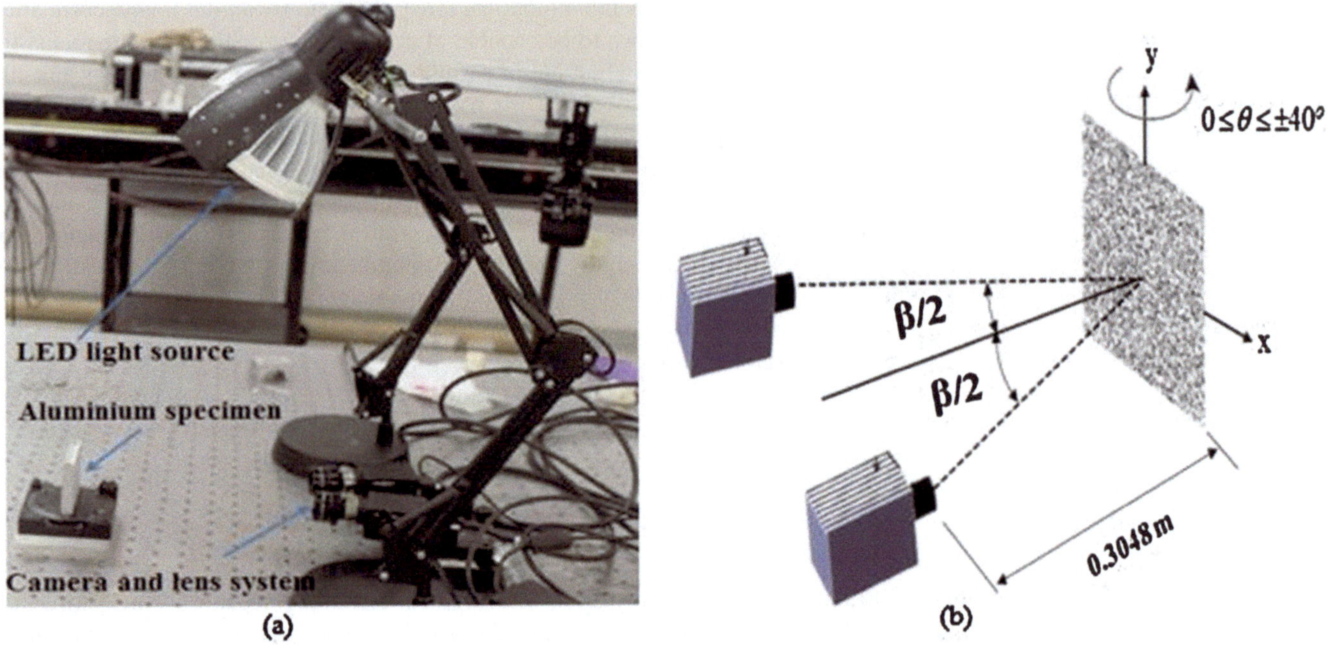

Fig. 50.1 Experimental setup for out of plane rotation (**a**) full experimental setup, (**b**) camera and sample positions

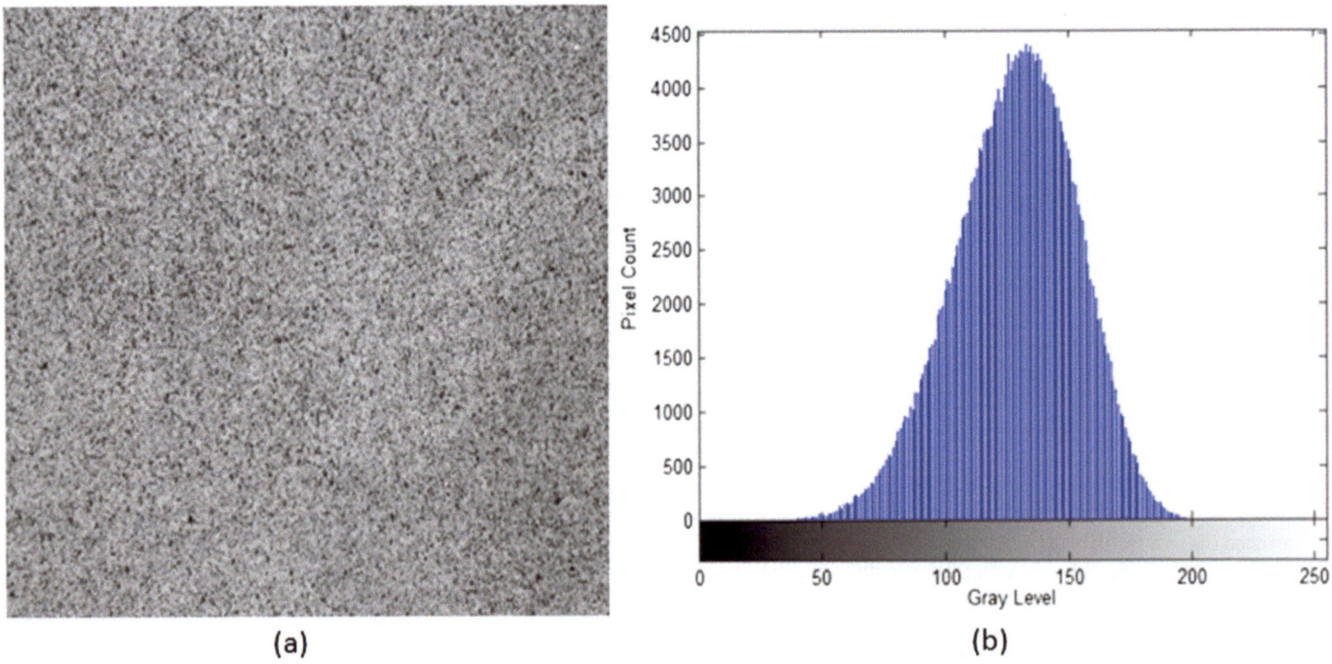

Fig. 50.2 Speckle Pattern (**a**) image, (**b**) *grey* level histogram for image shown

Table 50.1 Optical system components

	Stereovision system components
Cameras	Grasshopper3 GS3-U3-91S6M (8bits, 3376X2704)
Lenses	Schneider XENOPLAN 1.9/35-0511
Lighting	White LED Lighting

Table 50.2 Parameters used for analysis of images

StereoDIC analysis parameters	
Subset size	47 × 47 pixels^2
Step size	13 pixels
Software	VIC-3D
Strain filter size	5 × 5

Table 50.3 Calibration parameters for stereovision system

Parameter	Cam0 – SAC	Cam1 – SAC	Cam0 – LAC	Cam1 – LAC
Cx(pixel)	1676.743	1712.878	1656.124	1728.674
Cy(pixel)	1372.491	1432.081	1359.785	1430.952
$f\lambda_x$(pixel)	11,176.322	11,134.508	11,140.056	11,114.127
$f\lambda_y$(pixel)	11,175.219	11,133.544	11,138.959	11,112.591
Skew	−3.742	2.538	−1.635	−0.958
κ_1	−0.273	−0.225	−0.271	−0.223
κ_2	0	0	0	0
κ_3	0	0	0	0

Table 50.4 Camera 0→1 transformation parameter for stereovision system

Camera 0→1 transformation	Small angle calibration	Large angle calibration
Alpha(°)	0.288°	0.341°
Beta(°)	15.966°	15.815°
Gamma(°)	1.489°	1.482°
T_x(mm)	−77.473	−77.412
T_y(mm)	−0.992	−1.006
T_z(mm)	10.229	10.487
Baseline(mm)	78.1521	78.1256

To determine the effect of calibration procedures on the accuracy of stereovision system measurements in the presence of large rotations, the investigators performed a set of baseline experiments with two calibrations. Designating the first calibration process as the "Small Angle Calibration" (SAC), a total of 61 images of a translated and rotated planar dot pattern [4, 5] were acquired where the rotation angles in all directions were less than 10°. Designating the second calibration process as the "Large Angle Calibration" (LAC) process, a total of 142 images are used, where the first 61 images used for LAC are supplemented with 81 images as the same grid is rotated out of plane in 5° increments within the range $-40° \leq \theta \leq 40°$. Tables 50.3 and 50.4 show the calibration parameters obtained using the SAC and LAC process described above. The parameters shown include: image plane center location, (Cx, Cy), in pixels for each camera; the factors ($f\lambda_x$, $f\lambda_y$) in pixel for each camera in the horizontal and vertical directions; skew, the deviation from orthogonality between the row and column directions in the sensor plane; radial distortion coefficient κ_1, κ_2, κ_3;position of pinhole in camera 1 relative to camera 0 (t_x, t_y, t_z)(mm); relative orientation of camera 1 with respect to camera 0 (α, β, γ) (degrees), where α is the relative tilt, β the relative pan angle, and γ the swing angle.

50.3 Experimental Results

When using the SAC process, Fig. 50.3 shows the averaged surface strains ε_{xx}. ε_{yy} and ε_{xy} that are obtained in the central 1/3 portion of the specimen as rigid body out of plane rotation is applied. Though all the strains should be zero, only the strain ε_{yy} is near zero for all rotation angles. For the other in-plane strains, as shown in Fig. 50.3 the measured average normal

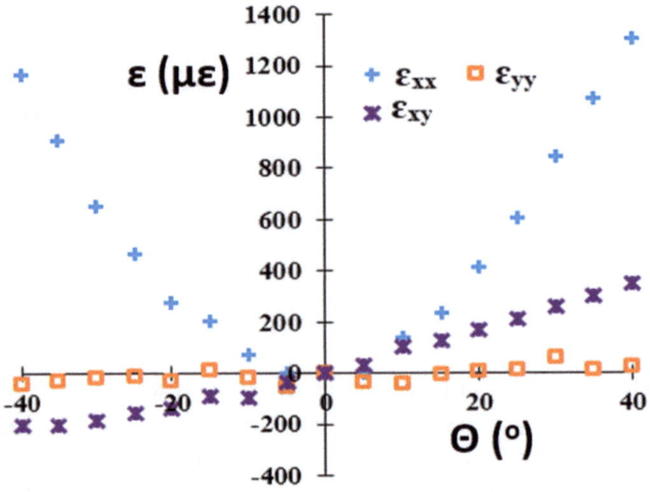

Fig. 50.3 Measured average in-plane strain as a function of out-of-plane rotation angle using the Small Angle Calibration process when specimen undergoes only rigid body out of plane rotation

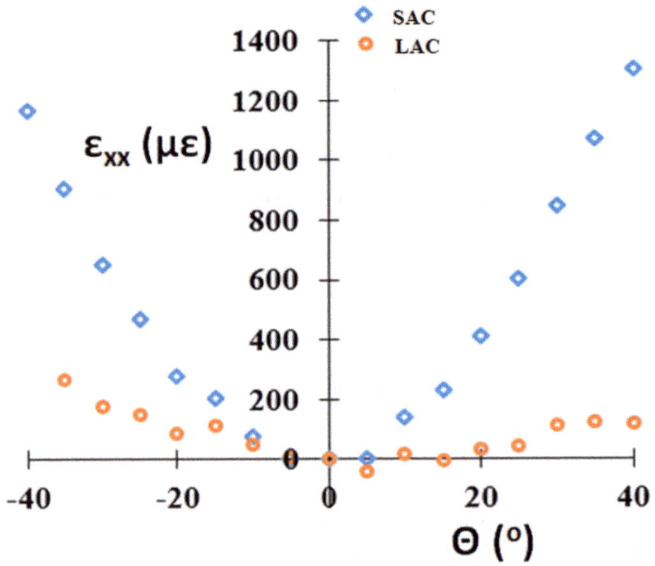

Fig. 50.4 Measured average εxx using both Small Angle Calibration and Large Angle Calibration procedures when specimen undergoes only rigid body out of plane rotation

strain ε_{xx} increases in a highly non-linear manner with rotation angle, reaching a maximum value $\approx$1400 $\mu\varepsilon$. In addition, the in-plane shear strain ε_{xy} is small but increases linearly with rotation angle.

Since the largest strain errors occur in the x-direction, Fig. 50.4 directly compares the experimental results for ε_{xx} obtained when using SAC and LAC procedures. As shown in Fig. 50.4, the use of several calibration grid images that are rotated out of plane in the range $-40° \leq \theta \leq 40°$ provides significant improvement in the measured ε_{xx} strain field.

Inspection of the camera parameters in Tables 50.3 and 50.4 shows that the changes in camera parameters are quite small, with all but skew changing in the 3rd significant figure or later. The results clearly demonstrate the importance of the calibration process for the accuracy of strain measurements obtained by stereo image correlation. It is noted that, in separate experiments, it was confirmed that mere out-of-plane translations of the calibration target do not sufficiently improve results, while acquiring images of the grid undergoing large rotations provides significant improvement.

50.4 Outlook and Conclusions

The influence of out of plane rotation in the range $-40^{\circ} \leq \theta \leq 40^{\circ}$ on strain measurements was investigated. The results illustrate that careful attention must be paid to the calibration process used for stereo DIC systems. Using only small rotations of the calibration target, errors in strain in excess of 1400 $\mu\varepsilon$ were obtained. Using large target rotations, the errors could be suppressed to around 200 $\mu\varepsilon$ in the experiments reported here. In our attempt to further suppress this bias, a promising calibration scheme using a speckle pattern for calibration has shown great potential, with bias levels around 50 $\mu\varepsilon$. The speckle-based calibration process will be the focus of further investigation.

References

1. Schreier, H., Orteu, J.J., Sutton, M.A.: Image Correlation for Shape, Motion and Deformation Measurements. Springer (2009)
2. Sutton, M.A., Wolters, W.J., Peters, W.H., Ranson, W.F., McNeill, S.R.: Determination of displacements using an improved digital correlation method. Image Vis. Comput. **1**(3), 133–139 (1983)
3. Kahn-Jetter, Z.L., Chu, T.C.: Three-dimensional displacement measurements using digital image correlation and photogrammic analysis. Exp. Mech. **30**(1), 10–16 (1990)
4. Luo, P.F., Chao, Y.J., Sutton, M.A., Peters Iii, W.H.: Accurate measurement of three-dimensional deformations in deformable and rigid bodies using computer vision. Exp. Mech. **33**(2), 123–132 (1993)
5. Luo, P.F., Chao, Y.J., Sutton, M.A.: Application of stereo vision to three-dimensional deformation analyses in fracture experiments. Opt. Eng. **33**(3), 981–990 (1994)
6. Sutton, M.A., Yan, J.H., Tiwari, V., Schreier, H.W., Orteu, J.J.: The effect of out-of-plane motion on 2D and 3D digital image correlation measurements. Opt. Lasers Eng. **46**(10), 746–757 (2008)

Chapter 51
Investigation of Optimal Digital Image Correlation Patterns for Deformation Measurement

G.F. Bomarito, T.J. Ruggles, J.D. Hochhalter, and A.H. Cannon

Extended Abstract Digital image correlation (DIC) relies on the visible surface features of a specimen to measure deformation. When the specimen itself has little to no visible features, a pattern is applied to the surface which deforms with the specimen and acts as artificial surface features. Since recent pattern application methods, e.g., micro-stamping [1] and lithography [2] allow for the application of highly customized patterns and because the accuracy and precision of DIC is dependent upon the applied pattern [3–6], an ideal pattern is sought for which the error introduced into DIC measurements is minimal. It is the goal of the present work to develop and refine an optimization technique to produce this type of DIC pattern.

The DIC literature has many examples of pattern quality metrics that aim to quantify the suitability of a pattern for DIC [e.g., 4–14]; however, most of these metrics have been developed with the intent of comparing pseudo-random patterns to each other. For that purpose, the quality metrics are well suited but typically do not generalize well to near-optimal patterns [15]. An example of this is the quality metric sum of square of subset intensity gradient [6–9], denoted SSSIG, which achieves a best value for a regular checkerboard pattern. Checkerboard patterns are ill-suited for DIC because they pose non-uniqueness problems. By using multiple pattern quality metrics simultaneously, Bomarito et al. [15] showed that these deficiencies can be avoided. They constructed an optimization method which centered around a quality metric Q which is a combination of two metrics:

$$Q = w_1 \hat{S} + w_2 (\hat{A}_2)^n \tag{51.1}$$

where $\hat{S}$ is derived from SSSIG and $\hat{A}_2$ is the secondary auto-correlation peak height similar to the metric developed by Stoilov et al. [12]. In the minimization of Q, the $\hat{S}$ term is intended to be the primary contributor and the $\hat{A}_2$ term is meant to act as a constraint that ensures the pattern retains some local uniqueness (i.e., that it avoids regular checkerboard-type patterns). w_1, w_2 and n are tuning variables that adjust the shape and relative intensity of the constraint.

In order to choose optimal values of w_1, w_2, and n, a study is performed which searches for the ideal balance of the two terms in Eq. (51.1). In the study, several patterns are generated ranging from near-checkerboard to completely random which are then subjected to a numerical experiment. The experiment consists of the following steps:

1. The pattern is magnified to a desired resolution of 5 pixels per speckle [16, 17].
2. The contrast of the image is reduced, simulating imperfect lighting conditions: i.e., the grayscale values are changed from 0.0 and 1.0 to 0.25 and 0.75.
3. Several numerical deformations are applied to the image.
4. A Gaussian blur is applied to the patterns to simulate more realistic transitions from dark to light pixels.

G.F. Bomarito (✉) • J.D. Hochhalter
NASA, Durability, Damage Tolerance, and Reliability Branch, Hampton, VA, USA
e-mail: geoffrey.f.bomarito@nasa.gov

T.J. Ruggles
National Institute of Aerospace, Hampton, VA, USA

A.H. Cannon
1900 Engineering, LLC, Clemson, SC, USA

M. Sutton, P.L. Reu (eds.), *International Digital Imaging Correlation Society*, Conference Proceedings of the Society for Experimental Mechanics Series, DOI 10.1007/978-3-319-51439-0_51

5. Zero-mean Gaussian noise with standard deviation of 2% is added to the image to simulate random error in the imaging process.
6. DIC is performed on each image and compared to the applied deformations Step 3.

The result of the experiment is a distribution of DIC measurement errors for each of the input patterns.

For patterns that are closer to a checkerboard, small regions of spurious results are observed under some deformations. This is caused by the confusion of very similar subsets, and indicates that the results of the pattern may not be fully trusted when the true deformation is not known *a priori*. However, the general trend was that the more checkerboard-like patterns had higher accuracy and precision in areas where the spurious results did not present themselves. Using these results, an appropriate ratio could be identified between the two terms in Eq. (51.1), which accompanied minimal error introduction while avoiding spurious results. The values of w_1, w_2, and n are then calibrated.

Using the updated values of w_1, w_2, and n, the pattern optimization space is then explored using multiple initial points. Through this exploration, confidence is increased that a pattern more nearly the global minimum is found. Using the same numerical experiment outlined above, the optimized pattern is shown to have DIC measurement errors consistent with the best regions of the near-checkerboard patterns but without any incidents of spurious results. The pattern performs approximately 25% better than a random pattern.

Keywords Digital image correlation • Pattern • Optimization • Quality metric • Designer patterns

References

1. Cannon, A.H., Hochhalter, J.D., Mello, A.W., Bomarito, G.F., Sangid, M.D.: Microstamping for improved speckle patterns to enable digital image correlation. Microsc. Microanal. **21**(S3), 451–452 (2015)
2. Kammers, A.D., Daly, S.: Small-scale patterning methods for digital image correlation under scanning electron microscopy. Meas. Sci. Technol. **22**(12), 125501 (2011)
3. Haddadi, H., Belhabib, S.: Use of rigid-body motion for the investigation and estimation of the measurement errors related to digital image correlation technique. Opt. Lasers Eng. **46**(2), 185–196 (2008)
4. Lecompte, D., Smits, A., Bossuyt, S., Sol, H., Vantomme, J., Van Hemelrijck, D., Habraken, A.M.: Quality assessment of speckle patterns for digital image correlation. Opt. Lasers Eng. **44**(11), 1132–1145 (2006)
5. Sun, Y., Pang, J.H.: Study of optimal subset size in digital image correlation of speckle pattern images. Opt. Lasers Eng. **45**(9), 967–974 (2007)
6. Pan, B., Xie, H., Wang, Z., Qian, K., Wang, Z.: Study on subset size selection in digital image correlation for speckle patterns. Opt. Express **16**(10), 7037–7048 (2008)
7. Pan, B., Lu, Z., Xie, H.: Mean intensity gradient: an effective global parameter for quality assessment of the speckle patterns used in digital image correlation. Opt. Lasers Eng. **48**(4), 469–477 (2010)
8. Sutton, M.A., Orteu, J.J., Schreier, H.: Image Correlation for Shape, Motion and Deformation Measurements: Basic Concepts, Theory and Applications. Springer Science and Business Media, New York (2009)
9. Wang, Y.Q., Sutton, M.A., Bruck, H.A., Schreier, H.W.: Quantitative error assessment in pattern matching: effects of intensity pattern noise, interpolation, strain and image contrast on motion measurements. Strain **45**(2), 160–178 (2009)
10. Hua, T., Xie, H., Wang, S., Hu, Z., Chen, P., Zhang, Q.: Evaluation of the quality of a speckle pattern in the digital image correlation method by mean subset fluctuation. Opt. Laser Technol. **43**(1), 9–13 (2011)
11. Triconnet, K., Derrien, K., Hild, F., Baptiste, D.: Parameter choice for optimized digital image correlation. Opt. Lasers Eng. **47**(6), 728–737 (2009)
12. Stoilov, G., Kavardzhikov, V., Pashkouleva, D.: A comparative study of random patterns for digital image correlation. J. Theor. Appl. Mech. **42**(2), 55–66 (2012)
13. Crammond, G., Boyd, S.W., Dulieu-Barton, J.M.: Speckle pattern quality assessment for digital image correlation. Opt. Lasers Eng. **51**(12), 1368–1378 (2013)
14. Bossuyt, S.: Optimized patterns for digital image correlation. In: Imaging Methods for Novel Materials and Challenging Applications, vol. 3, pp. 239–248. Springer, New York (2013)
15. Bomarito, G.F., Hochhalter, J.D., Ruggles, T.J., Cannon, A.H.: Increasing accuracy and precision of digital image correlation through pattern optimization. Opt. Lasers Eng. **91**, 73–85 (2017)
16. Winiarski, B., Schajer, G.S., Withers, P.J.: Surface decoration for improving the accuracy of displacement measurements by digital image correlation in SEM. Exp. Mech. **52**(7), 793–804 (2012)
17. Lecompte, D., Bossuyt, S., Cooreman, S., Sol, H., Vantomme, J.: Study and generation of optimal speckle patterns for DIC. In: Proceedings of the Annual Conference and Exposition on Experimental and Applied Mechanics, vol. 3, pp. 1643–1649. Springer, New York (2007)

Chapter 52
Characterization of Elastic-Plastic Fracture Behavior in Thin Sheet Aluminum

D.S. Dawicke and I.S. Raju

Abstract Composite overwrapped pressure vessels (COPVs) are used in space flight vehicles to contain pressurized gases. A COPV consists of a metallic liner to contain the gas or fluid and a composite overwrap for strength. The metallic liner must survive four lifetimes with the largest crack that could be missed during an inspection. Current light weight designs are considering liners that experience elastic-plastic behavior that violates the assumptions of current linear elastic fracture mechanics (LEFM) theory. Tests are being conducted to characterize the behavior of small cracks in thin aluminum coupons to both validate elastic-plastic fracture mechanics (EPFM) models and to develop test methods for experimental certification. The tests are conducted under strain control to replicate the conditions in a COPV liner. A digital image correlation (DIC) system is used to measure the applied far field strain and uses that strain to control a servo-hydraulic load frame. Small fatigue cracks (0.5–2.5 mm long) are cycled under a strain range that yields the material in both tension and compression. Two additional DIC systems are used to make high magnification measurements (about 100 pixels per mm) of the crack opening displacement on one side and the back face strain on the other side of the coupon.

Keywords Elastic-plastic • Fatigue • COPV • Strain-control • DIC

The fracture control of composite overwrapped pressure vessel (COPV) used in space flight applications requires that two qualification conditions be met: (1) safe-life and (2) leak before burst (LBB) [1]. The safe-life condition requires that the largest flaw in the liner that can escape detection in the operational NDE inspection with a 90/95 confidence and reliability level will not grow to failure or leakage in four lifetimes. The LBB condition requires a crack growing in the liner of a COPV at the maximum design pressure (MDP) will grow through the liner and leak the pressurized media without fracture occurring. The metallic liner can be either elastically responding or plastically responding depending on the thickness of the liner, strength of the composite, and MDP. Plastically responding liners offer the advantage of lower weight. The fracture control qualification requirements of COPVs with elastically responding liners can be addressed using conventional linear elastic fracture mechanics (LEFM) tools. However, no validated tools exist for analytically qualifying COPVs with plastically responding liners. Experimental techniques have been developed to qualify COPVs with plastically responding liners using uniaxial coupons [2]. The previously developed experimental techniques have been expanded to generate data that can be used to validate elastic-plastic fracture mechanics (EPFM) analytical tools for predicting safe-life in COPVs with plastically responding liners.

Fatigue crack growth tests were conducted on 2.28 mm thick uniaxial coupons of 6061-T6 aluminum. Small fatigue crack were introduced into the coupons using electrical discharge machining (EDM) to create small notches (0.5–2.5 mm long by 0.2 mm deep) and fatigue cycling under elastic conditions was performed until a fatigue crack was observed growing from the notch. The plastically responding liners experience both tensile and compressive yielding, so guide plates were used to prevent buckling of the thin coupons at compressive loads. The operational pressurization of a plastically responding COPV liner loads the liner to specific strain levels, so strain controlled cycling was used in the coupon testing. The characterization of crack growth behavior was performed by measurement of the crack opening displacement, measurement of the deformation on the opposite side of the crack, and post-test examination of the fracture surfaces.

D.S. Dawicke (✉)
AS&M, Inc, 107 Research Dr., Hampton, VA 23666, USA
e-mail: david.s.dawicke@nasa.gov

I.S. Raju
NASA Langley Research Center, Hampton, VA 23681, USA

M. Sutton, P.L. Reu (eds.), *International Digital Imaging Correlation Society*, Conference Proceedings
of the Society for Experimental Mechanics Series, DOI 10.1007/978-3-319-51439-0_52

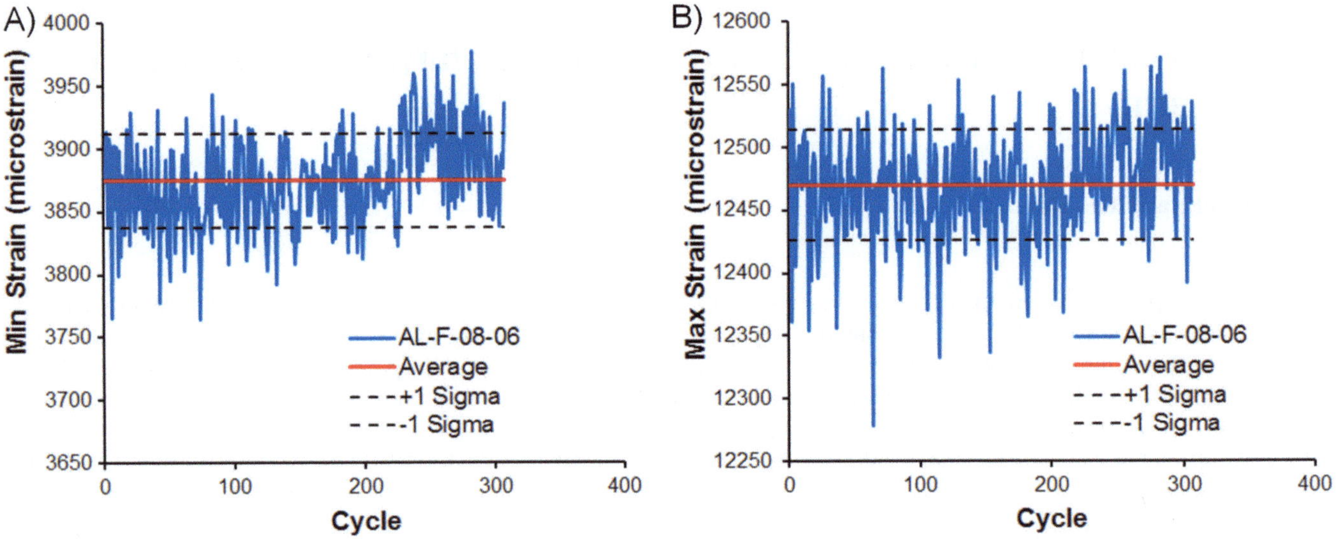

Fig. 52.1 Example of the peak and minimum strains achieved during the fatigue cycling. (**a**) minimum actual strains. (**b**) Peak actual strains

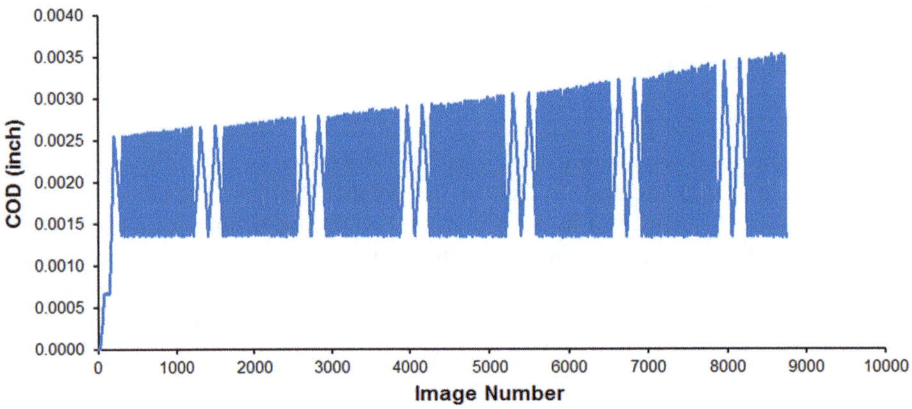

Fig. 52.2 Measured crack opening displacement (COD) for an elastic-plastic fatigue test

The strain controlled loading was obtained using a digital image correlation (DIC) system to measure the far field strain of the coupon and used the strain output to control a servohydraulic load frame. Holes were machined into the guide plate to allow the DIC cameras a clear line of sight to 12 mm diameter areas about 25 mm above and below the crack. The area of the test coupon behind the holes was painted with a white background and black speckles. Points were selected in both visible speckled regions and the strain ($\Delta L/L$) was calculated from measurement of the displacement between the two points (ΔL) and the undeformed distance between the two points (L). The DIC strain control resulted in peak and minimum applied cyclic strains within 200 microstrain of the desired levels, as shown in Fig. 52.1.

Two additional 12 mm holes were machined into the guide plates to view the crack and the area directly opposite the crack (back face). The regions of the test coupon behind the holes were painted white with black speckles that were 0.05–0.1 mm in diameter. The DIC camera resolution was about 100 pixels per mm. The center of the crack was located and points 0.6 mm above and below the center of the crack were identified. The crack opening displacement (COD) was defined as the relative displacement of the two points above and below the crack. The COD was measured as a function of the image number, as shown in Fig. 52.2. The applied load and far field strain were also recorded as a function of the image number.

An identical DIC system was placed directly opposite the crack to view the back face displacements during the elastic-plastic fatigue cycling. The plastic deformation at the maximum crack depth location will create a dimple where the material at the surface is pulled inward. This displacement (w) is small, but measurable, as shown in Fig. 52.3. The measurements of COD and back face dimple depth are correlated with crack length measurements obtained from post-test fracture surface examination. The resulting data will be used to validate EPFM predictions of crack growth behavior.

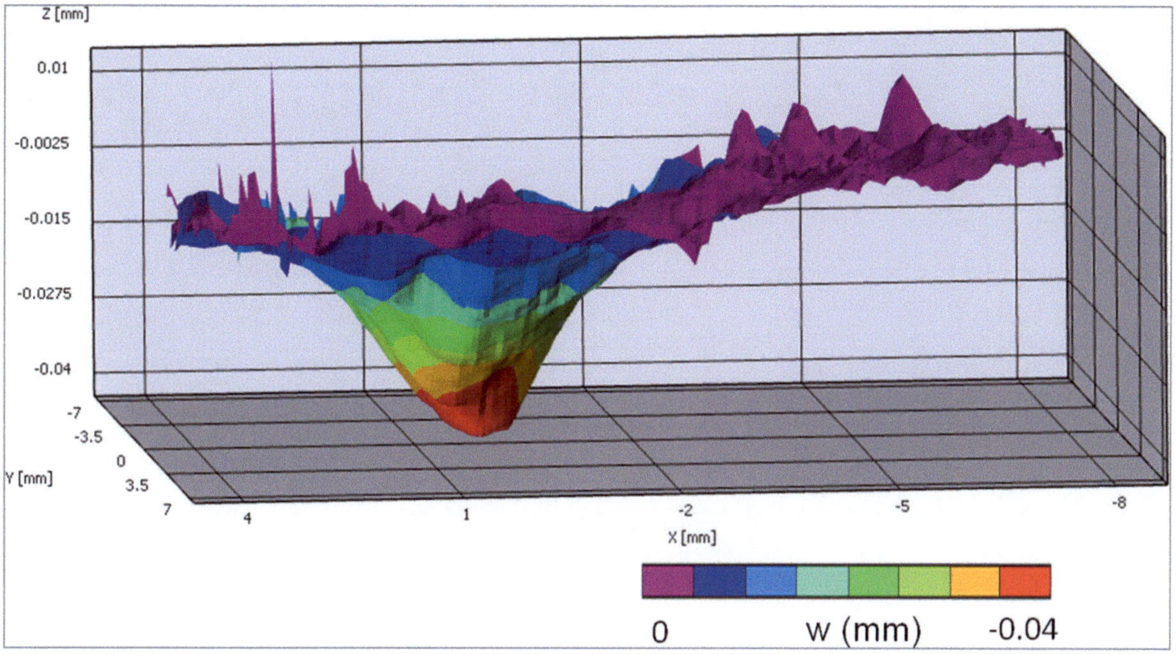

Fig. 52.3 Measurement of the back face dimple for an elastic-plastic fatigue test

References

1. ANSI/AIAA S-081A-2006 Space Systems-Composite Overwrapped Pressure Vessels (COPVs), 24 July 2006
2. Dawicke, DS; Lewis, JC: Fracture Test Methods for Plastically Responding COPV Liners. In: 50th AIAA/ASME/ASCE/AHS/ASC Structures, Structural Dynamics, and Materials, Conference, Palm Springs, CA, May 2009.

References

Chapter 53
Full-Field Structural Dynamics by Video Motion Manipulations

Yongchao Yang, Charles Farrar, and David Mascarenas

Abstract Structures with complex geometries, material properties, and boundary conditions, exhibit spatially local, temporally transient, dynamic behaviors. High spatial and temporal resolution vibration measurements and modeling are thus required for high-fidelity characterization, analysis, and prediction of the structure's dynamic phenomena. For example, high spatial resolution mode shapes are needed for accurate vibration-based damage localization. Also, higher order vibration modes typically contain local structural features that are essential for high-fidelity dynamic modeling of the structure. In addition, while it is possible to build a highly-refined mathematical model (e.g., a finite element model) of the structure, it needs to be experimentally validated and updated with high-resolution vibration measurements. However, it is a significant challenge to obtain high-resolution vibration measurements using traditional techniques. For example, accelerometers and strain-gauge sensors provide low spatial resolution measurements. Laser vibrometers provide high-resolution measurements, but are expensive and make sequential measurements that are time-consuming. On the other hand, digital video cameras are relatively low-cost, agile, and provide high spatial resolution, simultaneous, measurements. Combined with vision based algorithms (e.g., image correlation, optical flow), digital video camera based measurements have been successfully used for vibration measurements and subsequent modal analysis, based on techniques such as the digital image correlation (DIC) and the point-tracking. However, they typically require speckle pattern or high-contrast markers to be placed on the surface of structures, which poses challenges when the measurement area is large or inaccessible.

Keywords Structural dynamics • Full-field imaging • Video processing • Blind source separation • High-fidelity simulation

A new framework is first developed for the blind extraction and visualization of the full-field, high-resolution, dynamic parameters of an operating (output-only) structure from the digital video measurements using video motion manipulation and unsupervised machine learning techniques. It removes the need of structural surface preparation (markers or paints) associated with existing vision-based methods and can be implemented in a relatively efficient and autonomous manner with little user supervision and calibration. Videos of vibrating structures are first filtered using the complex steerable filters technique to extract local phases at each pixel, which are then blindly separated into individual vibration modes by a blind source separation (BSS) technique called complexity pursuit (CP). By magnifying the extracted modes, the constructed video then displays a magnified operating mode shape, which can be extracted by edge detection method. Thus, the proposed method is able to *blindly* and accurately extract modal frequencies, damping ratios, and full-field (as many points as the pixel number of the video frame) mode shapes of the output-only structure. See Fig. 53.1 for our experimental results of a vibrating cantilever beam and Fig. 53.2 of a vibrating building structure. Comparisons with traditional accelerometers results are shown in Table 53.1. The video demos including a building structure and the cantilever beam can be seen at http://www.lanl.gov/projects/national-security-education-center/engineering/research-projects/blind-modal-id.php.

Further, this high-resolution, full-field dynamic characterization framework enables a variety of applications that traditionally have not been possible. These include the ability to accurately localize minute, invisible, structural damage (see the experimental results in Fig. 53.3), and a new method enabling realistic video-space, high-fidelity simulations and visualizations/animations of structural dynamics (see the experimental results in Fig. 53.4).

Y. Yang (✉) • C. Farrar • D. Mascarenas
Los Alamos National Laboratory, Los Alamos, NM 87545, USA
e-mail: yyang@lanl.gov

M. Sutton, P.L. Reu (eds.), *International Digital Imaging Correlation Society*, Conference Proceedings of the Society for Experimental Mechanics Series, DOI 10.1007/978-3-319-51439-0_53

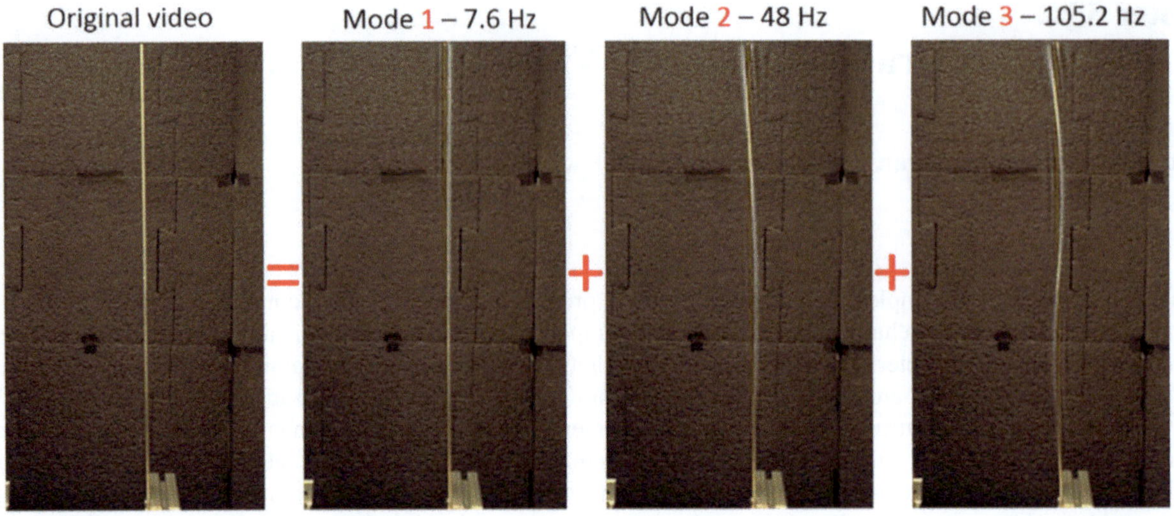

Fig. 53.1 Video-based, output-only, full-field dynamic extraction technique applied to an operating *cantilever beam* excited by a hammer impact. The original video is blindly and autonomously separated into three videos, each contains one individual vibration mode with very high spatial (pixel) resolution. *More Video Demos*: http://www.lanl.gov/projects/national-security-education-center/engineering/research-projects/blind-modal-id.php

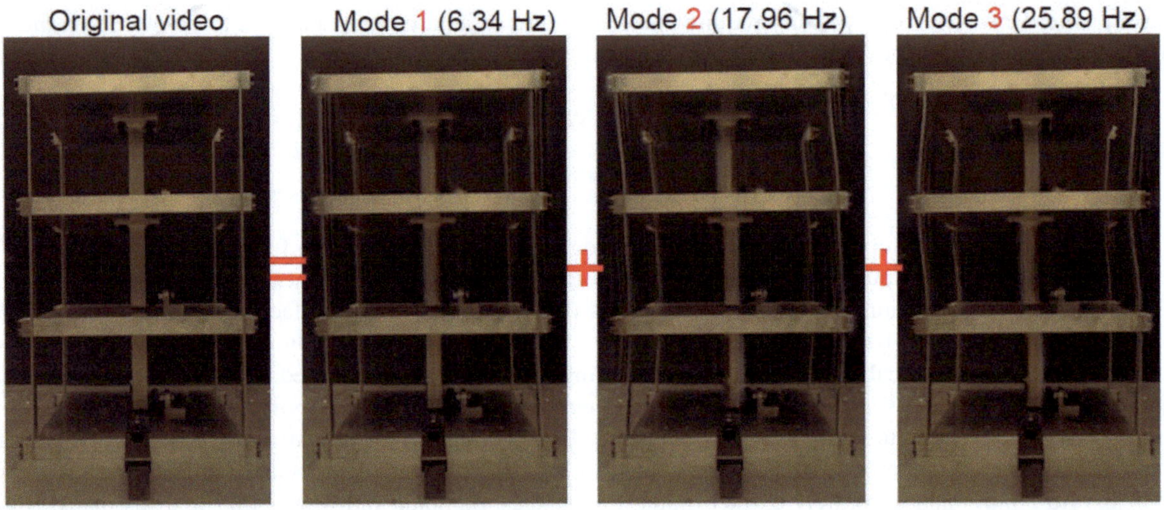

Fig. 53.2 Video-based, output-only, full-field dynamic extraction technique applied to an operating *building* excited by a hammer impact. The original video is blindly and autonomously separated into three videos, each contains one individual vibration mode with very high spatial (pixel) resolution. *More Video Demos*: http://www.lanl.gov/projects/national-security-education-center/engineering/research-projects/blind-modal-id.php

Table 53.1 The modal parameters of the 3-story structure estimated from the video measurements compared to those estimated from the accelerometers

	Frequency (Hz)		Damping ratio (%)	
Mode	Accelerometer	Video	Accelerometer	Video
1	6.27	6.31	0.28	0.27
2	17.88	17.92	0.30	0.30
3	25.82	25.92	0.25	0.25

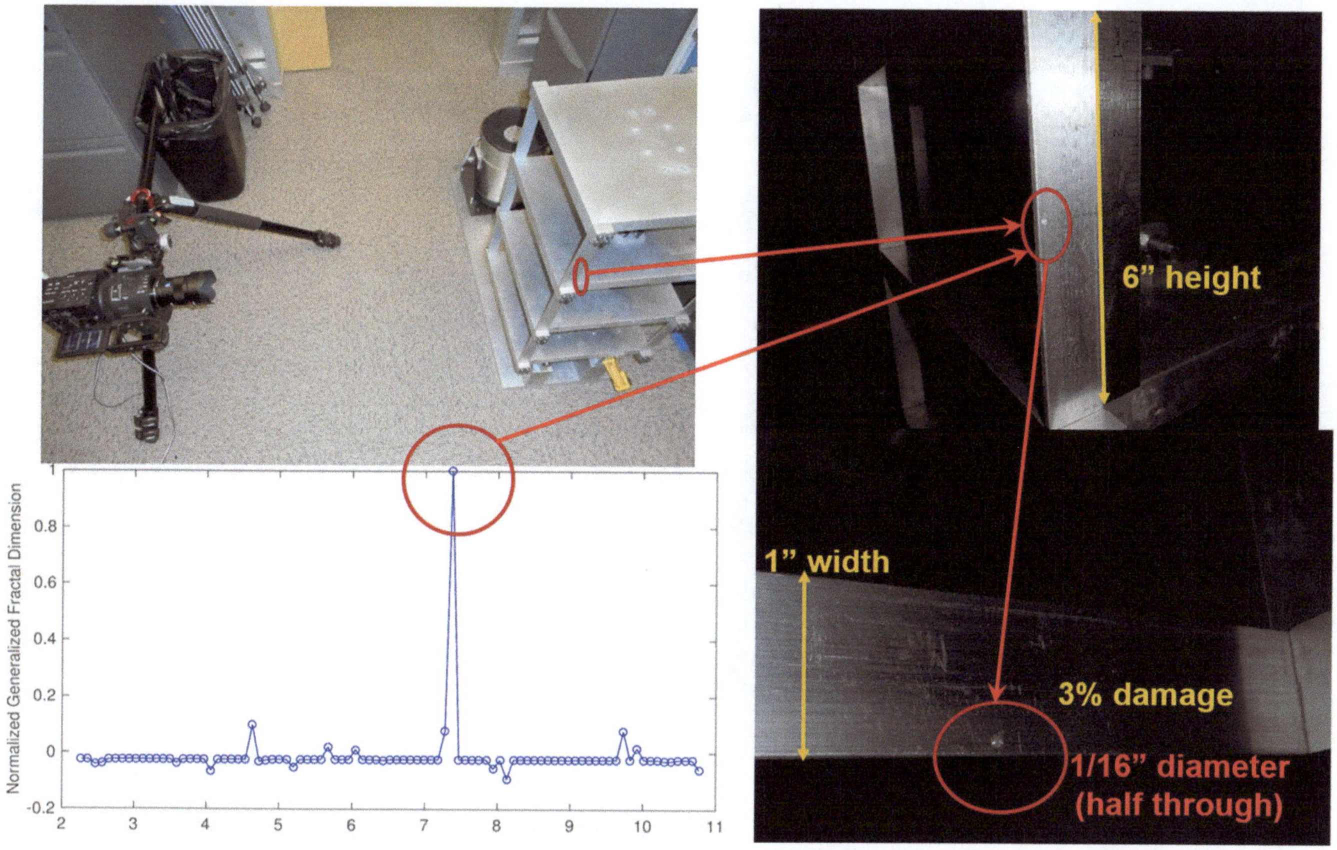

Fig. 53.3 The fractal dimension curves analyzing the full-field 1st mode shapes blindly extracted from the video measurements of the damaged *right top front column* of the 3-story *building structure* excited by an impact hammer. Damage is a hole drilled half through the thickness equivalent to a 3% cross section stiffness reduction back to the camera

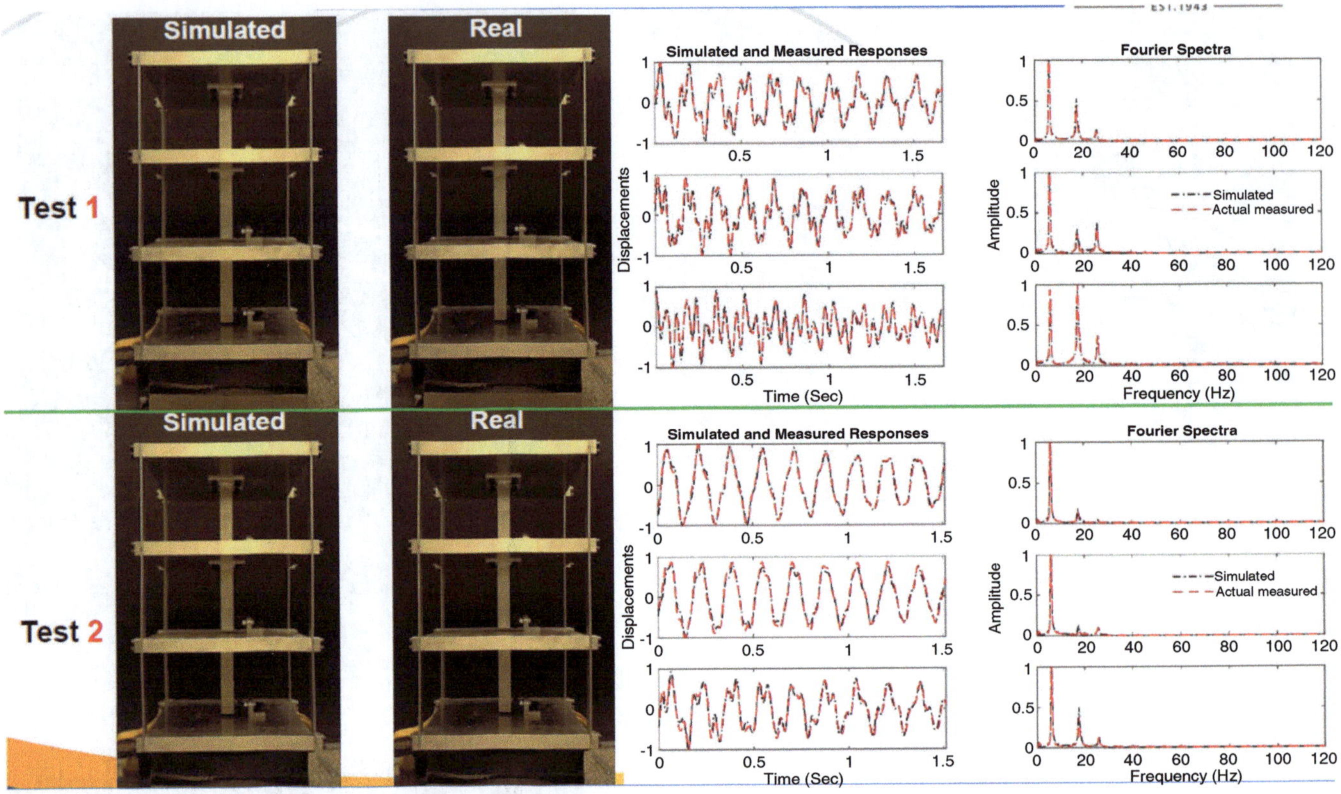

Fig. 53.4 Experimental results of the virtual, video-space, high-fidelity simulations and visualizations/animations of structural dynamics. The structure are subjected to impact tests. The simulated videos match the actual videos of the structural vibration testing, as seen in the right plots of the simulated and actual vibration in both time and frequency domains

Chapter 54
Coupled Experimental/Numerical Approach to Determine the Creep Behavior of Zr-4 Cladding Under LOCA Condition

D. Campello, N. Tardif, M.-C. Baietto, M. Coret, and J. Desquines

Abstract The thermo-mechanical behavior of Zircaloy-4 fuel rods under Loss-Of-Coolant Accident (LOCA) conditions is investigated. A custom experimental setup is dedicated to the high-temperature creep ballooning study of 90mm long cladding samples. Creep tests were performed under an inert environment (argon), for temperatures from 750 to 850°C and internal pressures ranging from 1 to 5 MPa. As the high-temperature creep of metals is strongly influenced by the temperature, the setup allow for a heterogeneous thermal distribution along the specimen. A unique test provides a rich database about the steady-state creep of the alloy. A first campaign is dedicated to bare Stress Relieved Annealed Zr-4.

Creep-rates are computed along a generatrix of the tube using 2-Dimensional Digital Image Correlation (2D-DIC) and are correlated to thermal distribution measurements allowed by Near Infra-Red Thermography (NIRT). As-received Zr-4 behavior laws are determined using Finite Element Model Updating (FEMU). Norton exponents and activation energies are determined for thermal-mechanical conditions of the performed tests.

Keywords 2D-DIC • NIRT • FEMU • Creep power-law • Zircaloy-4

54.1 Introduction

Loss-Of-Coolant of the primary loop is a complex multi-physical design basis accident. The study of the thermal mechanical and thermal hydraulics behavior of a fuel rod assembly is the context of PERFROI project [1]. It relies on both experimental and modeling activities about the flow blockage caused by fuel rods deformation, fuel fragment relocation and the potential loss of cladding integrity. In this framework, the ballooning behavior of fuel rod claddings submitted to high temperature and internal pressure is investigated.

Creep-tests and thermal ramps are usually performed in order to investigate the cladding behavior under LOCA conditions [2–5]. The setups proposed in this literature provide on a test a single temperature condition assuming thermal field is homogeneous when testing. A unique stress condition is often investigated. The number of the tests to be performed is consequent in order to accurately determine material parameters.

The study focuses on creep ballooning tests. Creep power-law in Eq. (54.1) is assumed to well model the secondary-creep behavior $\dot{\varepsilon}_{eq}$ of zirconium alloys [6] depending on the temperature T and the von Mises stress σ_{vM}. The multiplicative coefficient A, the Norton exponent n and the activation energy Q are widely investigated with experimental studies.

$$\dot{\varepsilon}_{eq}(\sigma, T) = A.{\sigma_{vM}}^{n}.e^{\frac{-Q}{R.T}} \tag{54.1}$$

These creep parameters depend on the temperature and applied stress conditions.

D. Campello (✉) • N. Tardif • M.-C. Baietto
Université de Lyon, CNRS, INSA-LYON, LaMCoS UMR 5259, 20 Avenue Albert Einstein, F69621 Villeurbanne, Cedex, France
e-mail: damien.campello@insa-lyon.fr

M. Coret
GeM (UMR 6183), Ecole Centrale de Nantes, 1 rue de la Noë, 44321 Nantes, France

J. Desquines
IRSN/PSN-REX/SEREX/LE2M, Saint Paul lez Durance, France

M. Sutton, P.L. Reu (eds.), *International Digital Imaging Correlation Society*, Conference Proceedings of the Society for Experimental Mechanics Series, DOI 10.1007/978-3-319-51439-0_54

54.2 Setup

The experimental setup initially designed by Tardif et al. [1] is upgraded to test Zircaloy-4 claddings under inert atmosphere using argon gas [8]. The thermo-mechanical loading applied to the specimen is represented in Fig. 54.1.

The tube is first heated using an induction system, and then three successive internal pressure levels are applied using argon gas. The bottom end effect is compensated. The bench is fitted to perform 2D-DIC [2] and NIR measurements [3].

Axial and hoop displacements fields are calculated in the ROI using a 2D-DIC algorithm. It is performed using Ufreckles software developed at LaMCoS by Réthoré [15]. The conservation of the optical flow is solved using a non-linear least-square method relying on a finite element basis. Performing 2D-DIC on tubular specimen is not common considering the 3D geometry. Uncertainties related to out-of-plane displacement (ballooning) have been estimated and are negligible with respect to the cladding deformation when steady-state creep rates are calculated. The equivalent steady-state creep-rates are calculated assuming material incompressibility and using the von Mises criterion.

The second digital image processing is dedicated to near infra-red measurements [9]. Standard cameras (with silicon detectors) are preferred to infra-red cameras in order to higher the image definition. The K_1 and K_2 parameters of the radiometric model detailed in Eq. (54.2) are calibrated using the data measured during the heating part of the test on the specimen itself.

$$T = \frac{K_1}{\ln\left(\frac{K_2}{I} + 1\right)} \tag{54.2}$$

54.3 Results

The steady-state creep rate and temperature axial distributions measured in the gauge length (z_o=20 mm) on a specimen heated up to 850°C are plotted in Figs. 54.2 and 54.3.

The thermal distribution clearly induces a variation of the creep-rates along the height of the specimen. The heterogeneous thermo-mechanical conditions are fully capitalized with the modeling of the tests. An iterative Gauss–Newton algorithm is then performed in order to lower the error between experimental creep-rate logarithms and calculated ones. A constitutive behavior law is determined using the data of a unique test. The measurements of nine experiments were processed in order to map the Norton exponent and the activation energy as represented in Fig. 54.3.

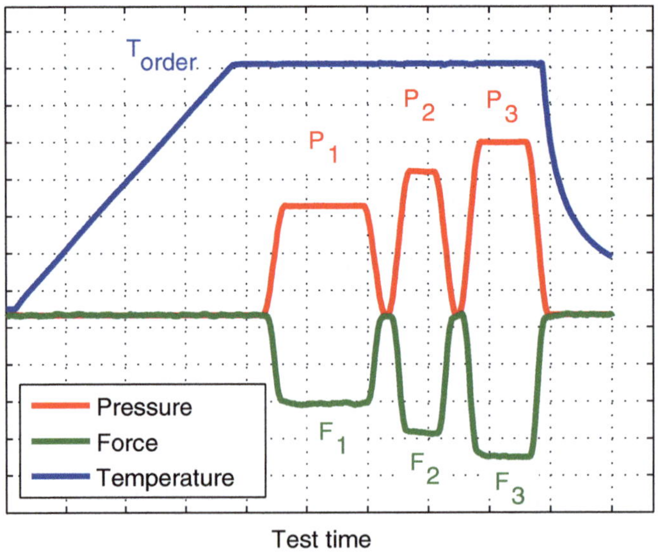

Fig. 54.1 Schematic thermal mechanical loadings applied to the specimen

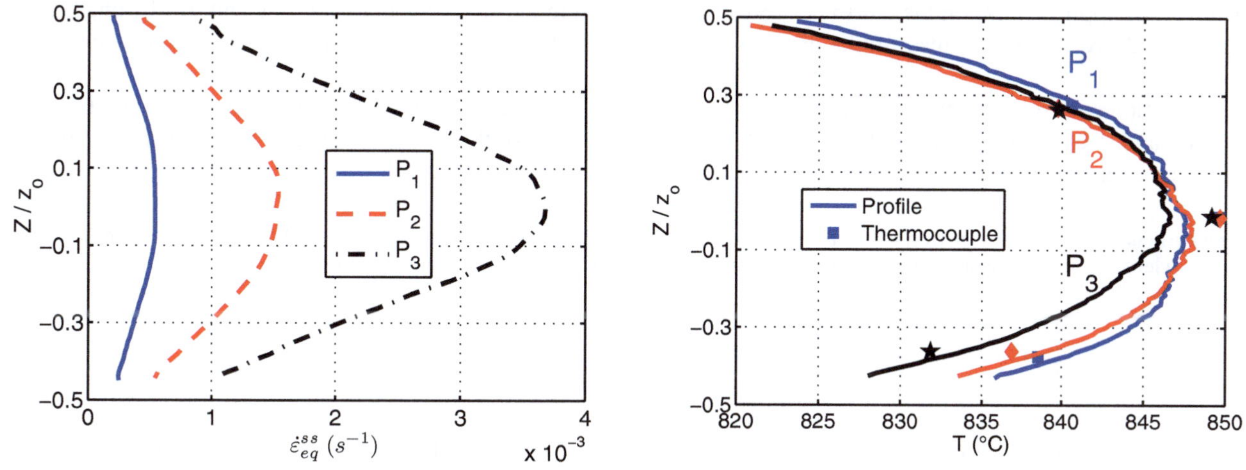

Fig. 54.2 Kinematics and thermal measurements on a tube heated up to 850°C resulting from 2D-DIC and NIRT

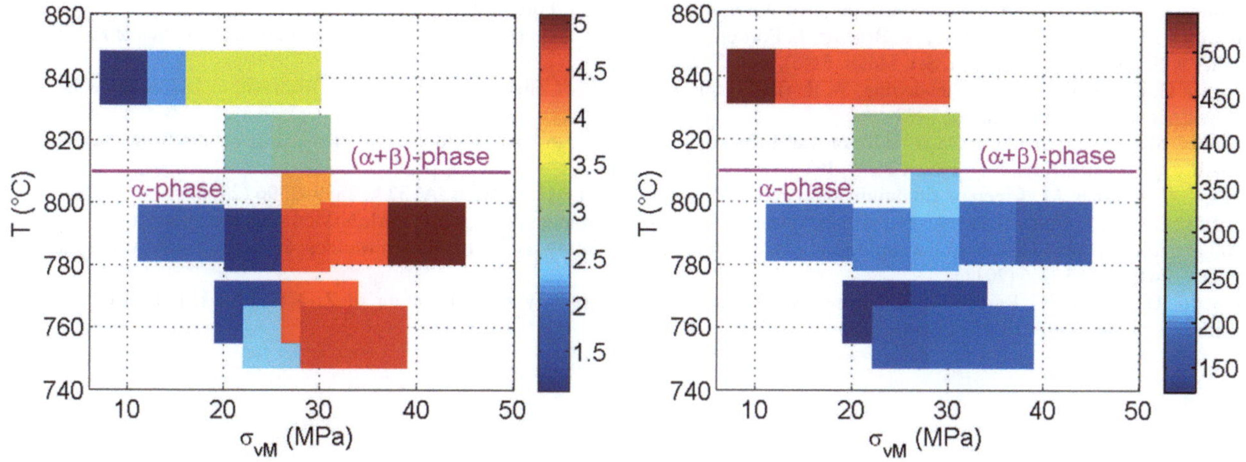

Fig. 54.3 Norton exponent and activation energy determined in the domain of the study using data of nine experiments

The results are in a good agreement between each other and are consistent with literature data. A stress of 25 MPa is determined in the α–phase domain of Zr-4 as a boundary between diffusion and dislocation creep mechanisms. The microstructure of the alloy is changing from α to α+β phase at a temperature of 820°C. As an effect, the determined activation energy is changing from around 180 kJ/mol to 300–400 kJ/mol in this domain.

54.4 Conclusion

A FEMU methodology was performed in order fully capitalize high-temperature creep ballooning tests data and determine the parameters a steady-state creep power-law. The ballooning structural effect are accounted for using FEM granting access to the true stress level and state over a 20 mm gauge length. Results of seven tests are detailed in this paper addressing to thermal conditions ranging from 750 to 850°C and von Mises stresses in a range of 7–48 MPa.

The effect of applied stress on the Norton exponent is evidenced in the study for both α-phase and (α+β)-phase domains. The determined stress exponent magnitudes are very consistent with literature data depending on the thermo-mechanical test conditions.

The activation energy parameter is very sensitive to the thermo-mechanical conditions and remains valid within a limited temperature and stress ranges to the authors' knowledge. The beginning of the (α+β)-phase domain is clearly identified with the activation energy increase.

Acknowledgments The study was performed in the framework of the PERFROI ANR project (No. ANR-11-RSNR-0017-01). The authors would like to acknowledge all the participants of the project: EDF, IRSN-SEREX, LEMTA, Ecole Centrale of Paris.

References

1. Repetto, G., Dominguez, C., Durville, B., Carnemolla, S., Campello, D., Tardif, N., and Gradeck, M., The R&D PERFROI project on the thermal mechanical and thermal hydraulics behaviors of a fuel rod assembly during Loss-Of-Coolant Accident. Proceedings of the Nureth16, Chicago, USA, 30 August–4 Sept 2015.
2. Busby, C., and Marsh, K., High temperature, time-dependent deformation in internally pressurized Zircaloy-4 tubing (LWBR Development Program), in Technical report, Colorado Cooperative Wildlife Research Unit, Fort Collins, USA, 1974
3. Kaddour, D., Frechinet, S., Gourgues, A., Brachet, J., Portier, L., Pineau, A.: Experimental determination of creep properties of zirconium alloys together with phase transformation. Scr. Mater. **51**(6), 515–519 (2004)
4. Rosinger, H. E., Bera, P. C., and Clendening, W. R., The steady-state creep of Zircaloy-4 fuel cladding from 940 to 1873K in Technical report, Atomic Energy of Canada Limited, 1978
5. Réocreux, M., de Martinville, E.S.: A study of fuel behavior in PWR design basis accident: and analysis of results from the PHEBUS and EDGAR experiments. Nucl. Eng. Des. **124**(3), 363–378 (1990)
6. Hayes, T.A., Kassner, M.: Creep of zirconium and zirconium alloys. Metall. Mater. Trans. A. **37A**, 2389–2396 (2006)
7. Tardif, N., Coret, M., Combescure, A.: Experimental study of the fracture kinetics of a tubular 16MnNiMo5 steel specimen under biaxial loading at 900 and 1000°C. Application to the rupture of a vessel bottom head during a core meltdown accident in a pressurized water reactor. Nucl. Eng. Des. **241**(3), 755–766 (2011)
8. Campello, D., Tardif, N., Baietto, M.-C., Coret, M., Desquines, J., Secondary creep behavior of Zr-4 claddings under LOCA conditions. Proceedings of the TOPFUE 2016, Boise, USA, 11–16 Sept 2015

Chapter 55
Comparison of DIC Methods of Determining Necking Limit of PLC Material

Junying Min, Thomas B. Stoughton, John E. Carsley, and Jianping Lin

Abstract Digital image correlation (DIC) techniques have been used in forming limit tests to obtain forming limit strains and many different methods have been proposed to determine the onset of localized necking in order to improve the accuracy of forming limit curves in recent ten years. In this work, necking limit strains of a PLC material (AA5182-O) was determined by the DIC strain rate contours as well as the existing DIC methods in the literature. The effect of PLC band propagation on identification the onset of localized necking was discussed.

Keywords Localized necking • Forming limit strain • DIC methods • Sheet metals • FLC

The Portevin-Le Chatelier (PLC) effect in 5000 series Al alloy sheets has been extensively reported in uniaxial tension [1] and pure shear [2] using DIC techniques, and the PLC behavior of AA5182-O in hydraulic bulge was recently reported by Min et al. [3] where the deformation was out-of-plane and biaxial, close to balanced-biaxial. The observed PLC behavior in hydraulic bulging with ringed or crescent shape bands are distinct from the "/" or "X"-shaped PLC bands observed in uniaxial tension. Nevertheless, PLC behavior of Al alloys under in-plane, plane strain deformation has rarely been studied in the literature. Before understanding the curvature changes in AA5182 specimens as well as how the PLC behavior affects detection of the onset of localized necking, the PLC behavior (also a plastic instability of sheet metals) in the AA5182-O specimen is described first.

Figure 55.1a presents the major strain contour (ε_1) of a 1.1 mm thick AA5182-O Marciniak specimen (with a width of 114 mm) just prior to fracture, which is measured by DIC techniques. This specimen exhibited plane-strain path in its central area. Figure 55.1b shows the evolution of the major strain rate ($\dot{\varepsilon}_1$) variation along a line perpendicular to the specimen fracture (namely, along line CD, the major strain direction on the AA5182-O specimen) in the time range from $\tau = 25.0$ s up to fracture. The evolution of $\dot{\varepsilon}_1$ along line CD for $\tau < 25.0$ s is not detailed here since it is far before necking. The light blue, yellow and red contours in Fig. 55.1b indicate the peak strain rates within PLC bands. For example, at $\tau = 27.9$ s, a PLC band with a peak strain rate ($\dot{\varepsilon}_{1P}$) of 7.0×10^{-3} s^{-1} nucleated at $X = -8.4$ mm as indicated by the lower white hollow arrow, and propagated continuously toward point O (Fig. 55.1a) to $X = -1.8$ mm within 0.7 s with an averaged propagation speed of 9.4 mm/s and then vanished. During its propagation, $\dot{\varepsilon}_{1P}$ increased to 6.1×10^{-2} s^{-1} at $\tau = 28.4$ s and then decreased until disappearance. Another example is the last set of PLC bands: a PLC band with a $\dot{\varepsilon}_{1P}$ of 1.0×10^{-2} s^{-1} nucleated at $X = 21$ mm when $\tau = 28.5$ s as shown by the upper white hollow arrow in Fig. 55.1b, and this PLC band moved continuously toward point O until $\tau = 29.7$ s. After that the band remained at $X = 0$ mm (point O) until fracture occurred. Similar to the previously described set of PLC bands, $\dot{\varepsilon}_{1P}$ increased to $\sim 6.0 \times 10^{-2}$ s^{-1} at $\tau = 29.3$ s during its initial propagation, and then $\dot{\varepsilon}_{1P}$ remained at $\sim 6.0 \times 10^{-2}$ s^{-1} until $\tau = 29.7$ s. From the temporal aspect of the PLC bands, the bands involved in Fig. 55.1b are Type A bands reported in uniaxial tension [1], which repetitively and continuously propagate along the gage section of tensile specimens.

For better understanding of the development of the last set of PLC bands and their spatial characteristics, the $\dot{\varepsilon}_1$ contours on the AA5182-O specimen surface calculated by DIC software from 29.3 to 29.9 s are presented in Fig. 55.2a–g, where the red-yellow-green contours indicate a nearly straight PLC band on the AA5182-O specimen. The tick on the X-axis indicates

J. Min (✉)
Chair of Production Systems, Ruhr-University Bochum, Bochum 44780, Germany
e-mail: junying.min@gmail.com

T.B. Stoughton • J.E. Carsley
General Motors Research and Development, Warren, MI 48095-9055, USA

J. Lin
School of Mechanical Engineering, Tongji University, Shanghai 201804, China

M. Sutton, P.L. Reu (eds.), *International Digital Imaging Correlation Society*, Conference Proceedings of the Society for Experimental Mechanics Series, DOI 10.1007/978-3-319-51439-0_55

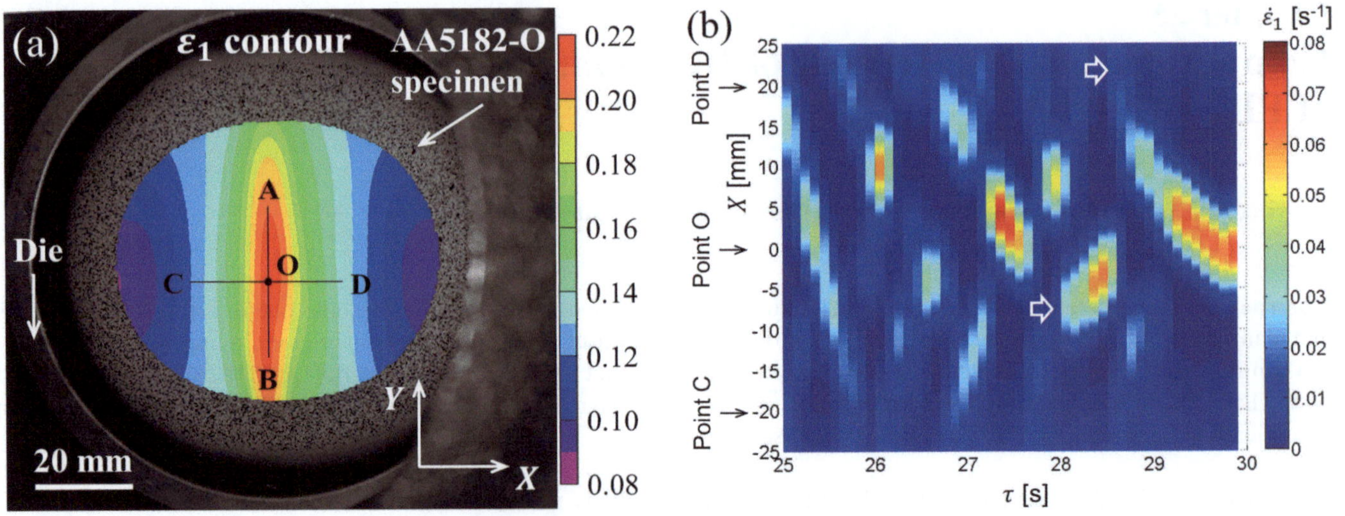

Fig. 55.1 (**a**) Major strain contour (ε_1) of the AA5182-O Marciniak specimen just prior to fracture. (**b**) Evolution of the major strain rate ($\dot{\varepsilon}_1$) along line CD (refer to Fig. 55.1a, *X*-axis) in the time range from 25 s up to fracture

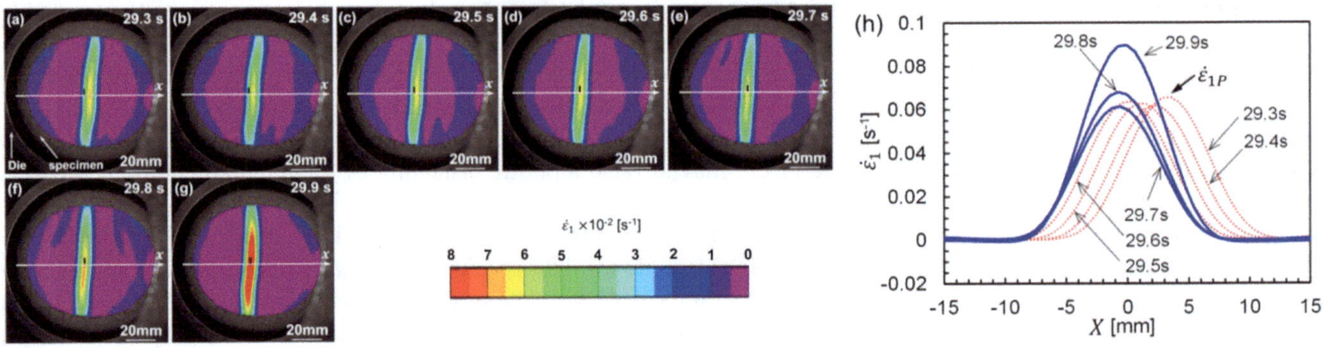

Fig. 55.2 Contours of the major strain rate ($\dot{\varepsilon}_1$) on the AA5182 specimen at (**a**) $\tau = 29.3$ s, (**b**) $\tau = 29.4$ s, (**c**) $\tau = 29.5$ s, (**d**) $\tau = 29.6$ s, (**e**) $\tau = 29.7$ s, (**f**) $\tau = 29.8$ s and (**g**) $\tau = 29.9$ s. (**h**) Profiles of $\dot{\varepsilon}_1$ along the *X*-axis in (**a**)–(**g**) show propagating PLC bands (*dotted*) and static localized necking bands (*solid*)

point O (Fig. 55.1a, $X = 0$ mm) for reference, and from Fig. 55.2a–e the PLC band propagated from right to left since the peak of the PLC bands moved from positive *X*-values to $X = 0$ mm. After $\tau = 29.7$ s, the peak of the bands remained at $X = 0$ mm until fracture occurred at this point. In particular, $\dot{\varepsilon}_{1P}$ increased to 9.0×10^{-2} s^{-1} just before fracture. The movement of the peak of PLC bands is clearly presented in Fig. 55.2h by plotting the $\dot{\varepsilon}_1$ profile along the *X*-axis. From both temporal and spatial aspects of the localized deformation bands shown in Figs. 55.1b and 55.2, it is believed that the PLC bands eventually developed to a localized necking band at $\tau = 29.7$ s in such plane-strain deformation, since beyond this moment the localized deformation band became static and $\dot{\varepsilon}_{1P}$ increased continuously and significantly before fracture. The difference between a PLC band and a localized necking band is that the former propagates and brings deformation to its in-plane adjacent material while the later leads to continuous (through-thickness) thinning at a local area.

Figure 55.3a–f detailed how the existing DIC methods determine the necking limit strains, and Fig. 55.3g compares the limit major strains (ε_{1L}) from the methods considered in Fig. 55.3a–f. From the overall comparison, the methods by Min et al. [4] and Martínez-Donaire et al. [5] identified the onset of localized necking at 29.7 s (refer to Fig. 55.3a, b) despite the serrations of due to PLC band propagation, which also matches exactly with the observation in Figs. 55.1b and 55.2. However, $\tau = 29.6$ s (Fig. 55.3c) and $\tau = 29.5$ s (Fig. 55.3d) were identified as the onset of localized necking by the Huang et al. [6] and Wang et al. [7] methods, respectively; and the method by Hotz et al. [8] determined the start of localized necking was at $z = 15.08$ mm corresponding to $\tau = 28.8$ s (Fig. 55.3e). The error of 0.1 and 0.2 s (i.e. one frame or two frames according to the used framing rate of 10 frames/s) of the methods by Huang et al. [6] and Wang et al. [7], respectively, are probably due to the fact that a differential method was used to calculate $\ddot{\varepsilon}_1$ (the second derivative of the major strain) and $\Delta\dot{Z}$ (the first derivative of surface height difference), where $\ddot{\varepsilon}_1$ and $\Delta\dot{Z}$ of the current frame was respectively calculated based on $\dot{\varepsilon}_1$ and ΔZ of the adjacent (before and after) frames, and $\ddot{\varepsilon}_1$ and $\Delta\dot{Z}$ were used to identify the onset of localized necking

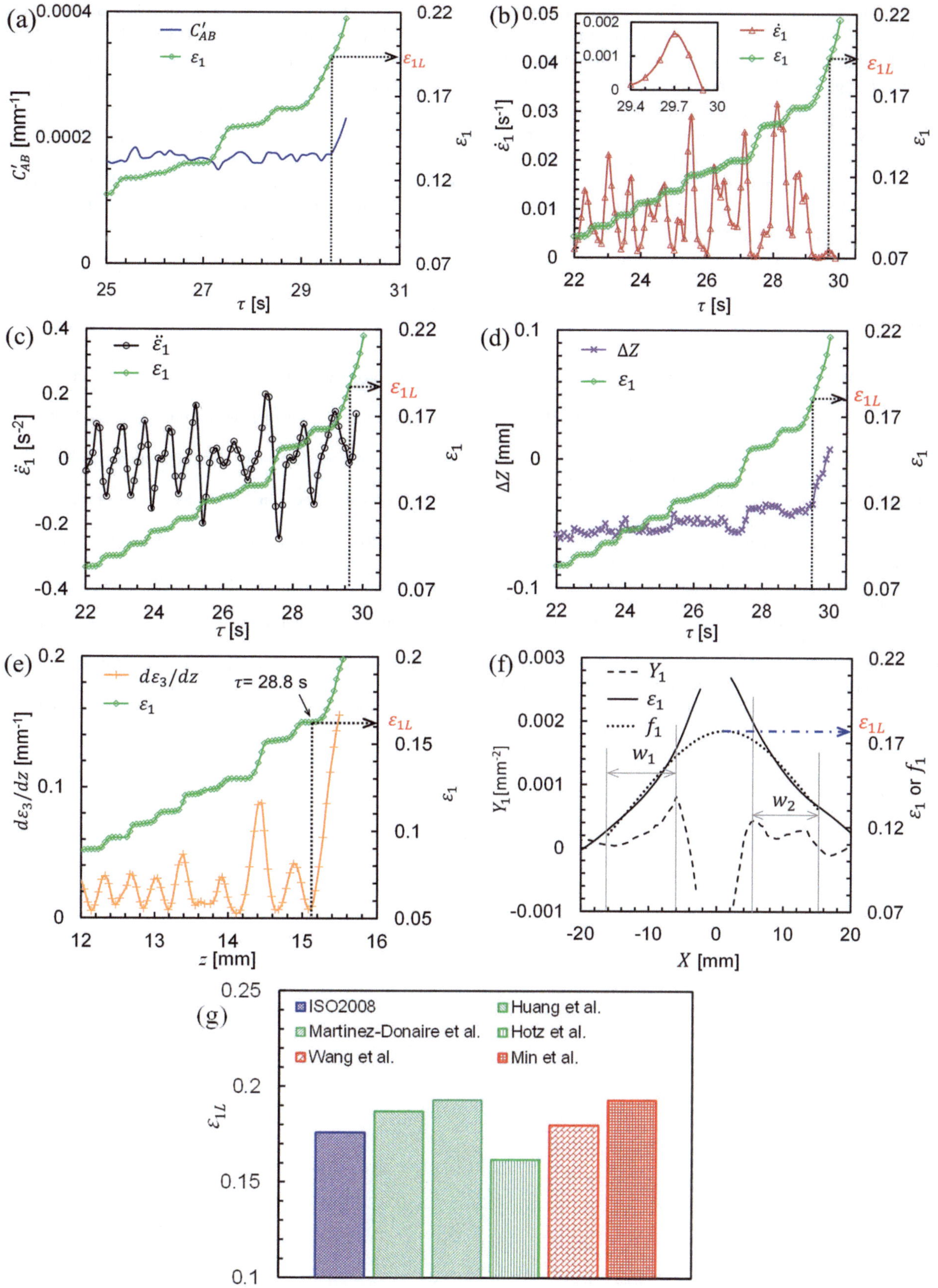

Fig. 55.3 Determination of necking limit of the AA5182-O specimen by (**a**) Min et al. [4], (**b**) Martínez-Donaire et al. [5], Huang et al. [6], Wang et al. [7], Hotz et al. [8] and (**f**) ISO standard [9]. (**g**) Comparison of necking limit strain (ε_{1L}) from the methods shown in (**a**)–(**f**)

in Huang et al. [6] and Wang et al. [7] methods, respectively. While in the method by Hotz et al. [8], 7 successive images were required to calculate $d\varepsilon_3/dz$ (z is the punch position), which may lead to larger time error; in addition, the Hotz et al. [8] method is based on the thinning rate at the neck center, and before the last set of PLC bands developed to a localized necking band, e.g. $\tau < 29.7$ s, point O ($X = 0$ mm in Fig. 55.2) was already in front of the PLC bands. This suggests that continuous thinning started as the front of PLC band propagated into point O; therefore, this method may not be applicable to materials exhibiting the PLC effect. Nevertheless, no differential calculation was used in the methods by Min et al. [4] and Martínez-Donaire et al. [5] and both methods predict exactly the same results for the AA5182-O specimen. Since the ISO standard method [9] is not time-dependent and only evaluates the strain profile after fracture (Fig. 55.3f), it is not able to provide a prediction of when localized necking occurred, but underestimates the necking limit strain as the Hotz et al. [8] method does according to the comparison in Fig. 55.3g.

References

1. Ait-Amokhtar, H., Fressengeas, C.: Crossover from continuous to discontinuous propagation in the Portevin-Le Chatelier effect. Acta Mater. **58**, 1342–1349 (2010)
2. Manach, P.Y., Thuillier, S., Yoon, J.W., Coër, J., Laurent, H.: Kinematics of Portevin-Le Chatelier bands in simple shear. Int. J. Plast. **58**, 66–83 (2014)
3. Min, J., Hector Jr., L.G., Carsley, J.E., Stoughton, T.B., Carlson, B.E., Lin, J.: Spatio-temporal characteristics of plastic instability in AA5182-O during biaxial deformation. Mater. Des. **83**, 786–794 (2015)
4. Min, J., Stoughton, T.B., Carsley, J.E., Lin, J.: A method of detecting the onset of localized necking based on surface geometry measurements. Exp. Mech. (2016). doi: 10.1007/s11340-016-0232-4
5. Martínez-Donaire, A.J., García-Lomas, F.J., Vallellano, C.: New approaches to detect the onset of localised necking in sheets under through-thickness strain gradients. Mater. Des. **57**, 135–145 (2014)
6. Huang G, Sriram S, Yan B. Digital image correlation technique and its application in forming limit curve determination. Proceedings of the IDDRG 2008 International Conference, Olofström, Sweden, 153–162, 2008
7. Wang, K., Carsley, J.E., He, B., Li, J., Zhang, L.: Measuring forming limit strains with digital image correlation analysis. J. Mater. Process. Technol. **214**, 1120–1130 (2014)
8. Hotz, W., Merklein, M., Kuppert, A., Friebe, H., Klein, M.: Time dependent FLC determination—comparison of different algorithms to detect the onset of unstable necking before fracture. Key Eng. Mater. **549**, 397–404 (2013)
9. ISO/DIS 12004-2:2008. Metallic materials—Sheet and strip—Determination of forming Limit curves. International Organization for Standardization 20087

Chapter 56
Large Field Digital Image Correlation Used for Full-Scale Aircraft Crash Testing: Methods and Results

Justin Littell

Abstract Since the summer of 2013, five full-scale crash tests have been conducted at NASA Langley Research Center's (LaRC) Landing and Impact Research Facility (LandIR) on deformable airframes for the evaluation of a variety of crashworthy concepts. Two tests were conducted on two CH-46E "Sea Knight" airframes as part of the Transport Rotorcraft Airframe Crash Testbed (TRACT) project and three tests were conducted on three Cessna 172 General Aviation aircraft as a part of the Emergency Locator Transmitter Survivability and Reliability (ELTSAR) project. Full field digital image correlation (DIC) data was obtained as a part of each aircraft's experimental suite, thus giving insight into the crash severity and airframe deformation which occurred for each test [1].

Keywords Crash testing • Large structures • Full field digital image correlation • Aircraft • Impact dynamics

56.1 Extended Abstract

Since the summer of 2013, five full-scale crash tests have been conducted at NASA Langley Research Center's (LaRC) Landing and Impact Research Facility (LandIR) on deformable airframes for the evaluation of a variety of crashworthy concepts. Two tests were conducted on two CH-46E "Sea Knight" airframes as part of the Transport Rotorcraft Airframe Crash Testbed (TRACT) project and three tests were conducted on three Cessna 172 General Aviation aircraft as a part of the Emergency Locator Transmitter Survivability and Reliability (ELTSAR) project. Full field digital image correlation (DIC) data was obtained as a part of each aircraft's experimental suite, thus giving insight into the crash severity and airframe deformation which occurred for each test [1].

In order to facilitate the acquisition of DIC data, a large field calibration procedure was performed using the fixed coded target array present on a large backboard, which can be seen in views behind each test article [2]. Monochromatic, synchronized 4 megapixel high-speed cameras filming at 500 Hz. were used for all data acquisition. Camera positions were typically 40 ft. from the test article, 90 ft. from the backboard and 30 ft. from each other. They were focused on the center of each test article, filming the impact from slightly different angles.

Two TRACT tests, called TRACT 1 [3] and TRACT 2 [4], were conducted for the evaluation of a variety of crashworthy experiments, which gave two opportunities to examine fuselage deformation on two separate test articles undergoing similar expected impact conditions. For the TRACT tests, the port side of each vehicle was first cleaned and then painted with a base coat of white paint. One inch diameter dots were applied by hand all along the fuselage in a random pattern, achieving approximately 50% coverage of black/white contrast. Figure 56.1 shows the painted airframe used in TRACT 1.

Camera lenses were chosen such that a 50 ft. horizontal field of view (FOV) was present, which extended about 5 ft. beyond the airframe length. This FOV allowed for the viewing of a small amount of vehicle slide out, post impact. TRACT 1 impacted a soil surface at 25 ft./s vertical and 33.0 ft./s horizontal velocities at a nose up pitch of 2.5°. There was approximately 8 ft. of post-test slide out. Figure 56.2 shows the lateral deformation of the airframe during four discrete times during the impact.

TRACT 2 was conducted on a second CH-46 airframe under similar test conditions. One major difference between the two tests was the soil for TRACT 2 was much softer due to a higher moisture content. Because of this, the airframe "stuck" into the soil more than TRACT1, with a slide out only measuring approximately 4 ft. With roughly half of the slide out, the

J. Littell (✉)
Research Aerospace Engineer, Structural Dynamics Branch, NASA Langley Research Center, 12 W. Bush Rd MS 495, Hampton, VA 23681, USA
e-mail: Justin.D.Littell@nasa.gov

M. Sutton, P.L. Reu (eds.), *International Digital Imaging Correlation Society*, Conference Proceedings of the Society for Experimental Mechanics Series, DOI 10.1007/978-3-319-51439-0_56

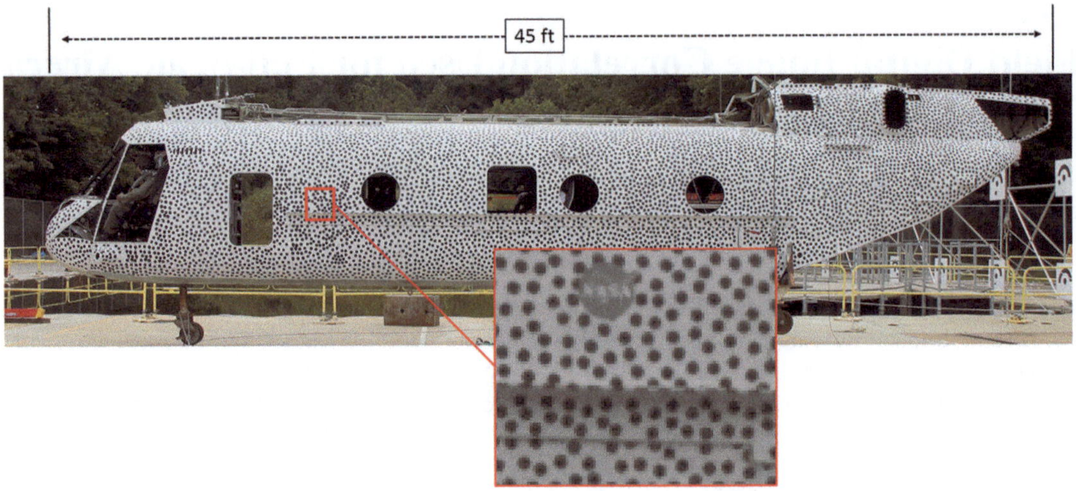

Fig. 56.1 TRACT 1 fuselage with dot pattern highlighted

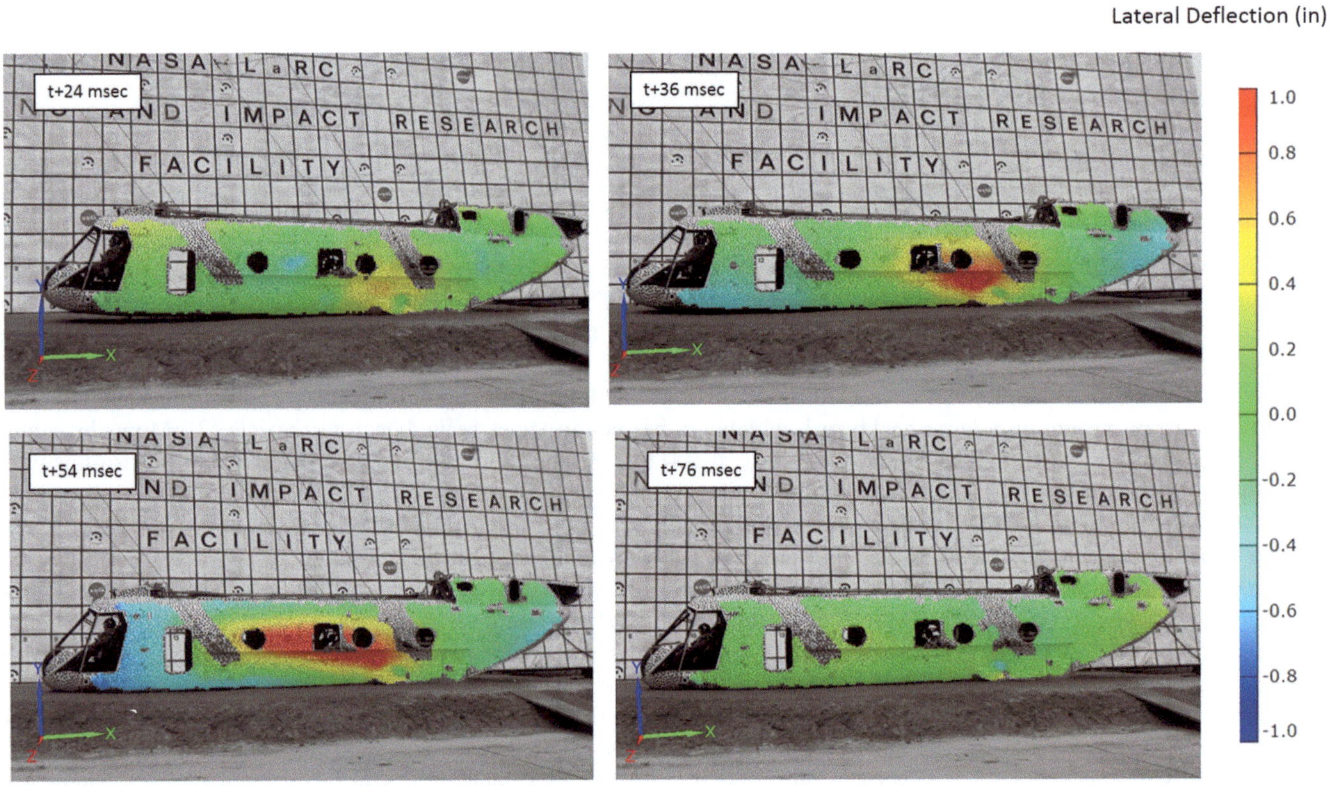

Fig. 56.2 TRACT 1 fuselage lateral deformation

horizontal deceleration was twice as high which caused failure in the lower floor section in the front cabin area. Figure 56.3 shows the lateral deflection comparisons between TRACT 1 and TRACT 2, with the failure region highlighted. Figure 56.4 shows detailed strain analysis of the failure region in TRACT 2.

The second series of tests focuses on the crash response of three Cessna 172 aircraft as a part of the ELTSAR Project [5]. There were two main factors in determining the camera setup for the ELTSAR tests—the size of the vehicle and the post-test impact response. The horizontal field of view was chosen to be approximately 40 ft. in order to balance data resolution based on two conflicting factors: a smaller (relative to TRACT) test article, but larger (relative to TRACT) expected impact velocities and larger post-test slide out. Other factors made DIC difficult; some an artifact of the test configuration, requiring

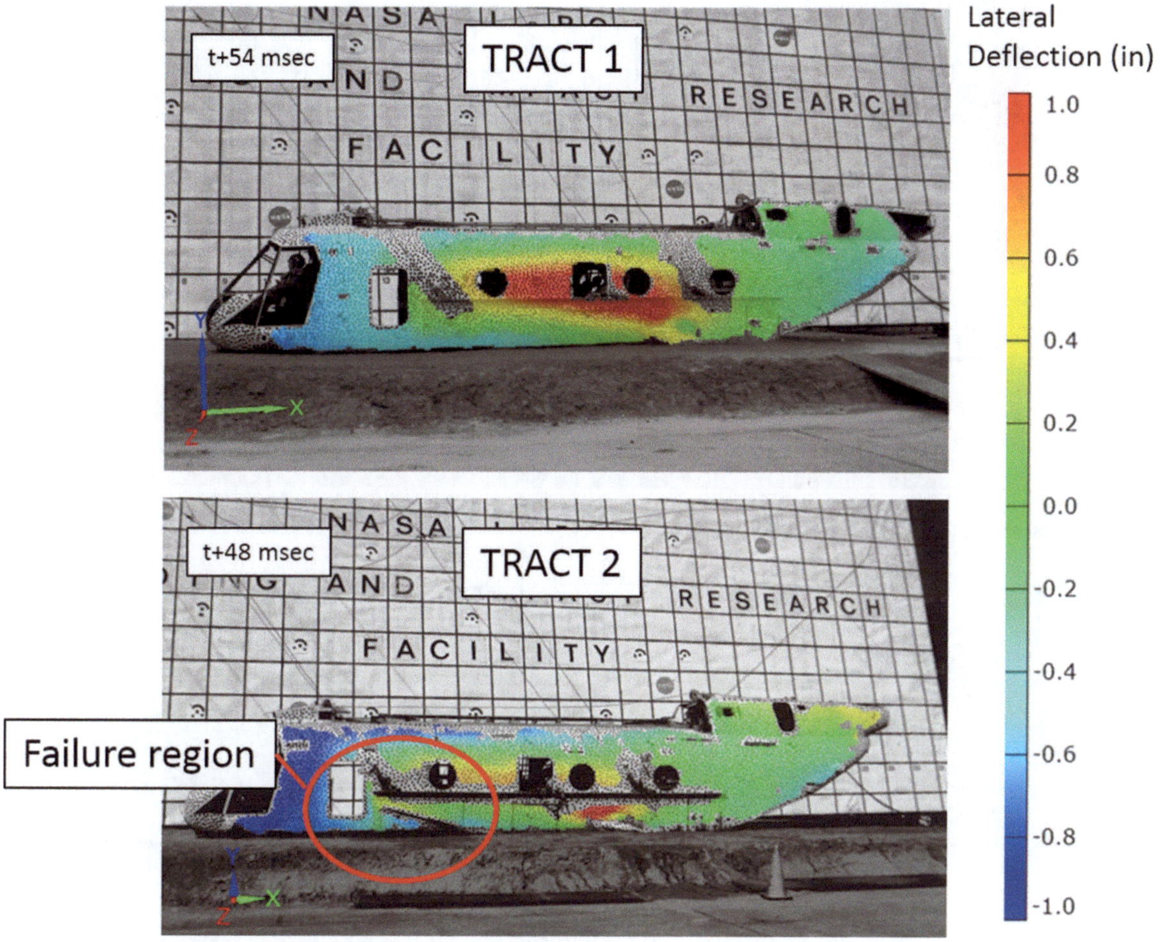

Fig. 56.3 TRACT 1 vs. TRACT 2 lateral deformation

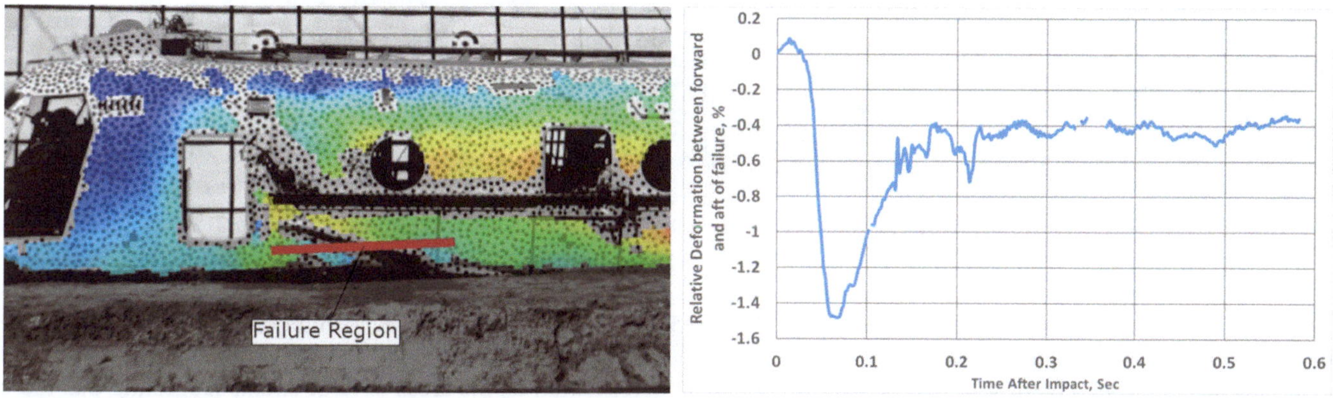

Fig. 56.4 TRACT 2 strain across failure region

data to be masked in order to remove artificial distortions due to test rigging and data cabling, and others an artifact of the Cessna 172 design. Figure 56.5 shows the Test 2 Cessna 172 test article used in the ELTSAR test series.

Two-dimensional DIC was acquired on ELTSAR Test 1, while three-dimensional DIC was acquired on the ELTSAR Tests 2 and 3. Each test represented a different crash condition with different model year aircraft, so direct comparisons were not made between tests. However, each test represented a scenario for an impact, and data gathered represented three discrete data points. The first represented a hard landing onto a rigid surface such as a highway or runway at 60.2 ft./s horizontal and

Fig. 56.5 ELTSAR Cessna 172 test article (Test 2 shown)

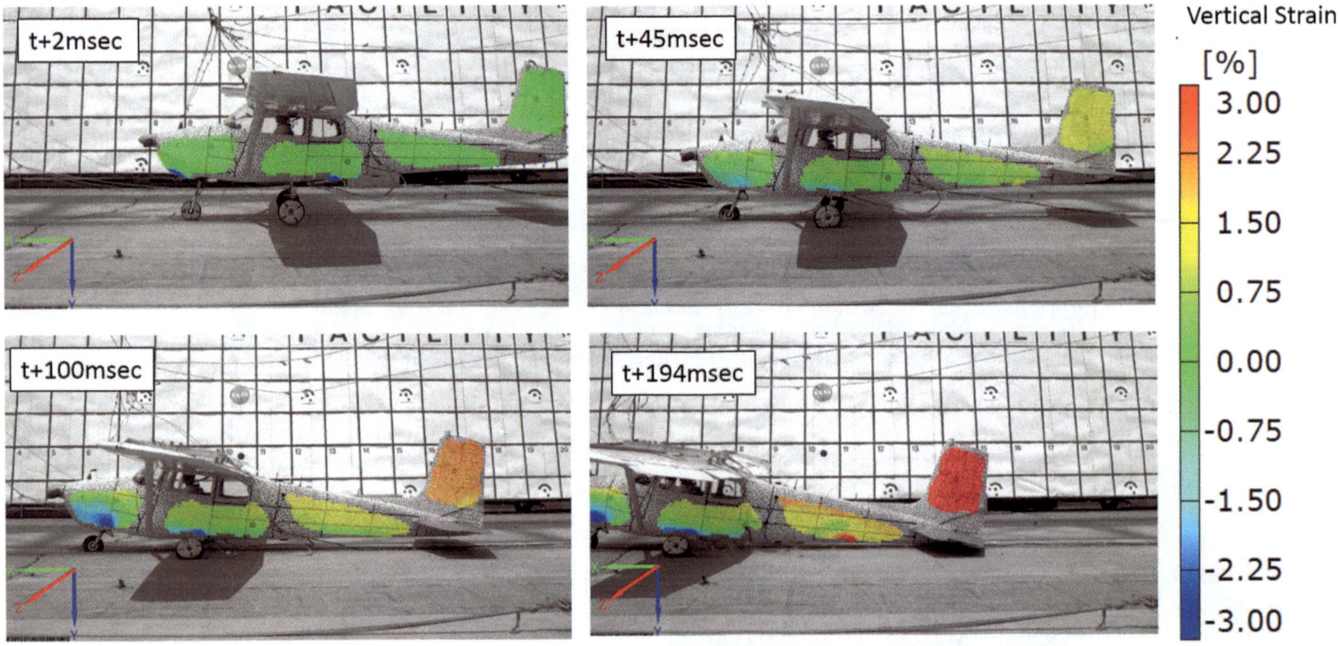

Fig. 56.6 Vertical strain time history for ELTSAR Test 1

23.0 ft./s vertical velocities at a pitch angle of 1.5°. Figure 56.6 shows the global vertical strain on the aircraft during the impact. Note that the airplane leaves the FOV before the rebound so it is unclear from the data whether the aircraft sustained a purely elastic impact. Post-test inspections show permanent deformation on the areas of large strain, suggesting that these areas sustained plastic impact.

The second and third ELTSAR tests were conducted with a soil impact surface, replicating a controlled flight into terrain (CFIT) scenario. In Test 2, the aircraft was pitched nose down, while in Test 3, the aircraft was pitched nose up. Both test were conducted at velocities in the range of the aircraft's stall speed. Figure 56.7 shows strains in Test 2 and Test 3 during the impact.

Tests 2 and 3 show the aircraft exhibiting very high deformations with corresponding to high strains. Distinct areas which show large deformations include near the door opening and near the engine cowling for Test 2, and on the landing gear and vertical stabilizer attachment points for Test 3.

Through the development of the large field calibration techniques, optimization of camera placement, lenses and FOV choices, large field DIC results can be obtained from a variety of test articles under a wide range of test conditions. This

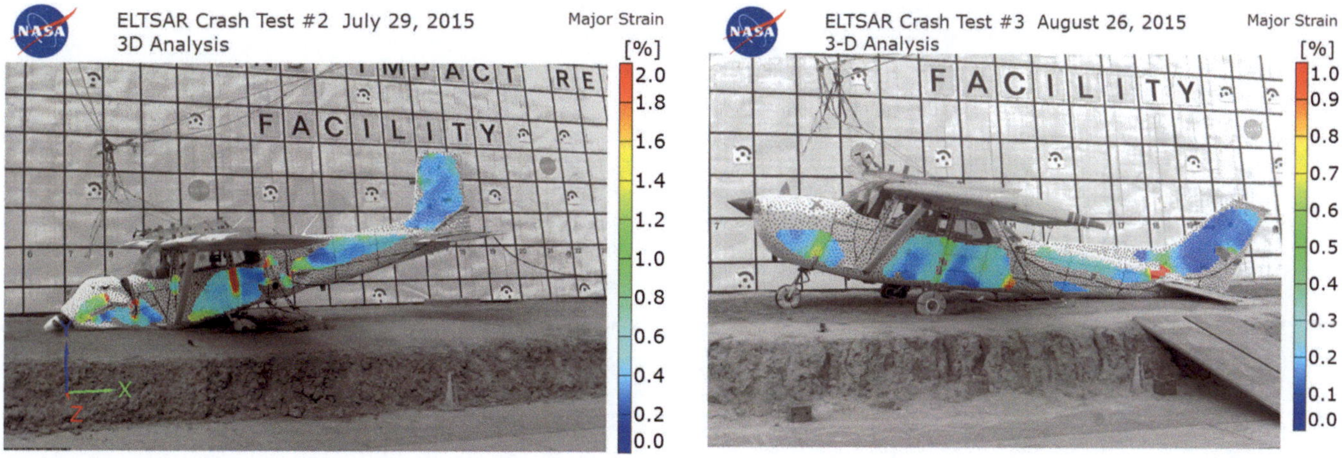

Fig. 56.7 ELTSAR Test 2 (left) and Test 3 (right) major strain during impact

presentation presents five sample results obtained from five full-scale crash tests on real airframes under realistic crash conditions. Overall the DIC data give previously unknown insight into airframe deformation and strain which is able to be processed, interrogated, presented and discussed.

References

1. Littell, J.D.: Experimental photogrammetric techniques used on five full-scale aircraft crash tests. NASA TM-2016-219168. March 2016
2. Littell, J.D.: Large field photogrammetry techniques in aircraft and spacecraft impact testing. In: Proceedings from the Society of Experimental Mechanics Annual Conference. Indianapolis, Indiana, 7–10 June 2010
3. Annett, M.S., Littell, J.D., Jackson, K.E., Bark, L.W., DeWeese, R.L., McEntire, B.J.: Evaluation of the first transport rotorcraft airframe crash testbed (TRACT 1) full-scale crash test. NASA TM-2014-218543. October 2014
4. Annett, M.S., Littell, J.D.: Evaluation of the second transport rotorcraft crash testbed (TRACT 2) full scale crash test. In: Proceedings from the American Helicopter Society 71st Annual Forum, Virginia Beach Virginia, 5–7 May 2015
5. Stimson C.M., Littell, J.D. Emergency locator transmitter survivability and reliability study. NASA TM. In Pub

Chapter 57
Desirable Features of Processing Dic Data with a Stress Function

A. Alshaya, W.A. Samad, and R.E. Rowlands

Abstract Structural and mechanical components often contain holes or notches, thus introducing stress concentrations which can control member integrity. While DIC is an extremely popular experimental technique, there are difficulties in recording quality DIC data at and near edges, rendering it problematic to obtain reliable stress or strain concentration factors. Moreover, traditional displacement-based strain/stress analyses require knowing both in-plane displacements. For linear-elastic isotropic or orthotropic materials, using a stress function enables one to obtain both in-plane displacements, hence complete strains and stresses, from measured information of only *one* of the two displacement fields. This can be advantageous since it is not uncommon for one or the other of the in-plane displacements to be of inadequate quality and/or extremely small. Analyzing isotropic materials involves real variables while one represents the stress function in complex variables for orthotropic materials.

Keywords Digital Image Correlation • Hybrid Method • Stress Concentration • Stress Analysis

Figure 57.1 compares the vertical strain distribution as one approaches the edge of an elliptical hole in a vertically-loaded aluminum plate. DIC-based results include those processed using either a stress function (Hybrid DIC) or commercial software. Each experimental analysis of Fig. 57.1 was conducted with the hole either unfilled or filled with a very compliant material. Figure 57.2 compares the vertical strain distribution in the neighborhood of a side notch in an orthotropic ($E_{11}/E_{22} = 3.7$) composite plate. The DIC-based results again include those processed using either a stress function (Hybrid DIC) or commercial software. In both of these cases, the two in-plane displacement components, and the three components of in-plane strain and stress were obtained full-field from only one of the two in-plane DIC-recorded displacement fields.

The cited results illustrate just some of the stress analysis advantages of combing DIC with a stress function. Compared with commercial software, results herein demonstrate significantly improved reliability in magnitudes of stress concentrations in components. Inherent with synergizing displacement information with a stress function is the strong, fundamental mechanic's base by which the measured data are processed. In addition to necessitating recording only one of the in-plane displacement fields, the approach smooths the measured data and provides individual stresses/strains full-field, including at edges of geometric discontinuities (evaluate stress concentration factors); thus making it possible for fatigue analyses and strength theories to be applied.

A. Alshaya (✉) • R.E. Rowlands
University of Wisconsin-Madison, Madison, WI 53711, USA
e-mail: alshaya@wisc.edu

W.A. Samad
Rochester Institute of Technology, Dubai, UAE

M. Sutton, P.L. Reu (eds.), *International Digital Imaging Correlation Society*, Conference Proceedings of the Society for Experimental Mechanics Series, DOI 10.1007/978-3-319-51439-0_57

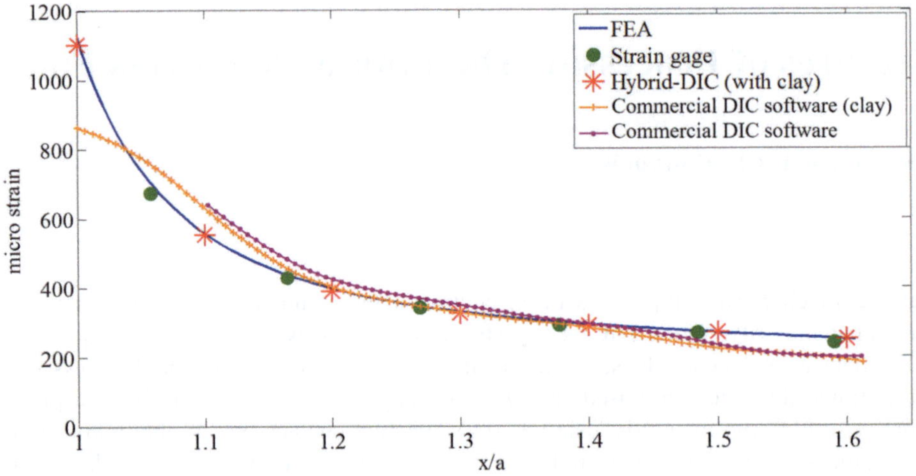

Fig. 57.1 Vertical strain ϵ_{yy} along horizontal axis approaching boundary of elliptical hole in aluminum plate from FEA, strip-gage, Hybrid-DIC and commercial software—with and without clay-filled hole

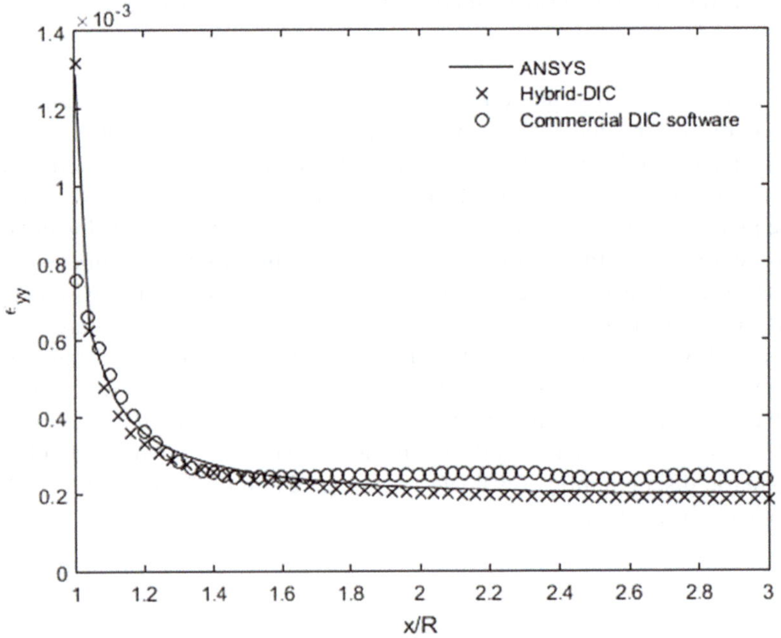

Fig. 57.2 Vertical strain ϵ_{yy} along horizontal axis approaching boundary of edge notch in orthotropic composite from ANSYS, hybrid-DIC and commercial software (Other)

Chapter 58
Characterization of Deformation Localization Mechanisms in Polymer Matrix Composites: A Digital Image Correlation Study

Jay Patel and Pedro Peralta

Abstract Polymer matrix composites (PMCs) are attractive structural materials due to their high strength to low weight ratios. However, due to their low shear strength, failure can occur due to kink bands that develop on compression via plastic microbuckling. This phenomenon has been modeled extensively; however, experimental measurements of the strain fields leading to and developing inside these bands are scarce, as the sample preparation for digital image correlation (DIC) requires a microscale speckle pattern, which is hard to obtain using inks as PMCs they absorb them. In this study, two dimensional DIC is used to measure strains inside kink bands during 3-point bending of Dyneema HB80 samples. A novel micro-scale speckle pattern along with precision optics have been used to measure strains within the kink band, which had a width of approximately 600 micrometers. The speckle pattern was created by dropping copper particles through fine mesh sifters on a thin adhesive film that was applied and cured on the side surface of beam samples. This led to speckle dots that have a minimum of 15–20 μm mean diameter along with an inter-spacing $\leq$10 μm through the kinkband width. Images were captured during sample loading and processedusing ARAMIS™ software. Measurements indicate a mixture of large shear and normal strains ($\sim$30%) in Dyneema HB80 inside the kink band.

Keywords Polymer matrix composites • Plastic microbuckling • Digital image correlation • Micro-scale speckle pattern • Kink band • Deformation localization

58.1 Introduction

Polymer Matrix Composites (PMCs) are attractive materials due to their high strength-weight ratio and controlled anisotropy. The use of PMCs in demanding applications in aerospace and defense industries makes it essential to understand their complex failure mechanisms under compression and bending. Many researchers [1, 2] have studied these failure mechanisms, including plastic microbuckling, and significant insight has been gained into PMC behavior. However, the same literature has not shed light on understanding kink band evolution under stress-gradients, such as those present during bending. This can be achieved, experimentally, by using high-resolution DIC to quantify the evolution of displacement and strain fields inside the kink band in the presence of stress-gradients. This investigation is focused on Dyneema HB80—Trademark of DSM Dyneema, NL. This is a cross-ply laminate [0/90] made of ultra-high molecular weight polyethylene (UHMWPE) fibers and polyurethane (PU) matrix. The traditional method of creating a speckle pattern by airbrush and water based inks does not work on these materials as they absorb them leading to smearing; hence, a new technique was developed to create a micro-scale speckle. Details on the procedures used during the experiments are described next.

J. Patel (✉) • P. Peralta
Mechanical and Aerospace Engineering, Arizona State University, Tempe, AZ, 85287-6106, USA
e-mail: jkpatel1@asu.edu

M. Sutton, P.L. Reu (eds.), *International Digital Imaging Correlation Society*, Conference Proceedings of the Society for Experimental Mechanics Series, DOI 10.1007/978-3-319-51439-0_58

Table 58.1 Sample dimensions (in mm) for 3 point bending tests (aspect ratio used followed [3])

Total beam length	203.2
Supported beam length	152.4
Beam width	12.38
Beam thickness	10.57

58.2 Methodology

58.2.1 Three Point Bending Test and DIC

Samples of Dyneema HB80 and Spectrashield were cut from plates using a water-jet to conduct 3-point bend tests to study the kink band characteristics. Sample dimensions are shown in Table 58.1, which result in a 1–20 thickness to span ratio, which is acceptable as per ASTM standards [4].

A micro-scale speckle pattern was created on these samples to match the length scale of the kink band, which was expected to be 600 μm [1, 2]. In this method, a thin adhesive film, 3M™ Adhesive Transfer Tape 9485PC, with 100 μm thickness was applied with on the side surface of beam samples. For better adherence, the film was cured for 24 h at room temperature. After curing the film, Cu particles were applied on top of the film by dropping them from a fine mesh sifter. As a result, the fine particles adhered to the film and formed a micro-scale speckle pattern size of dots that had a minimum of 15–20 μm mean size diameter along with an inter-spacing $\leq$10 μm through the kink band width.

Understanding the sequential events of kink band formation and evolution can be difficult, since the process can be sudden; hence, all tests were conducted under displacement control to insure stability. An appropriate ASTM standard [3] was followed as closely as possible. Tests were performed at room temperature using a computer-controlled, screw-driven, electro-mechanical load frame with a constant displacement rate of 2 mm/min. Samples were rested on knife-edge roller supports), a radius of 5 mm, as accepted by the standard [3] and load was applied through a roller of 5 mm radius [3] in the middle of the span. Load vs. displacement data were captured digitally during the tests.

58.2.2 Data Acquisition for DIC

A 5 megapixel monochrome CCD camera with a ½ inch CMOS sensor equipped with a Rodenstock™ 35 mm lens was used, along illumination from 1400 lumen fluorescent CFL light bulbs. The manipulation of the focal length and aperture of the camera-lens assembly was a key factor to resolve the speckle through the band width. An aperture stop of *f*/4 was found appropriate and when the system was in focus resulted in 15 × 15 mm field of view, large enough to capture enough speckle dots. The picture had 2560 pixels horizontally across the 15 mm, i.e., about 100 pixels across the expected kink band width of 600 μm. Figure 58.1a shows the resulting speckle pattern. The band was expected to develop on either side of the load point application, so the large field of view helped with this issue. However, a translation stage to move the camera along 3 axes was also built to allow studying kink band formation and propagation during loading. This entire assembly was placed in front of the load frame, and photographs of the sample were taken at different values of load during testing. At some instances, the test was paused to reposition the camera-lens assembly to follow the kink band. All these pictures were taken before the test was paused and after the load-displacement curve returned to its original path after the loading was resumed. The ARAMIS™ software used these pictures to calculate the strains within and around the kink bands.

58.3 Results and Discussion

Load-displacement curves were found to have load oscillations, which were associated to the formation of kink bands. Tests were carried out along with DIC to find the strains inside the kink band during its evolution up to two subsequent peaks (pointed out with two red circles) using the stages shown in Fig. 58.1b for Dyneema HB80. The drops in Fig. 58.1b correspond to the point of the pause of the test as the camera-lens assembly was moved to follow the propagating band. After re-starting the test, the data acquisition continued recording the corresponding loads and displacements. Images were captured at zero load and also as load increased, with particular emphasis at the first two subsequent peak loads to visualize the development of the kink band from the compression side of the beam.

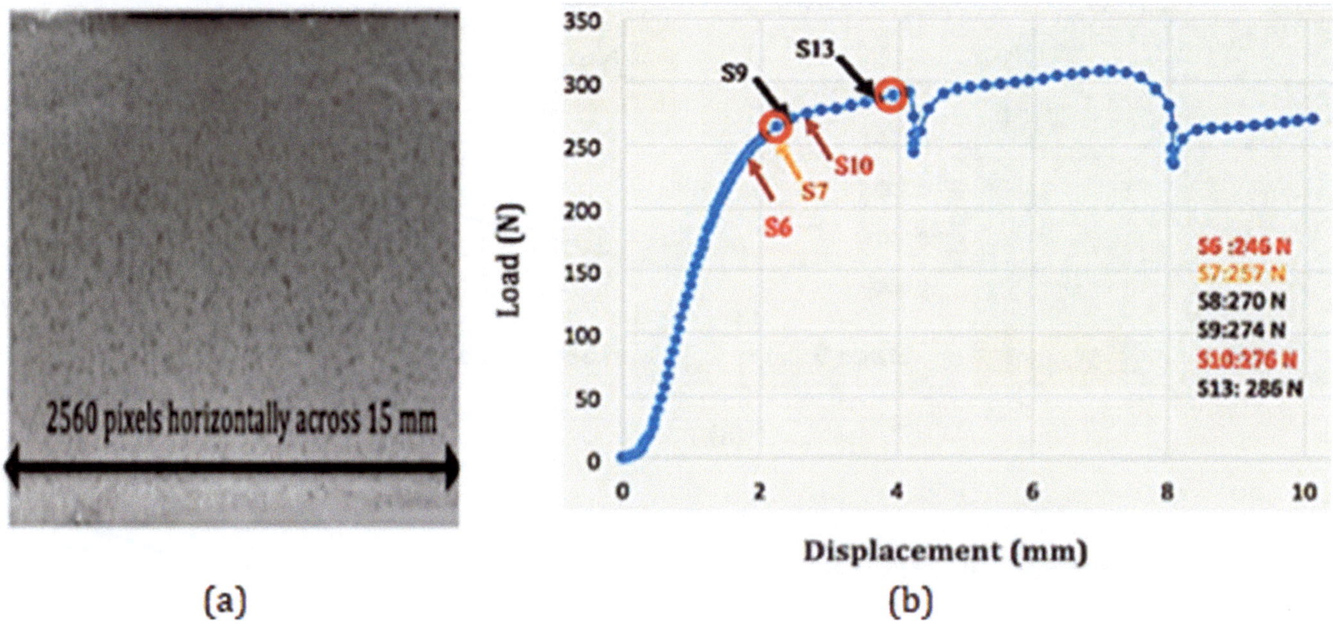

Fig. 58.1 (**a**) Speckle pattern used here (**b**) Load-displacement curve for Dyneema HB80. Different stages of the DIC measurement are marked with arrows and were done at the loads shown in the inset

58.3.1 DIC Results

The series of images captured during testing of the samples were processed using DIC software to determine the displacements and strains within and around the kink band. Variables of interest extracted from the DIC results include shear strains (%) inside the kink band and around it, as the band evolved. Different sections were created to determine the corresponding values of strains. The resolution of the DIC is based on the number of facets within the interested region. The resulting images had 100 pixels across the expected kink band width of 600 μm. In this case, at least 4 facets with one-facet, size of 21 × 21 pixels were used to solve for displacements and strains across the kink band. Figure 58.2a shows a kink band in a Dyneema HB80 sample. The coordinate system has been rotated to be parallel and perpendicular to the band to extract actual strains inside it. The shear strain for section 0 was approximately 8%. As the band propagated with further bending, the shear strain increased for the same section as shown in Fig. 58.2b.

The strain in section 0 of stage 12 increased by 300%. The strains shown here are the peak values within the band extracted from profiles in Fig. 58.2a. Note that, the shear strains in all stages increased as the band evolved with further beam displacement. This is because with further bending, the plies rotate (ply rotation angle dominates) cooperatively near the vicinity of the load point application. The behavior is similar to Dyneema HB26 as reported in [2]. The shear strains in these bands have been found to be as a function of ply rotation and kink band inclination angles [1, 2].

58.4 Conclusions

Oscillations on load-displacement curves indicate the presence of kink bands triggered by plastic microbuckling. The DIC results show that the bands nucleate at the compression side and propagate into the sample carrying a mixture of large shear and normal strains (~30%), while also decreasing its bending stiffness. The shear strains in all stages increased as the band evolved with further beam displacement. This is because with further bending, the plies rotate (ply rotation angle dominates) cooperatively near the vicinity of the load point application. Dyneema HB80 has a kink band width of approximately 560 μm.

Acknowledgments Support from BAE Systems, which provided composite samples for this research, and the Athletics Research Grant, operated by GPSA at Arizona State University (ASU), which provided partial support to carry out finite element simulations, is gratefully acknowledged. Moreover, JP offers sincere thanks to the School of Matter, Transport and Energy department at ASU for providing resources for research.

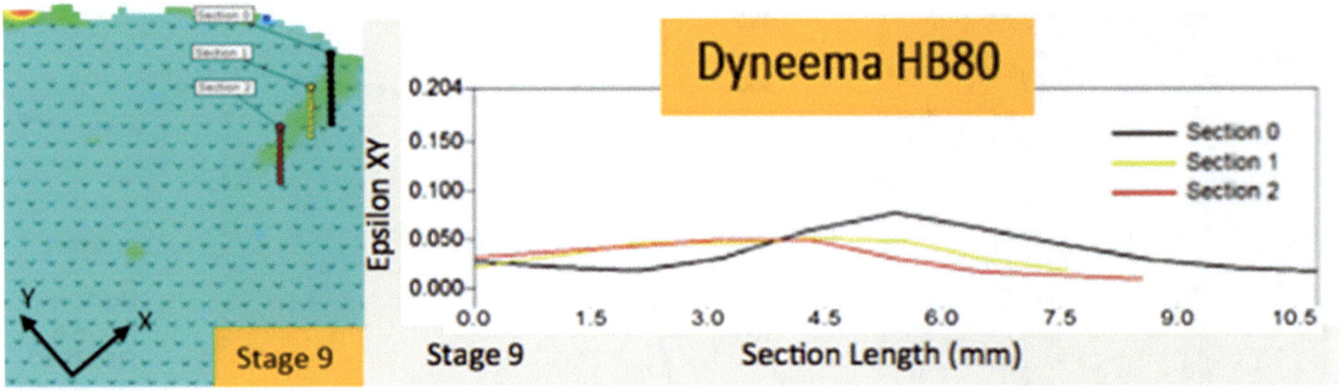

(a)

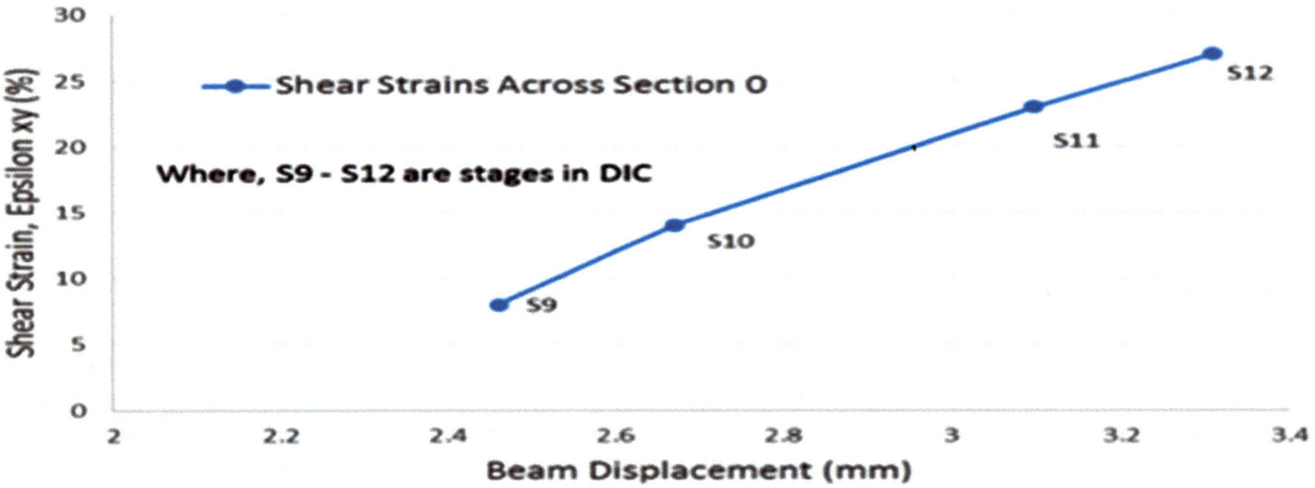

(b)

Fig. 58.2 (**a**) Kink band in rotated coordinate system along with stages and Shear strains across sections 0 (*black*), 1(*yellow*) and 2 (*maroon*) (**b**) shear strain (%) vs. beam displacement (mm) in DyneemaHB80

References

1. Budiansky, B., Fleck, N.A.: J. Mech. Phys. Sol. **41**(1), 183–211 (1993)
2. G Liu, MD Thouless, VS Deshpande, NA Fleck, Collapse of a composite beam made from ultra-high molecular-weight polyethylene fibres, Master's Thesis, Cambridge University, 2013
3. ASTM D7264/D7264M - 07: Standard Test Methods for Flexural Properties of Polymer Matrix Composite Materials

Chapter 59
Experimental Investigation on Macroscopic Fracture Behavior of Wood Plates Under Tensile Load Using Digital Image Correlation Method

Tzu-Yu Kuo, I-Feng Cheng, and Wei-Chung Wang

Abstract Wood is a kind of composite material in nature. Variations of latewood distribution and grain orientation are due to the climate and environment changes. The fracture behavior of the wood is difficult to observe and predict. In this paper, Japanese cedar (*Cryptomeria japonica*), a kind of conifers, was used to prepare the tensile test specimen to investigate the macroscopic fracture behavior under tensile load. During the tensile test, both the three-dimensional digital image correlation (3D-DIC) and two-dimensional digital image correlation (2D-DIC) methods were employed simultaneously. The 3D-DIC method was used to observe the specimen from the beginning of the tensile test to failure. The 2D-DIC method was employed to record the fracture mechanism by using a high-speed camera. According to the 3D-DIC results, higher longitudinal and radial strains occurred on the location of fewer latewood of the specimen. Based on the captured images of the fracture process obtained from the high-speed camera, both mode I and mode II fracture behaviors were observed. It is obvious that the mode I fracture occurred at the very beginning of the equivalent defect on the specimen. Then, the crack extended along the boundary of earlywood and latewood. Later on, the mixed mode I and II fracture occurred passing the latewood. Finally, the mode I fracture occurred on the crack passing the latewood.

Keywords Japanese cedar • Fracture behavior • Gray orientation • Digital image correlation • High-speed camera

59.1 Material and Experimental Procedure

In this study, Japanese cedar (*Cryptomeria japonica*) produced in Hsinchu County, northern Taiwan was used as the specimen material. By following the specification of CNS 450 O2001 [1], the dimensions of all specimens used in this paper are 180 mm × 25 mm × 3mm (longitudinal, radial, tangential). The tensile specimen was placed in a room with temperature controlled at 25°C and relative humidity was set at about 60% for more than one week. As shown in Fig. 59.1a, the front view and top view are the longitudinal-radial (LR) surface and the radial-tangential (RT) surface, respectively. The commercially available digital image correlation (DIC) software, VIC-2D and VIC-3D (Correlated Solutions, Inc., USA), were used to analyze the surface deformation of the LR surface of the test specimen under tension. Experimental apparatus of the tensile test is shown in Fig. 59.1b. The wood specimen was placed on the tensile fixture and tensile load was measured by a load cell. The LED light source (Skier Sunray 200 Cube) was placed on the right side of the two cameras. During the tensile test, three-dimensional digital image correlation (3D-DIC) images were captured from the VIC-3D system with one shutter per second. Once the specimen started to fracture, two-dimensional digital image correlation (2D-DIC) images were captured from a MEMRECAM NX-7 high speed camera system (nac Image Technology, Inc., Japan) with 6000 frames per second. A tensile test machine HT-2402 (Hung Ta Instrument Co., Ltd., Taiwan, R.O.C.) with a capacity 20 kN was used. The speed of the tensile test was set at 1 mm/min and force data were recorded every per second.

59.2 Results and Discussions

By using the 3D-DIC method, the strain distribution of the specimen during the tensile test can be obtained. Strain concentrations in both radial and longitudinal strain distributions can be observed in Figs. 59.2a, b respectively. However, the frame rate of the CCD cameras employed for general purposes in the 3D-DIC system are not fast enough to investigate the

T.-Y. Kuo (✉) • I.-F. Cheng • W.-C. Wang
Department of Power Mechanical Engineering, National Tsing Hua University, Hsinchu, Taiwan, Republic of China
e-mail: asmit.com@msa.hinet.net

M. Sutton, P.L. Reu (eds.), *International Digital Imaging Correlation Society*, Conference Proceedings of the Society for Experimental Mechanics Series, DOI 10.1007/978-3-319-51439-0_59

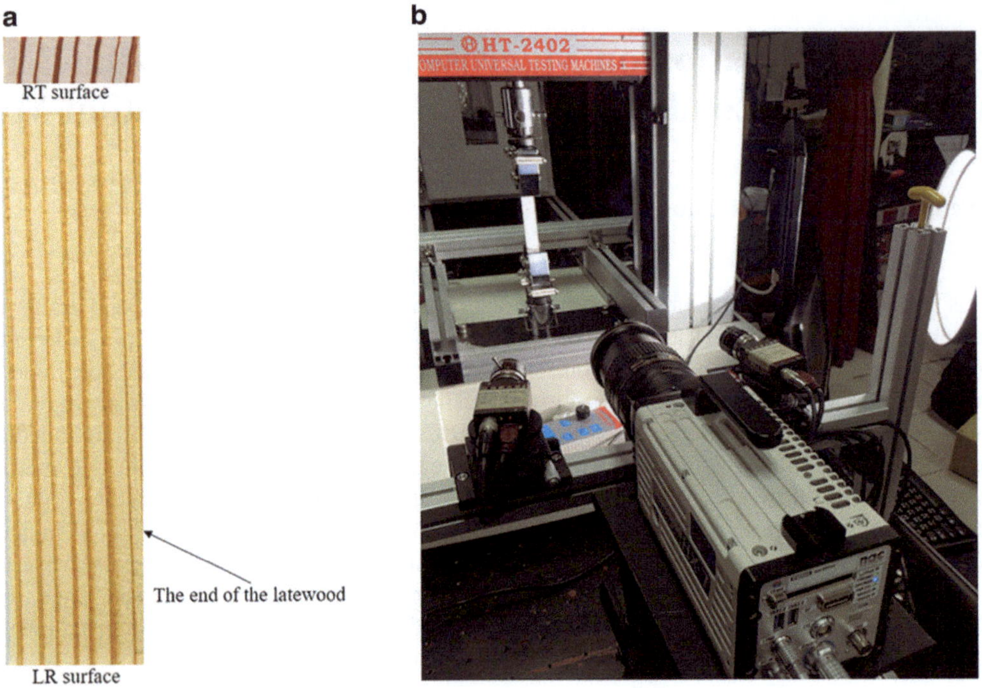

Fig. 59.1 Test specimen and experimental setup. (**a**) Test specimen, (**b**) Experimental apparatus of tensile test

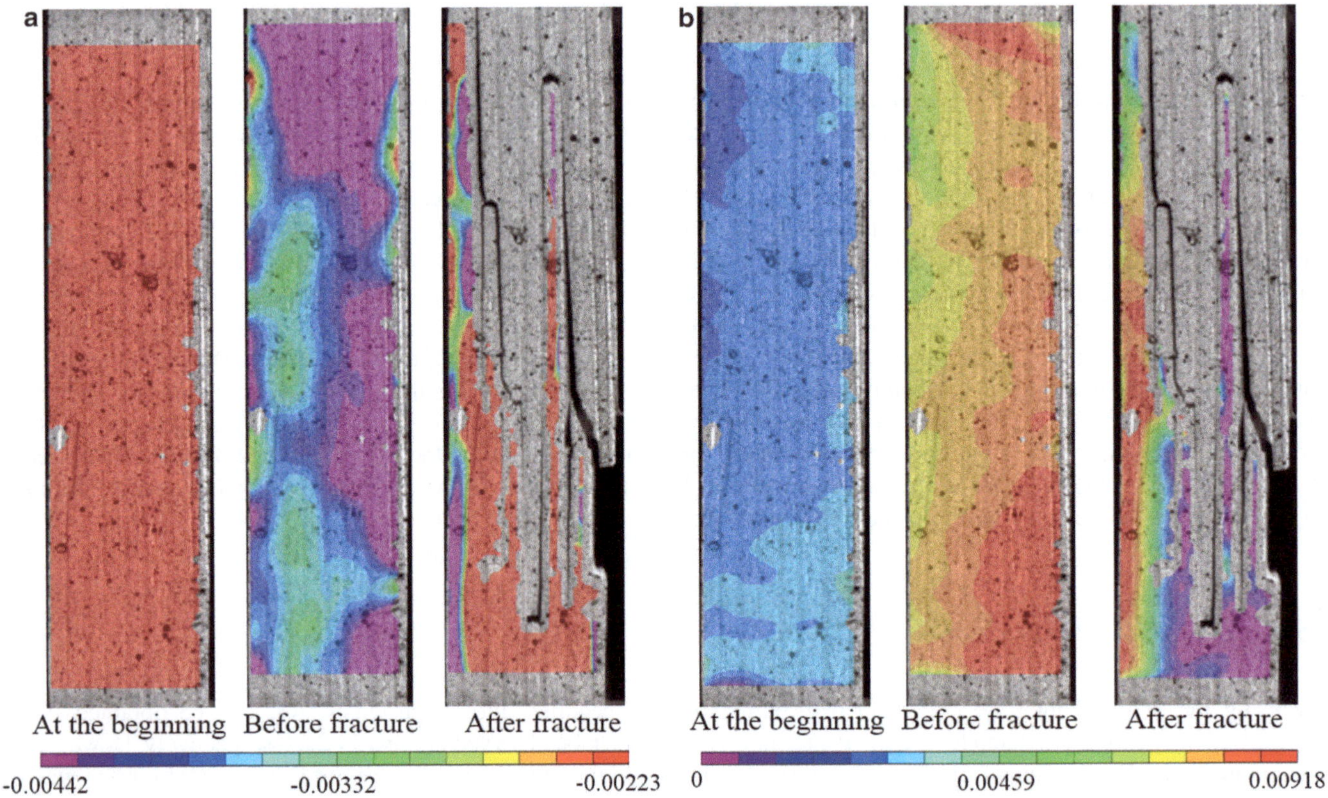

Fig. 59.2 3D-DIC results of surface strain in radial and longitudinal directions. (**a**) Radial direction, (**b**) Longitudinal direction

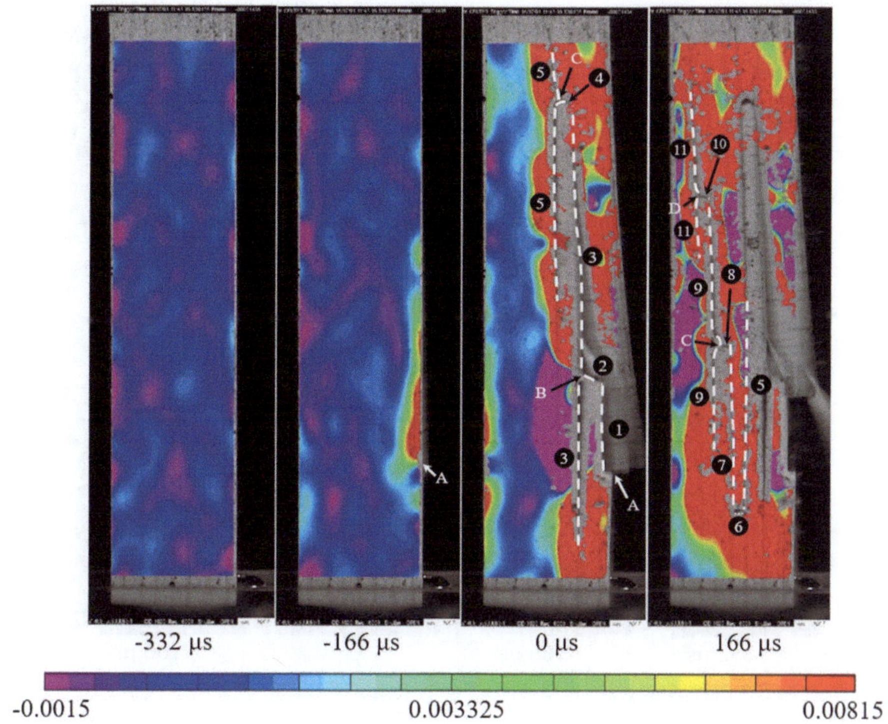

Fig. 59.3 2D-DIC results of surface strain in longitudinal directions

fracture process of the wood specimen. Therefore, the 2D-DIC method was employed to record the fracture mechanism by using a high-speed camera. By observing strain distributions obtained by 2D-DIC method at different time intervals shown in Figs. 59.3 and 59.4, the fracture mechanism of the wood specimen can be investigated. The time 0 μs was set at the instant when the high speed camera was triggered to record catastrophic failure. Numbers 1–11 marked in Figs. 59.3 and 59.4 denote the development of the crack propagation. Firstly, as depicted in Fig. 59.3 captured at -332 μs, the region of the discontinuous longitudinal strain was because of the formation of micro cracks of the wood fiber accompanied with sharp and loud sounds before the catastrophic failure. Later on, the longitudinal strain concentrated at point A in Fig. 59.3 captured at −166 μs indicates that mode I fracture of the specimen started there. This is not unexpected. As shown in Fig. 59.1a, the latewood ends at point A. It is obvious that the wood specimen started to fracture from the vicinity of point A. After the mode I fracture, the crack extended along the boundary between latewood and earlywood until crossing the earlywood to reach the bifurcation point B as shown in Fig. 59.3. The crack was then split by shearing from point B and propagated sequentially to the next two bifurcation points C and D [2]. In comparison with the longitudinal strain distribution shown in Fig. 59.3, as shown in Fig. 59.4, shear strain concentration also appears in the vicinity of point A before the specimen fails. Besides, the large shear strain near the crack indicates that mode II fracture dominates the debonding process between latewood and earlywood [3]. Finally, since most of the latewood and earlywood fractured on the right part of the specimen, the latewood on the left is under large tensile strain and fractured in mode I.

59.3 Conclusions

The DIC method is an efficient technique to investigate the full-field strain distribution. By using general purpose CCD cameras, the strain concentration in the wood specimen can be observed during the tensile test. Moreover, fracture behavior investigation is possible by using the high speed camera to capture images during the fracture process. According to the strain measurement results, there are high correlations between the fracture behavior of the wood specimen and the latewood distribution. Therefore, by using the high speed camera in the DIC method, the macroscopic fracture behavior of the wood specimen can be readily analyzed.

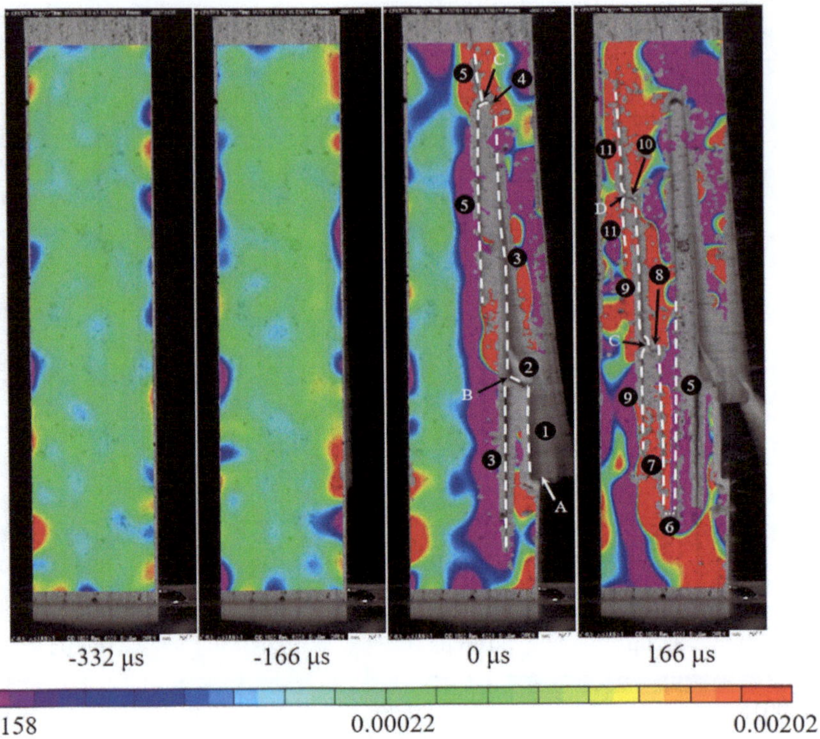

Fig. 59.4 2D-DIC results of surface shear strain

References

1. CNS 450 O2001, Wood-sampling Methods and General Requirements for Physical and Mechanical Tests, 2013
2. Bodig, J., Jayne, B.A.: Mechanics of Wood and Wood Composites. Krieger Publishing, USA (1993)
3. Anderson, T.L.: Fracture Mechanics: Fundamentals and Applications, 3rd edn. CRC Press, USA (2005)

Chapter 60
DIC Data-Driven Methods Improving Confidence in Material Qualification of Composites

Guillaume Seon, Andrew Makeev, Yuri Nikishkov, and Brian Shonkwiler

Abstract This work presents initial results of research activities focused on utilizing the Digital Image Correlation technique for measurement of the complex deformation of composite materials towards the development of consolidated common material qualification processes. These common processes will enable reduced material qualification test matrices potentially accommodating substitution of resin types and other modifications of the material systems. In particular, data-driven methods based on full-field DIC strain measurements and finite element analysis developed at the University of Texas Arlington Advanced Materials and Structures Lab, are verified on carbon-fiber reinforced/untoughened and toughened polymer-matrix composite material systems. The DIC data-driven methods allowed verification of simplifying assumptions, increasing our confidence in material allowables. The methods also allowed identifying standing challenges in lamina characterization, including inconsistencies of certain ASTM standard test fixtures and analysis methods, preventing accurate measurement of lamina characteristics key to laminate analysis.

Keywords Digital image correlation • Data-driven modeling • Inverse method • Polymer-matrix composites • Mechanical properties

60.1 Extended Abstract

Fiber-reinforced composite materials are increasingly used in aircraft structures to reduce weight and improve efficiency. A major challenge, delaying the implementation of recently developed higher-performance materials that offer improved mechanical strength and stiffness at a lower weight, is the lack of efficient common material qualification. The current standard practices are too costly and time-consuming. Due to low confidence in material allowables, the entire material qualification process, including numerous test methods and large test matrices, must be repeated for every seemingly minute change in the composite system.

This work presents initial results of research activities focused on utilizing the Digital Image Correlation (DIC) technique for measurement of the complex deformation of composite materials towards the development of consolidated common material qualification processes. These common processes will enable reduced material qualification test matrices potentially accommodating substitution of resin types and other modifications of the material systems.

Over the past five years at the University of Texas Arlington Advanced Materials and Structures Lab (AMSL), we have developed data-driven methodologies that take advantage of the full-field strain measurement capability of DIC to enable efficient and accurate measurement of multiple lamina material properties in a single experiment. In particular, we demonstrated that all key lamina properties in a composite material system could be measured based only on two test methods, illustrated on Figs. 60.1 and 60.2.

Figure 60.1, shows a test setup, test specimen configurations and example of DIC-based strain measurements for the first method utilizing a simple short-beam shear (SBS) specimen. The SBS method allows assessment of tensile and compressive modulus, Poisson's ratio, nonlinear shear stress-strain curve and shear strength in the plane of loading using only one experiment [1]. Figure 60.2 shows a test setup with three synchronized stereo camera systems for our second test method utilizing a small plate-twist (SPT) rectangular specimen. The SPT method enables simultaneous assessment of non-linear shear stress-strain relations in all three principal material planes, including 2–3 interlaminar behavior [2, 3].

It is worth noting that due to early tensile failure of 90°-transverse and interlaminar SBS specimen, the SBS method does not allow for measuring nonlinear interlaminar shear material properties in the 2–3 material plane. Accurate interlaminar

G. Seon (✉) • A. Makeev • Y. Nikishkov • B. Shonkwiler
Department of Mechanical and Aerospace Engineering, University of Texas at Arlington, 500 West First Street, Arlington, TX 76019, USA
e-mail: seon@uta.edu

M. Sutton, P.L. Reu (eds.), *International Digital Imaging Correlation Society*, Conference Proceedings of the Society for Experimental Mechanics Series, DOI 10.1007/978-3-319-51439-0_60

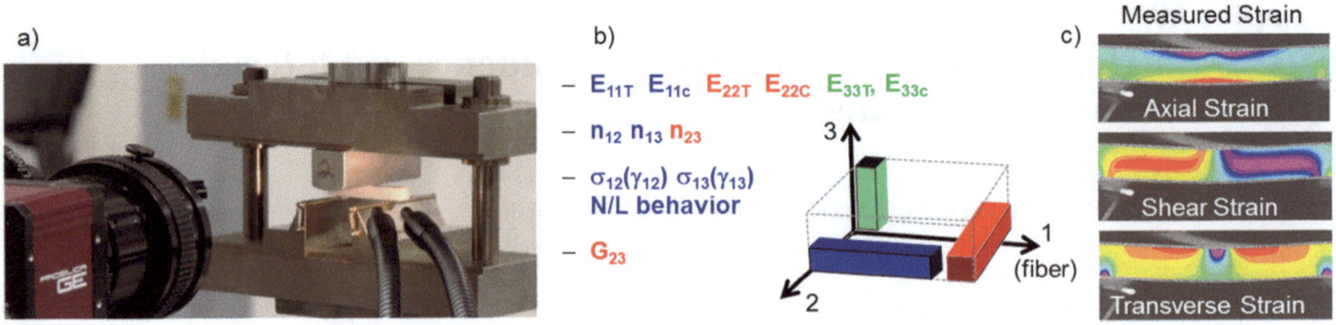

Fig. 60.1 A short-beam shear (SBS) method to measure multiple stress-strain properties of composites in a single experiment. (**a**) test setup, (**b**) specimen configurations, and (**c**) measured strains (DIC)

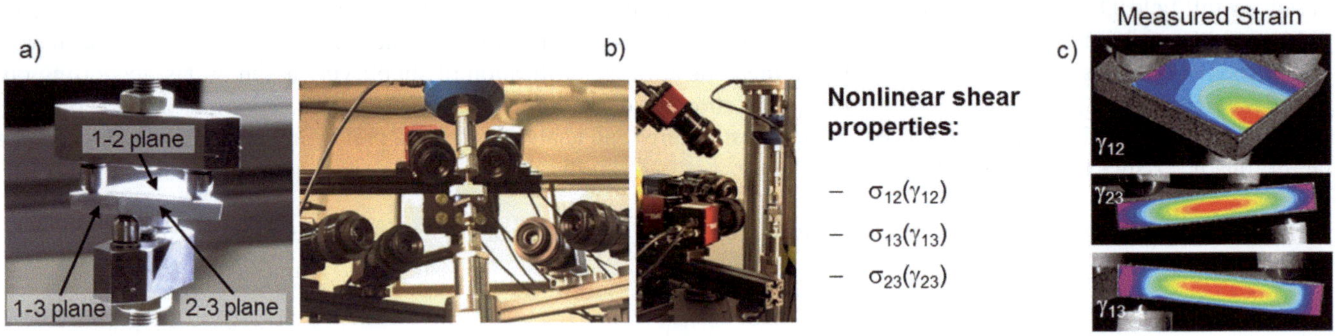

Fig. 60.2 A small plate-twist (SPT) method to simultaneously measure shear stress-strain properties in all three principal material planes. (**a**) SPT test setup, (**b**) three synchronized stereo camera systems for DIC based strain measurement, and (**c**) DIC-measured shear strains

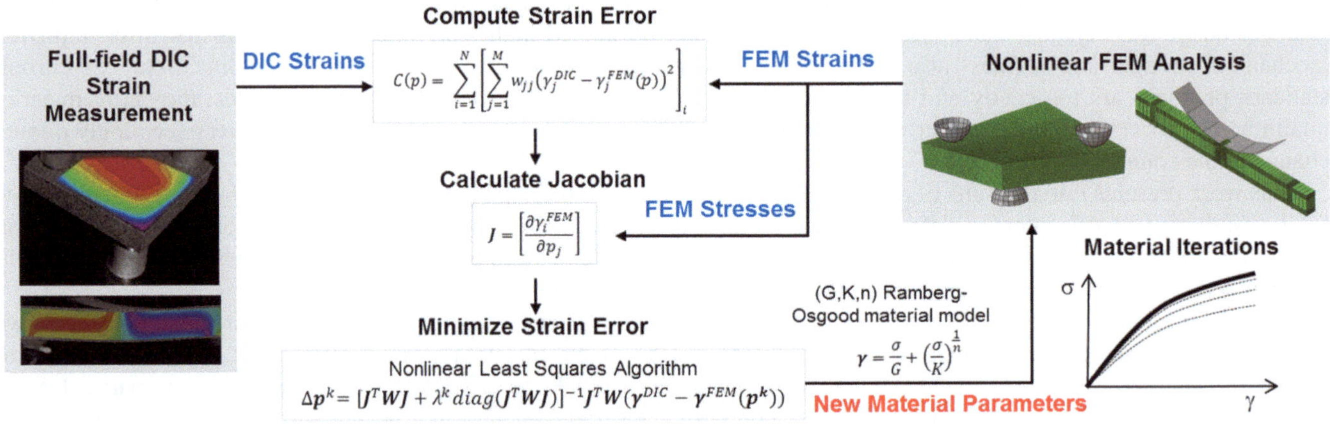

Fig. 60.3 Iterative FE model updating procedure used in the DIC data-driven methods

constitutive and strength properties in both 1–3 and 2–3 material planes might be essential for enabling analysis accurately predicting initiation and progression of delamination failure as well as its interaction with matrix cracking in multi-directional composite laminates. To this date and to the best of the authors' knowledge, the SPT method is the only test method allowing full accurate characterization of the ILS stress-strain behavior [2].

Both the SBS and SPT methods use elaborate experimental setups and advanced measurement techniques as well as data-driven analysis to verify simplifying assumptions commonly used for composites and to increase confidence in allowables. The data-driven analysis combines full-field DIC strain measurement and iterative finite element (FE) model updating for assessment of material parameters, following a procedure illustrated in Fig. 60.3.

Data-driven analysis in the SBS method allows correct characterization of the shear non-linear behavior of typical composite material systems used in the aerospace industry that exhibit significant shear non-linearity at high strain levels. Data-driven modeling is also required in the SPT method for assessing the shear stress-strain curves in 3D, since there is no closed-form solution for evaluation of surface stresses due to complexity of deformation exhibited by the SPT specimens.

The SBS and SPT analysis methods have been previously verified only on a Hexcel IM7 carbon-fiber/8552 toughened-epoxy matrix 350°F curing composite material system. Due to well-recognized susceptibility of polymeric composites to matrix-dominated failures, a more comprehensive verification of the analysis methods using a broad range of the matrix types is required for improving confidence in the analysis methods.

In this work, the SBS and SPT data-driven methodologies are verified on carbon-fiber reinforced/untoughened and toughened polymer-matrix composite material systems. The DIC data-driven methods allowed verification of simplifying assumptions, increasing our confidence in material allowables. The methods also allowed identifying standing challenges in lamina characterization, including inconsistencies of certain ASTM standard test fixtures and analysis methods, preventing accurate measurement of lamina characteristics key to laminate analysis.

References

1. Makeev, A., He, Y., Carpentier, P., Shonkwiler, B.: A method for measurement of multiple constitutive properties for composite materials. Compos. Part A Appl. Sci. Manuf. **43**(12), 2199–2210 (2012)
2. Makeev, A., Seon, G., Cline, J., Shonkwiler, B.: In quest of methods for measuring 3D mechanical properties of composites. Compos. Sci. Technol. **100**, 105–112 (2014)
3. Seon, G., Makeev, A., Cline, J., Shonkwiler, B.: Assessing 3D shear stress–strain properties of composites using Digital Image Correlation and finite element analysis based optimization. Compos. Sci. Technol. **117**, 371–378 (2015)

Chapter 61
DIC Applications Highlights from China

Darren Yang, Roger Lou, Belinda Chen, and Tim Schmidt

Abstract China has become a significant market for commercial digital image correlation systems, and usage is continuing to accelerate. The authors have been providing training and application support for more than 100 installations across the country. This paper reviews some of the most widely seen applications, as well as a few special cases.

China produces more cars annually than any other country, so it is natural that there are many applications in the automotive industry, as well as related areas such as steel production. DIC is used to characterize new alloys and to determine forming limit curves for sheet metal. Nearly every steel manufacturer in China is equipped with a commercial DIC system. In the aerospace industry, DIC is used for various testing requirements, from basic static materials mechanical tests to determine material behavior; high rate strains; from component testing (i.e. aircraft panel test) to entire airplane strength tests. In the civil engineering industry, we have done DIC tests such as concrete tensile and compression tests, concrete wall shear tests, large concrete platform with different steel fibers bending tests, and whole building strain tests under earthquake simulation. And in the biomechanics industry, we use DIC to study implants mechanics features after they are implanted into the human body and grow together with human bones. Last but not least, in order to assure the accuracy of DIC, we have done quite a few comparison experiments between DIC and extensometers or strain gauges, and the results are quite satisfactory.

Keywords Digital image correlation (DIC) • Material properties • Validation • FEA • Aircraft

This paper shows a closer look at several of the applications that will be presented. Figure 61.1 shows a typical application of DIC in materials parameters determination testing through tensile testing. Another typical material testing requirement that we see very often with our customers is the forming limit curve (FLC) test. The ARAMIS DIC system carries out Nakajima tests in accordance with the ISO 12004 standard to calculate FLC curves. It's quite convenient to do that with ARAMIS since it has a specific mode for FLC calculation. In China, almost every steel manufacturer uses DIC to do FLC calculation. With the GOM ARAMIS system, it's very fast and simple to obtain all primary mechanical properties of the material. The experiment we have done for Wuhan Iron and Steel as shown in Fig. 61.2 helps to verify the precision and repetition of ARAMIS system, as well as the ease of use for material property measurements.

The ARAMIS DIC system can be synchronized with testing machines, and the controller of ARAMIS system can capture force from the testing machine simultaneously, thus stress-strain curves can be calculated. The two graphs show stress-strain results of different specimens, with linear fits in the elastic range automatically computed and overlaid. As we can see, they show very good repetition of ARAMIS. The Young's modulus, Poisson Ratio, N value and R value can also be obtained from the integrated tensile test evaluation script.

Since so far, there is no international or national standard for DIC testing, although DIC is already widely known, we often meet demand for comparison testing with extensometers, from customers and potential users who are interested in DIC, yet also skeptical about its accuracy.

Therefore, we have done quite a few tests, and the results from these two methods match each other quite well. And the advantage of ARAMIS over an extensometer is that it continues to measure after the extensometer releases, until the specimen breaks. An example of this is shown in Fig. 61.3.

On Nov 2nd, 2015, China's first domestically produced large passenger aircraft, the C919, officially rolled off the final assembly line in Shanghai. DOM is very honored to be involved in part of its static tests, including torsion tests of fuselage

D. Yang • R. Lou • B. Chen
DOM 3D, Unit 318, Tower 1, German Center, 88 Keyuan Road, Shanghai 201203, China

T. Schmidt (✉)
Trilion Quality Systems, 500 Davis Drive, Suite 200, Plymouth Meeting, PA 19462, USA
e-mail: schmidt@trilion.com

M. Sutton, P.L. Reu (eds.), *International Digital Imaging Correlation Society*, Conference Proceedings of the Society for Experimental Mechanics Series, DOI 10.1007/978-3-319-51439-0_61

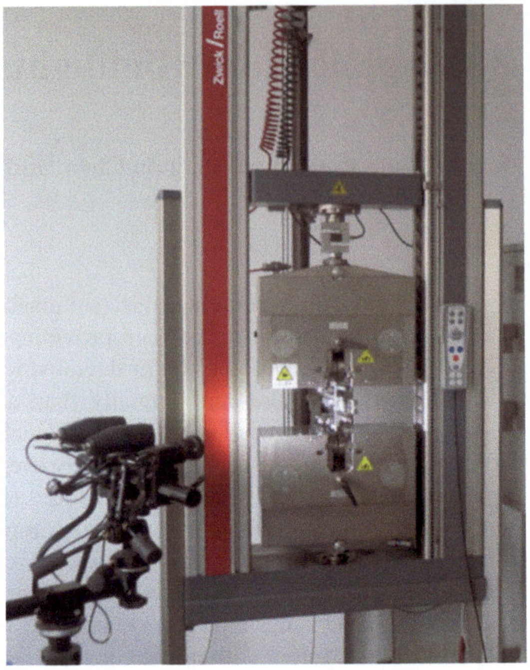

Fig. 61.1 The ARAMIS 3D DIC system measuring full-field strains off-axis at a load frame, while also capturing force data used to generate stress-strain curves and determine material properties

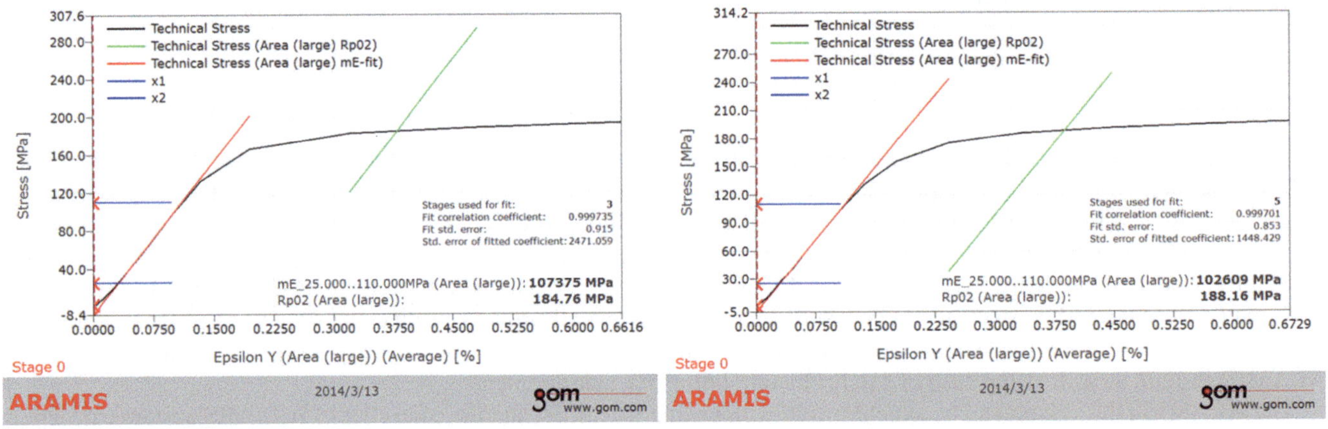

Fig. 61.2 The ARAMIS DIC tensile test evaluation script automatically plots and analyzes linear fits within the specified elastic range, for material property determination

panels. The purpose of the tests is to measure 3D full field displacement and strain of the panel over the whole process of torsion loading, until ultimate failure. COMAC is interested to know when buckling happens, how deformation develops before buckling, and what are the relatively weak areas in the panel structure (Fig. 61.4).

In the past, COMAC always used strain gauges to do torsion tests. They have had three primary difficulties with this. First, it was difficult to specify the areas where the largest strains take place, for gauge placement. They developed an FEA model to predict these areas, but this is not completely dependable, nor complete.

The second issue is that the gauges measurement could only get correct data for the elastic deformation period. When the panel starts to buckle, strain gauges fail. Therefore, they could only get testing data up to 55% of the design load with strain gauges, because once the panel was loaded to 60% of the rated load, the specimen began to buckle.

The third major limitation with strain gauges is that they couldn't obtain adequate testing data, therefore they were not really able to verify the accuracy of their FEA model. However, the ARAMIS DIC obtains data at thousands of points, and we could import FEA data into ARAMIS software, and compare it with real testing data, so as to improve accuracy of FEA model. An example of this from a composite component is shown in Fig. 61.5.

Experimentally, stochastic patterns were sprayed on the specimen surface, and point markers were also used, in order to indicate the rib reinforcement locations at the back side of the specimen. The measuring area was about 1140 × 1000 mm, and stepped loading was applied in 5% increments up to 70% of the rated load, and was held for 10 s. After 70%, continuous

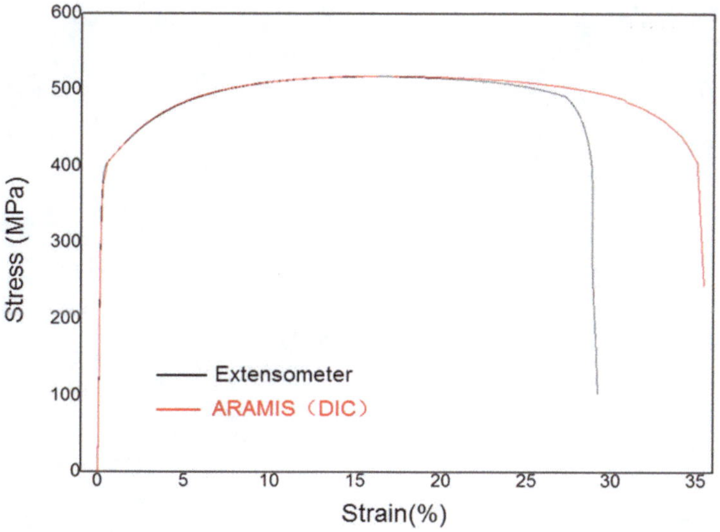

Fig. 61.3 Comparison of DIC data to extensometer, showing excellent matching

Fig. 61.4 On Nov 2nd, 2015, China's first domestically produced large passenger aircraft, the C919, officially rolled off the final assembly line in Shanghai

loading was applied until the specimen broke. Thus pictures were taken at specific time intervals from 0 to 70% of rated loading. After 70%, pictures were taken continuously. Figure 61.6 shows resultant displacements at increasing load. COMAC said that the results matched their FEA data, indicating that their model was accurate.

At the beginning of 2017, COMAC will do complete aircraft static testing. DOM is looking forward to be involved in the major, critical wing tests.

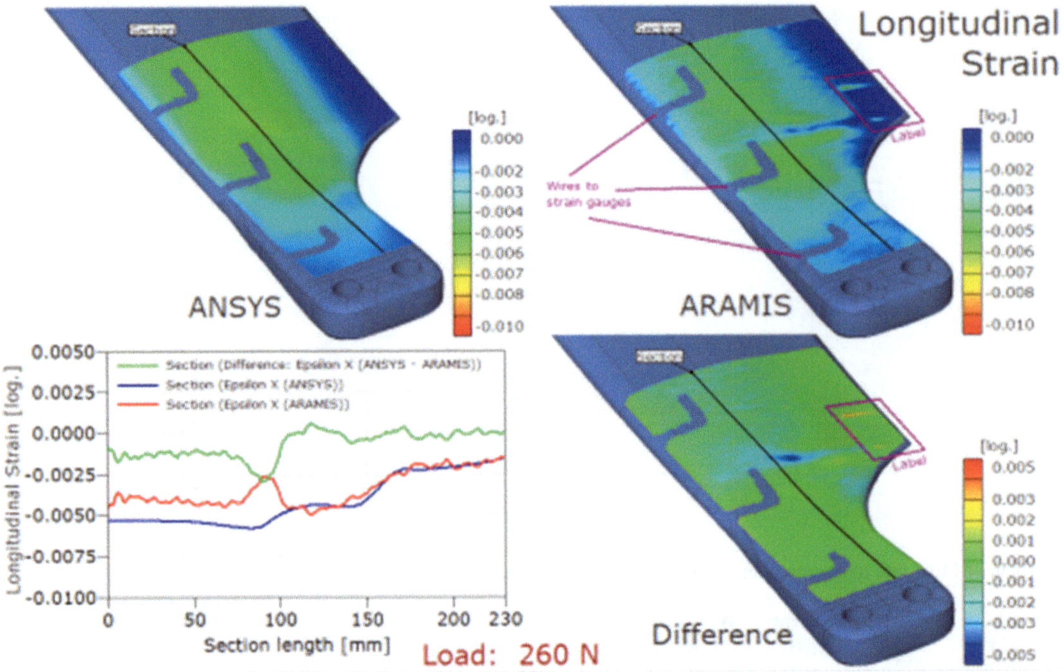

Fig. 61.5 Direct import of FEA into the ARAMIS DIC software enables significantly better FEA validation and improvements compared to sparse strain gauges

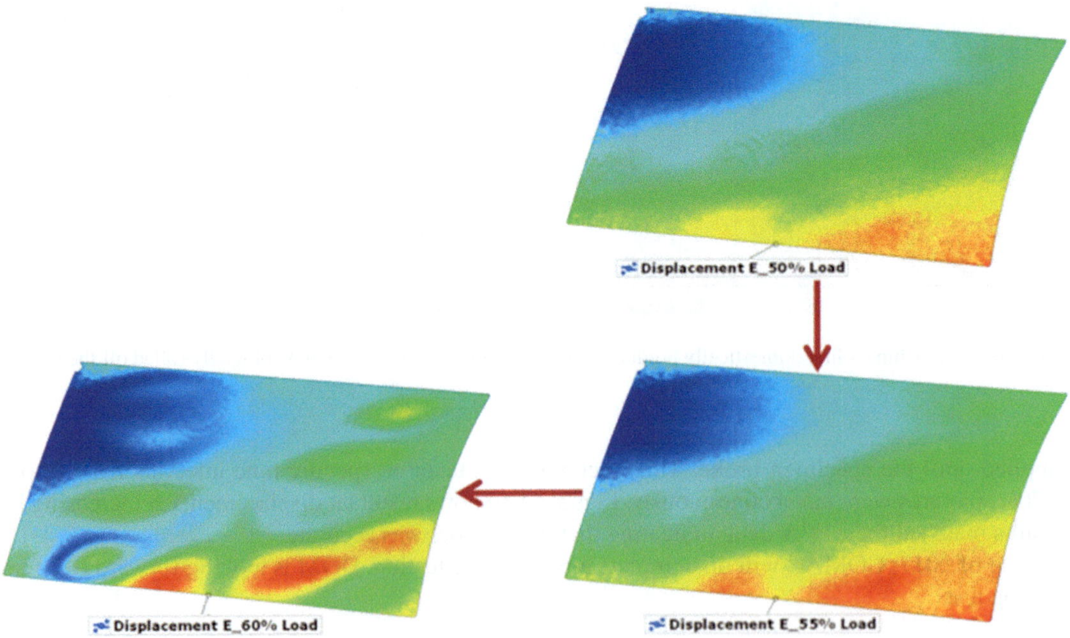

Fig. 61.6 Resultant displacement of a fuselage panel section at increasing load increments

Chapter 62
High-Speed DIC on Inside Perma-Gel During Ballistic Peneration

Matthew Grimm, Rory Bigger, and Christopher Freitas

Abstract Perma-Gel® is a melt-castable, reusable synthetic ballistic gelatin used as a soft tissue simulant in research, forensics, and shooting sports. It can also be used to study the temporary wound cavity that forms during a penetrating injury from a ballistic projectile such as a small arms bullet or a blast fragment. This cavity is typically imaged with high-speed video cameras and its measured manually in each video frame. The dynamic cavity size alone, however, does not directly provide data on the strain that material nearby, but outside of, the wound cavity is subject to. In a recent project, SwRI® determined that it would be feasible to embed a dot pattern inside of the Perma-Gel block and use DIC to directly measure the strain field surrounding the wound cavity. Successfully and reliably embedding a dot pattern on a flat plane in the middle of the gel block proved to be challenging, in part because casting on top of a set layer with the pattern tended to melt the bottom layer and distort the pattern. After some modifications, a successful technique was developed. The method developed and example applications of it and the resulting DIC strain fields measured from a variety of projectiles and impact speeds are presented. The data demonstrate that DIC is an effective technique for measuring dynamic response of ballistic penetration in Perma-Gel, and can provide new insights that may be difficult or impossible to achieve with other measurement techniques.

Keywords Ballistics • Perma-gel • High-rate • Wounds • Shear

62.1 Introduction

Engineers, scientists, and hobbyists have long used ballistic gelatin investigate and compare wound cavities caused by projectiles. Ballistic gelatin is a natural gelatin calibrated to correlate to swine muscle tissue [1]. Since it is a natural product, it has a very limited shelf life: essentially, it must be prepared, shot, and studied within a period of a few hours before it must be discarded. Recently, several shelf-stable alternatives to ballistic gelatin have come onto the market including Perma-Gel®, a melt-castable, reuseable synthetic gelatin (or "gel"). Southwest Research Institute® (SwRI®) has been using Perma-Gel on a number of recent projects to study both temporary (dynamic) and permanent wound cavities. For one of these projects, DIC was used to directly measure the strain field surrounding the wound cavity. A dot-pattern was embedded on a plane in the middle of the block. The shot-line of the projectile was lined up with this plane. DIC could then be performed on the recorded digital images of the deforming plane. The data is used to support validation of a finite-element model of the gel. In the past, measurements have been made with a few pressure sensors embedded into the gel, but, with DIC, full-field measurements can now be used to fully validate the model results.

62.2 Block Preparation

The gel block was cast in a 300 × 150 × 150 mm transparent polycarbonate box with one open side. The general preparation process is shown in Fig. 62.1. First, the box was filled half-way with molten gel (1). After cooling, a dot pattern was applied with a permanent marker or black spray paint and a stencil (2). The box was then filled the rest of the way with molten gel (3). After cooling, the bottom of the box was removed and the block and box could be rotated and mounted for testing (4).

M. Grimm • R. Bigger (✉) • C. Freitas
Southwest Research Institute (SwRI), 6220 Culebra Road, San Antonio, TX 78238, USA
e-mail: rory.bigger@swri.org

M. Sutton, P.L. Reu (eds.), *International Digital Imaging Correlation Society*, Conference Proceedings of the Society for Experimental Mechanics Series, DOI 10.1007/978-3-319-51439-0_62

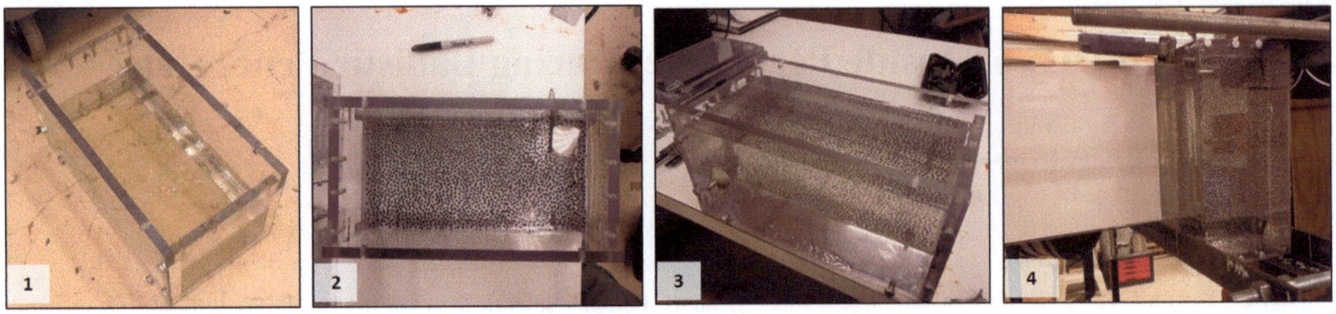

Fig. 62.1 Process of embedding dot pattern inside of Perma-Gel block

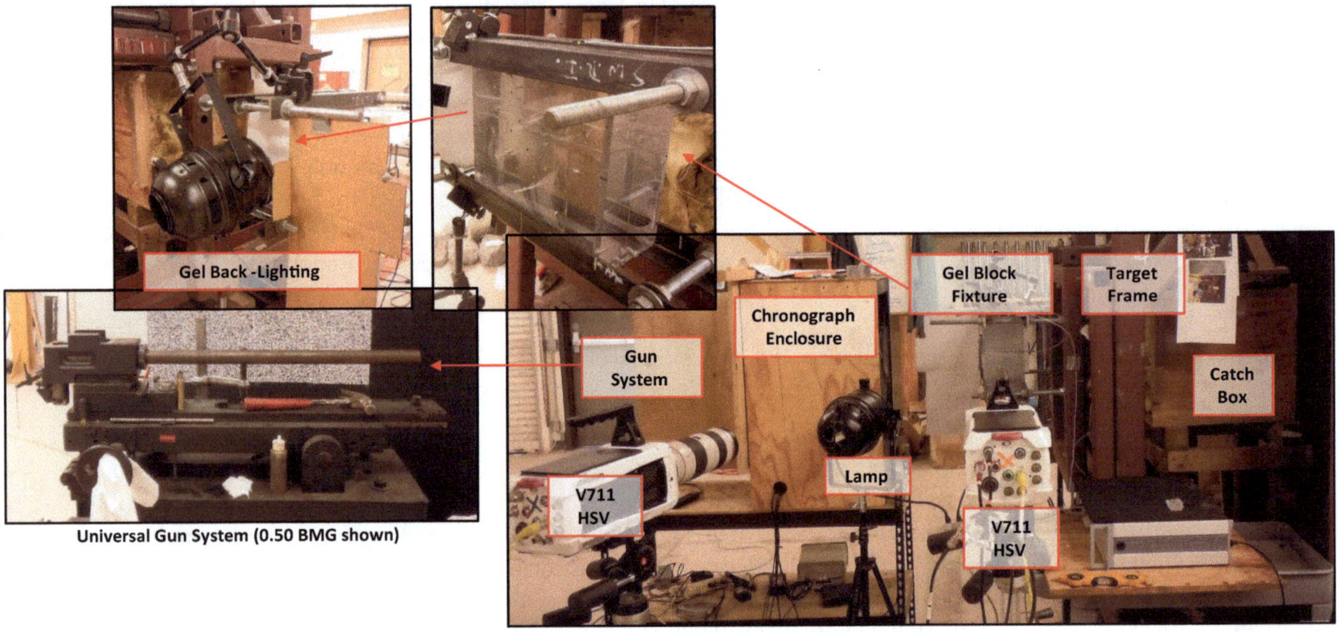

Fig. 62.2 Ballistic range configuration

Several initial attempts were made at embedding a planar dot pattern that produced poor results. The first attempt was made using glass beads for a pattern. During deformation, the gel separated from the beads, and the DIC software could not correlate to the original image. In a second attempt, a dot pattern was applied with gel and then molten gel was poured on top and left to cool at room temperature. This did not work well, because out-of-plane air bubbles in the poured top layer interfered with imaging of the pattern. Finally, good data was achieved by letting the block outgas in an autoclave at elevated temperature after filling halfway (between steps 1 and 2 in Fig. 62.1). This removed air bubbles between the wall of the box and the dot pattern, allowing for clean imaging of the pattern and full-field strain measurements with DIC.

62.3 Test Setup

The ballistic range setup is shown in Fig. 62.2. A variety of aerodynamic and non-aerodynamic projectiles were fired from a universal gun system. With a universal gun, the barrel and breech are one unit that may be interchanged to fire different caliber cartridges using a common receiver and firing system. The gun was fired via lanyard from an adjacent room for safety. The projectiles travel through two infrared light screens connected to a universal counter to form a chronograph for velocity measurement. Cartridges were hand-loaded with different powder masses to study the wound dynamics at different impact speeds. After passing through the chronographs the projectiles impact the gel block. A catch box was placed behind the gel block to stop any projectiles were not stopped by the gel. The gel block itself is mounted in a fixture attached to a 2-axis target frame that translates vertically and horizontally to align the dot pattern for DIC with the shot line of the gun.

The gel was backlit with a single 1000W PAR64 lamp. Since this lamp gets hot and the gel melts, it was turned on remotely just before the gun was fired. A single Phantom v711 high-speed camera was used to collect data at nominally 50,000 Hz for 2D DIC measurements. The single-lamp backlight prevented glare from the polycarbonate and was sufficient to use ~5 μs exposures needed to prevent motion blur. Another v711 was used to document the impact, but not for DIC. The cameras are triggered using a custom trigger signal generation circuit that pulses when the projectile passes through the second light screen.

62.4 Example Result

Contours of Von-Mises strain from DIC are shown in the top row of Fig. 62.3 for an impact with a 64 grain right circular cylinder (RCC). Overall DIC data quality is good. Strain is measurable up to the edges of the wound cavity. As the would cavity expands, strain magnitude increases. Elevated strain can also be observed at significant distance away from the wound cavity. This indicates potential damage to tissue from the temporary wound that may be 10–100 mm away from the path of the projectile and not clearly observable in tests where high-speed video imaging of the wound cavity alone is used. Using data gathered from tensile and compressive tests of Perma-Gel samples, a viscoplastic material model was developed. This model was implemented in the LS-DYNA finite element simulation software to model the ballistic tests. The simulation results for the 64 grain RCC are shown in the bottom row of Fig. 62.3. Overall, there is good agreement between the model and the DIC data from the tests. The DIC data provided a unique full-field measurement validation metric that would have otherwise been limited to a few pressure and/or strain gauges embedded in the block at fixed location.

Strain-time histories at specific points of interest are easily extracted from the DIC data. Fig. 62.4 shows such a history for the "P2-2" location (lower right callout in the DIC images in Fig. 62.3). The strain for different projectiles and impact speeds can be compared this way. In the case of Fig. 62.4, a similar response is seen between an aerodynamic projectile and an RCC at much lower velocity. The higher-velocity RCC causes significantly higher strains with a longer duration.

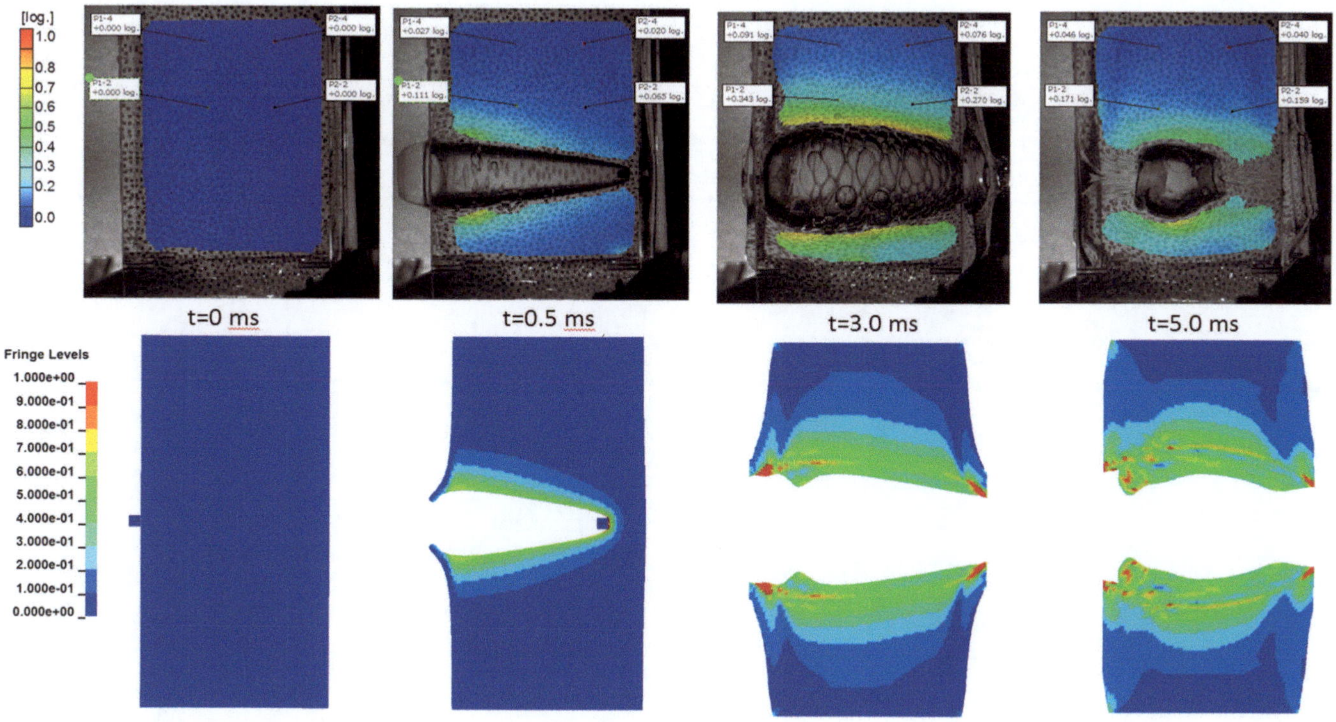

Fig. 62.3 DIC results compared to finite-element simulation for 64 grain RCC projectile

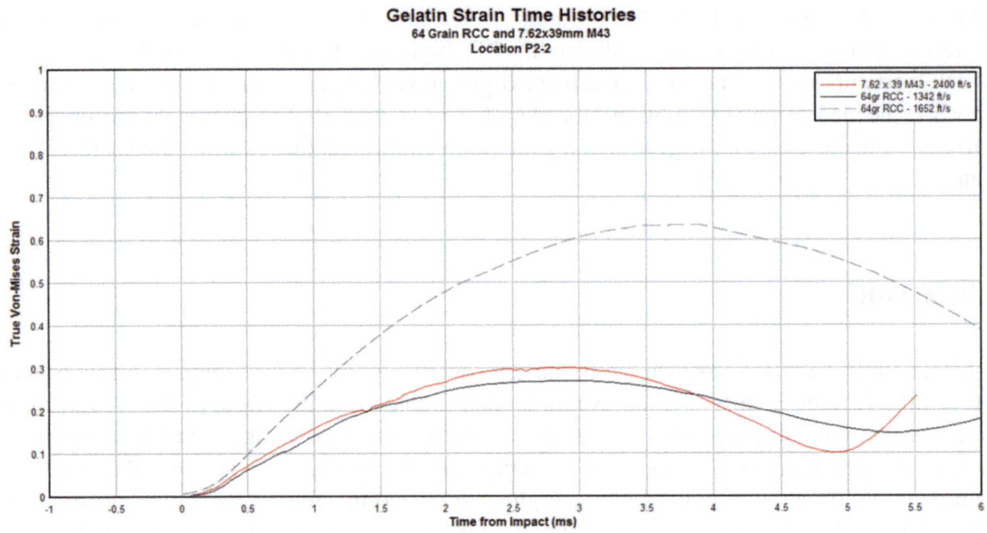

Fig. 62.4 Strain-time histories from DIC at the P2-2 location (Fig. 62.3 lower-right callout) for 3 impacts: a 7.62 mm M43 aerodynamic projectile at 732 m/s (*red*) and 64 grain RCC's at 409 and 504 ft/s (*solid black and dashed* respectively)

Reference

1. Fackler, M.L., Malinowski, J.A.: Ordnance Gelatin for Ballistic Studies. Detrimental Effect of Excess Heat Used in Gelatin Preparation. Am. J. Forensic Med. Pathol. **9**(3), 218–219 (1988)

Chapter 63
High-Speed DIC on Flat Panels Subjected to Ballistic Impacts

Rory Bigger, Christopher Freitas, and James Mathis

Abstract Over the past several years, Southwest Research Institute® (SwRI®) has developed and applied DIC to measure back-face deflection and strain response on hundreds of flat panel test articles subjected to ballistic impacts from both small arms fire and bird-strike. The data collected has been used to develop new composite panel layups and validate material models to a level that no previous measurement technique was able to achieve. Getting consistent, high-quality DIC data during a ballistic impact test, however, presents some unique challenges not present in lower-rate testing methods. Proper pattern application techniques to avoid spalling during impact and protection of the expensive stereo high-speed camera pair must be considered and are addressed here. For small arms tests, the required image exposure time to avoid motion blur is on the order of 1 μs, requiring appropriate illumination methods. Bird strike presents a separate set of challenges including an outdoor environment, larger measurement area, and biohazard. The methods used to address these will also be presented. Materials tested include metals, transparent plastics, and composites including carbon, aramid, and polyethylene. Target sizes range from 6 × 6 to 36 × 36 in. and data rates from 10,000 to 50,000 Hz. Comparison of the DIC data with traditional measurement techniques such as strain gauges and calipers was also made. Agreement proved to be good, prompting the most recent series of tests to be run with DIC as the exclusive data collection method.

Keywords Ballistics • Birdstrike • Armor • Aerospace • High-rate

63.1 Introduction

The Engineering Dynamics Department at Southwest Research Institute® (SwRI®) has been using DIC since 2010, primarily for measurements of high-speed events such as ballistic impacts on armor panels and bird impacts on aerospace structures. SwRI maintains two high-speed stereo DIC systems with Phamtom v710 and v711 cameras that have been used to make DIC measurements at rates up to 100,000 Hz for large-caliber impact tests. These cameras have a fixed pixel-per-second throughput rate of approximately 7 Gpx/s. At full 1280 × 800 pixel sensor resolution they record at 7500 Hz. By reducing the recorded area of interest, high rates are possible, up to 100,000 Hz. Rates of 10,000 Hz (for bird strike) to 50,000 Hz (small arms fire) are typical. Making reliable and accurate measurements with DIC for highly-dynamic events, often in less-than-ideal environments, presents a unique set of challenges when compared to lower rate testing (i.e. universal test machine applications). When care is taken in the experimental setup and application, however, good-quality results have been achieved that are not possible with other forms of instrumentation. Examples from ballistic impact of body armor and bird-strike testing will be presented.

63.2 Ballistic Impacts on Body Aromor

A key measurement quantity in ballistic testing of body armor is deflection of the back face of the armor sample. The amount of deflection provides information both on how close to failure the armor is and potential blunt trauma to the wearer that will occur even if the bullet is stopped. Although a single high-speed camera perpendicular to the shot line can provide a basic 2D measurement, the full-field deflection measureable with stereo DIC is preferable. SwRI uses DIC for deflection measurement on body armor regularly, including a program where over 150 armor coupons made of different composite

R. Bigger (✉) • C. Freitas • J. Mathis
Southwest Research Institute (SwRI), 6220 Culebra Road, San Antonio, TX 78238, USA
e-mail: rory.bigger@swri.org

M. Sutton, P.L. Reu (eds.), *International Digital Imaging Correlation Society*, Conference Proceedings of the Society for Experimental Mechanics Series, DOI 10.1007/978-3-319-51439-0_63

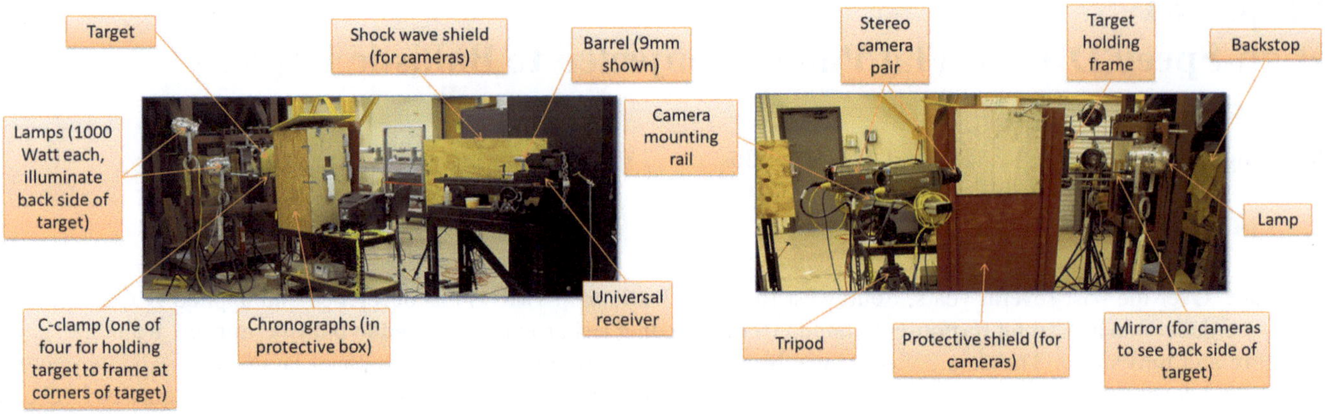

Fig. 63.1 Typical ballistic range setup for target back-face deflection measurement with DIC

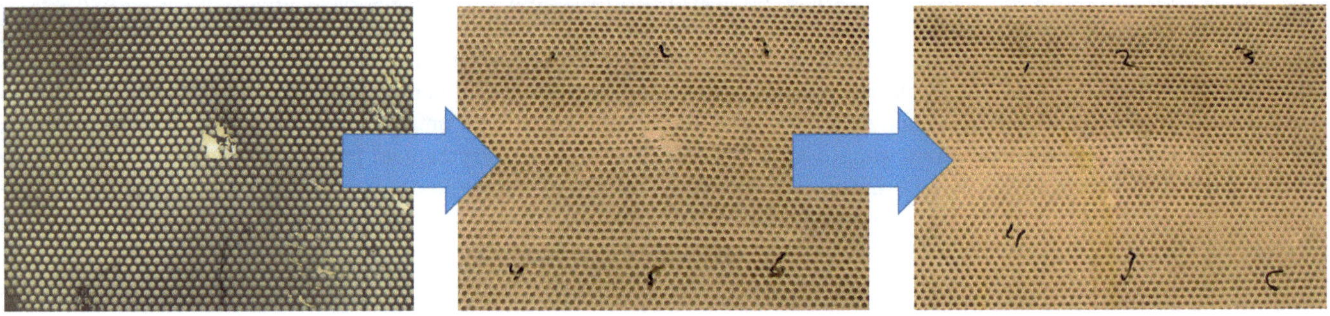

Fig. 63.2 Examples of poor dot pattern (*left and center*) showing spall and appropriately applied pattern (*right*)

panels were tested [1]. Small arms fire testing on body armor samples often involves shooting very close to the ballistic limit of the armor, or performing V50 tests [2], where there is significant probability of the projectile completely penetrating the target. Therefore, the DIC cameras risk damage if placed behind the target. SwRI addresses this risk by placing the cameras so that the back of the armor is observed through a mirror. A typical range setup is shown in Fig. 63.1. Since the mirrors are regularly destroyed by residual projectile material that fully penetrates targets during experiments, an inexpensive, readily available mirror source is required. SwRI has investigated a variety of mirrors and found that 300 × 300 mm frameless mirrors for craft projects, purchased from a local home improvement store for approximately $2 each provide sufficient image quality for accurate measurements.

Small arms projectiles are typically launched at ~1 km/s. Thus, very small exposure times are required to avoid motion blur. SwRI has found that for many small arms impacts, exposures of 1–2 μs are needed to avoid motion blur that prevents the DIC software from finding facet matches. In addition, the deflection of the armor is on the order of 100 mm, necessitating lens apertures of f/2.8–4 to keep the dot pattern in focus throughout the event. Although the light sensitivity of modern digital high-speed cameras has greatly improved recently, significant external lighting is still needed. SwRI typically uses 2–4 1000 W PAR64 incandescent lamps for DIC during ballistic testing. Since these lamps generate significant heat, switches on extension cords are typically used to turn on the lamps remotely just before the gun is fired. SwRI also uses flash lamps on occasion. The flash duration is usually shorter than the deformation events therefore, multiple flash lamps are sequenced using a digital delay generator. All DIC hardware is triggered using infrared light screens that are also used to measure projectile speed.

Dot pattern application on the back face of modern armor panels has proven to be challenging. Many polyethylene-based armors have a smooth surface that causes glare. Matte paint is typically used to cut down glare, but it is difficult to get paint to adhere to the panels. Typically, it spalls off of the back surface during the event. SwRI has found that spall is worst when paint is applied as a continuous sheet with no strain relief. Cracks form in the paint and it can be ejected from the panel during projectile impact. A non-continuous "dusting" of white paint is used to reduce glare, and black dots applied minimally with a stencil. This may still result in spall, necessitating light sanding of the target surface prior to dot pattern application. When only the outer resin layer is sanded, effect on performance is negligible. This improvement process is shown in Fig. 63.2.

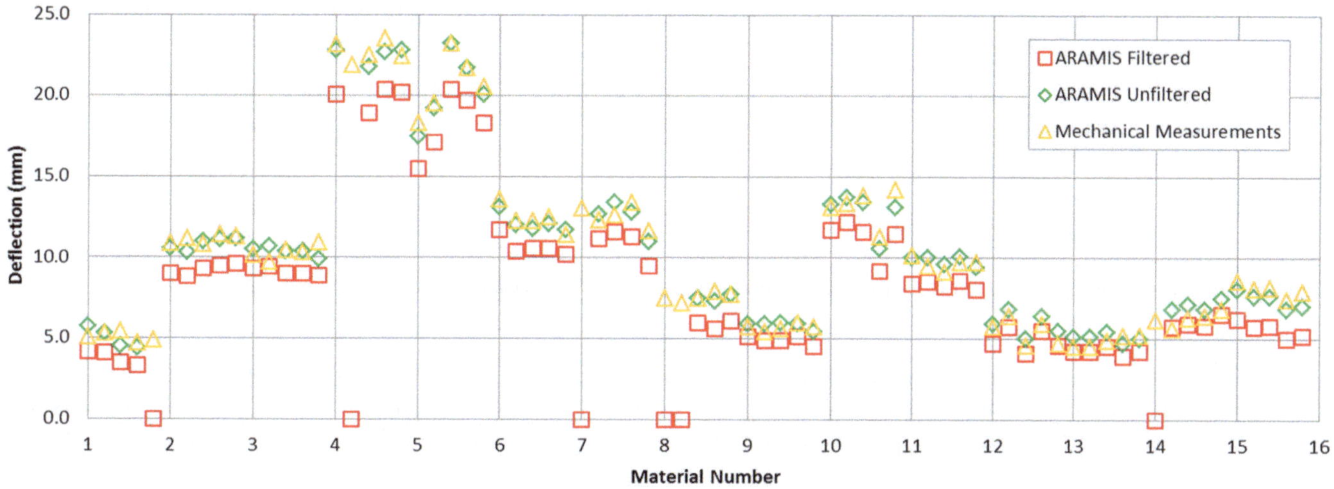

Fig. 63.3 Comparison of DIC and hand measurements (with calipers) of final deflection of a variety of armor panels, 9 mm projectile vs. various armor materials

Fig. 63.4 Typical DIC setup for a bird-strike test: in this case the target is a flat aluminum plate

A concern when imaging through a mirror is the accuracy of the DIC deflection measurements. During one test program [1], extensive measurements were made of final deflection using DIC after the test (replacing the mirror if necessary) and by hand using calipers. Results are shown in Fig. 63.3. It was found that spatially filtering the DIC results lead to aliasing of the peak and a reduction in deformation. The unfiltered results were mostly noise free and matched the hand measurements generally within ±1 mm, which SwRI considers acceptable for ballistic impact tests.

63.3 Bird-Strike Testing of Aerospace Components

SwRI uses DIC for bird-strike tests regularly. Historically, bird-strike tests have been used primarily for FAA certification, but are increasingly being used to provide validation data for computer simulations. DIC provides validation data that was previously unattainable through other means. A typical DIC setup for bird-strike is shown in Fig. 63.4. The cameras are placed as far away as practical from the measurement surface, using telephoto lenses, in order to minimize the amount of bird residue that is deposited on the cameras during the test. Through experience, SwRI has determined that leaving the cameras unprotected and cleaning their surfaces with disinfectant after the test is preferable to placing them in protective enclosures that are cumbersome and reduce the amount of light available through glass and/or polycarbonate shields.

SwRI experimented with moving the bird gun indoors, but this was problematic due to cleaning requirements. The outdoor test environment poses challenges such as lighting from the sun that shifts throughout the day and temperature/humidity variations. As much as possible, SwRI shades the DIC pattern and illuminates with external lighting as necessary, even on sunny days, to avoid problems when facets move in and out of shadows. The latest Phantom cameras have proven capable of dealing well with weather extremes. The large measurement area necessitates the use of calibration crosses rather than panels.

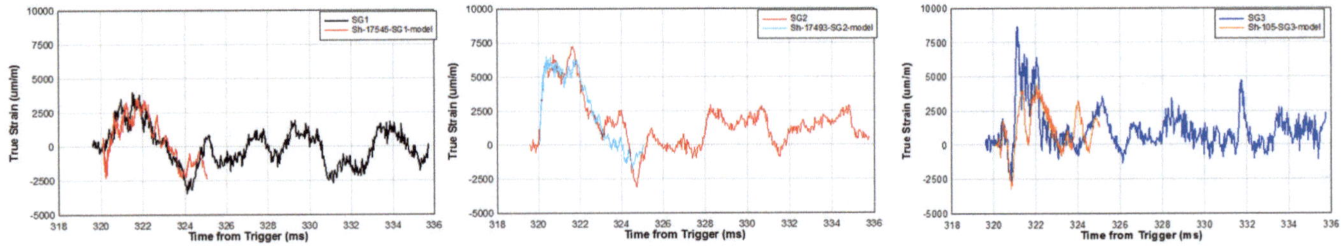

Fig. 63.5 Strain data from bird-strike test DIC at several locations compared to LS-Dyna simulation run prior to the test

Working with crosses in enclosed areas like cockpits is challenging, but a high-quality stereo camera bar can be moved in and out of the area without losing calibration. With these methods in place, good quality data that matches simulations well can be recorded (Fig. 63.5). Recently, SwRI has replaced strain gauges entirely with DIC on some bird-strike test programs.

References

1. Freitas, C.J., Bigger, R.P., Grimm, M.G., Mackiewicz, J.: Resin/fabric composites dynamic back face deflection characteristics due to ballistic impact, CAMX Conference Proceedings, Orlando, FL, October 13–16, 2014
2. V_{50} Ballistic Test for Armor, Department of Defense Test Method Standard MIL-STD-662F, January 1987

Chapter 64
Microscopic Height Change on the Surface of Polycrystalline Pure Titanium Plate Under Cyclic Tension

Naoya Tada

Abstract A cyclic tensile test was carried out using a plate specimen of commercial pure titanium on the stage of digital holographic microscope. The microscopic deformation of grains was observed and their height distribution was measured on the specimen surface. Each grain showed different height change from small elastic loading due to microscopic inhomogeneity, and the height change increased with the tensile load. It was found that the height change under elastic condition was correlated with that under plastic condition.

Keywords Titanium • Microscopic inhomogeneity • Elastic deformation • Plastic deformation • Height distribution

64.1 Introduction

Since industrial metals are usually composed of many crystal grains with anisotropic elasticity and plasticity, inhomogeneous microscopic deformation occurs by an external mechanical loading. The inhomogeneous deformation brings about various damages such as localized plastic deformation, microcracking, necking, and so on. Therefore, it is very important to investigate the microscopic elastic and plastic deformation behaviours. Recently, commercial pure titanium has been widely used in aerospace, chemical, and biomedical industries for its lightweight, high corrosion resistance, high strength, high heat resistance and good biocompatible properties. However, pure titanium takes hexagonal closed-pack structure and it is known that the structure has a large anisotropy. In this study, a cyclic tensile test was carried out using a plate specimen of commercial pure titanium on the stage of digital holographic microscope. The microscopic deformation of grains was observed and the height distribution was measured in a fixed area on the specimen surface.

64.2 Experimental Procedures

64.2.1 Specimen and Tensile Test

A plate of commercial pure titanium (CP-Ti) with the purity of 99.5 wt.% was used. The shape and size of specimen is shown in Fig. 64.1. Tensile test was carried out using a compact type tension-compression material testing machine with a load capacity of 2 [kN]. The machine was setup on the stage of the digital holographic microscope and the height distribution on the surface of specimen was acquired during tension hold and the following unloaded condition. The tensile load was increased and decreased manually by turning a handle, and kept constant during the measurement. The maximum load was increased step by step, as shown in Table 64.1. Tensile load and load point displacement were measured by load cell and dial gauge, respectively.

N. Tada (✉)
Graduate School of Natural Science and Technology, Okayama University, 3-1-1 Tsushima-naka, Kita-ku, Okayama 700-8530, Japan
e-mail: tada@okayama-u.ac.jp

M. Sutton, P.L. Reu (eds.), *International Digital Imaging Correlation Society*, Conference Proceedings of the Society for Experimental Mechanics Series, DOI 10.1007/978-3-319-51439-0_64

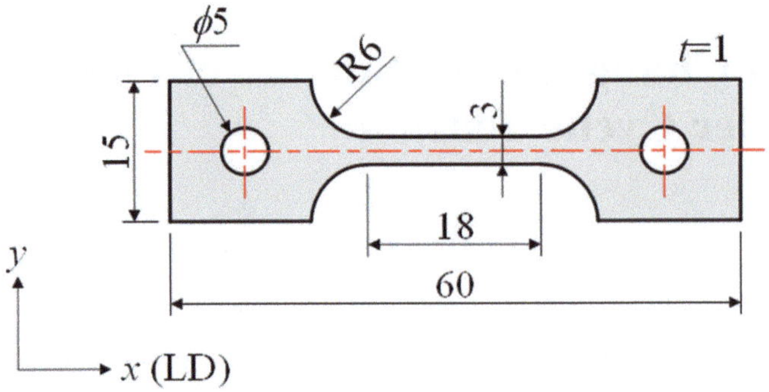

Fig. 64.1 Shape and size of plate specimen

Table 64.1 Load at each step

Step	Load P[N]
0 (before test)	0
1	317
2	0
3	423
4	0
5	635
6	0
7	741
8 (after test)	0

64.2.2 Digital Holographic Microscope and Digital Height Correlation Method (DHCM)

Height distribution on the surface of specimen was acquired by a reflection configured digital holographic microscope R1100 manufactured by Lyncee Tec, Switzerland. The location of evaluation area at different loading steps was identified by the digital height correlation method (DHCM) [1–3] which determines the same location with reference to the natural undulations of the order of nano-meter.

64.2.3 Observation Area and the Average Height of Grain

Specimen surface in the observation area after the tensile test is shown in Fig. 64.2, where there were 81 grains as were numbered and the average grain diameter by the planimetric procedure was 78.1 [μm]. A square area of 816 × 816 [μm^2] was set to measure the height distribution at each load step.

The average height h_m was evaluated for 61 grains out of 81 grains in the observation area using

$$h_{\mathrm{m}} = \frac{1}{n^2}\sum_{i=1}^{n}\sum_{j=1}^{n} h\left(x_i, y_j\right) \tag{64.1}$$

where h_m is the average height in a region of 21 × 21 pixels set around the center of grain. The red squares in Fig. 64.2 show the evaluation areas.

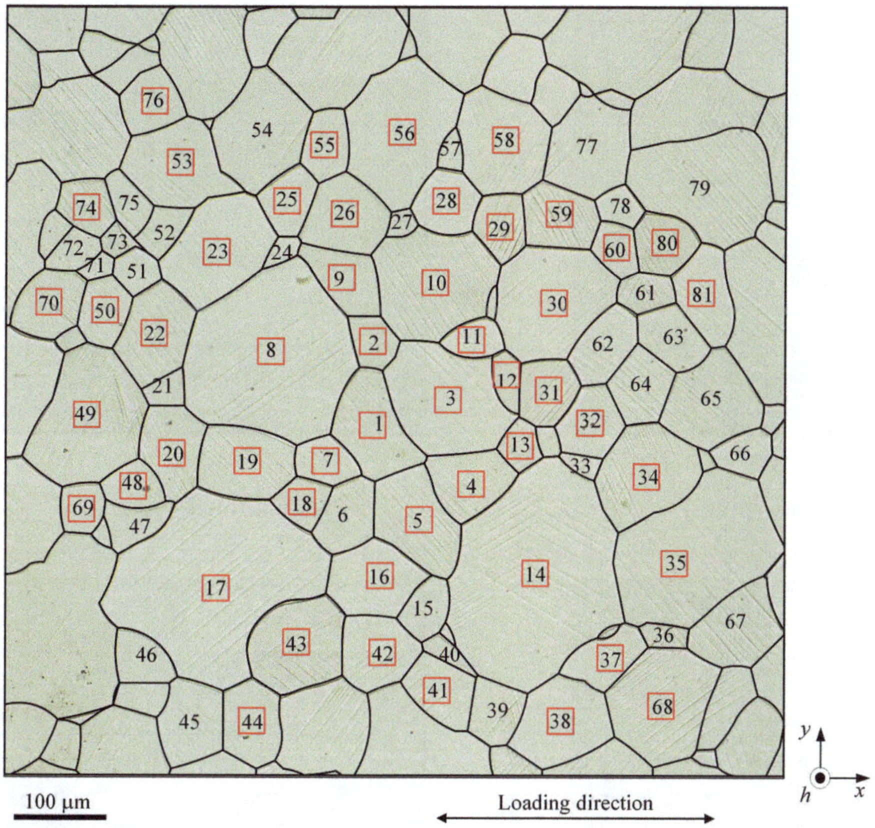

Fig. 64.2 Specimen surface after the tensile test. *Red squares* show the evaluation area for average height of grain

64.3 Results and Discussions

64.3.1 Height Distribution in Observation Area

Height distributions in observation area on the surface of specimen at four tension-hold load steps are shown in Fig. 64.3. Considering the visibility, the heights of 3 by 3 pixels are averaged and unified into one pixel. The first ($P = 317$ [N]) and third ($P = 423$ [N]) load steps were macroscopically elastic conditions, and the fifth ($P = 635$ [N]) and seventh ($P = 741$ [N]) steps were plastic conditions estimated by the macroscopic deformation curve. While little change is observed between the 0th ($P = 0$ [N]) and the third ($P = 423$ [N]) load steps in Fig. 64.3, a large change is observed at the fifth ($P = 635$ [N]) and the seventh ($P = 741$ [N]) load steps. Relatively large undulations at the fifth and seventh load steps reflect the inhomogeneous microscopic plastic deformation on the surface of specimen. The microscopic inhomogeneity at the first and third load steps is not clear in Fig. 64.3. It will be discussed in the next section.

64.3.2 Change in Average Height of Grain Under Elastic and Plastic Conditions

Change in the average height by a unit load, $\Delta h_{\mathrm{m}}/\Delta P$, is shown in Fig. 64.4 for different combinations of load steps where Fig. 64.4a shows the relationship between $\Delta h_{\mathrm{m}}/\Delta P$ at the first load step ($P = 317$ [N]) and that at the third load step ($P = 423$ [N]) and Fig. 64.4b shows the relationship between $\Delta h_{\mathrm{m}}/\Delta P$ at the third load step ($P = 423$ [N]) and that at the seventh load step ($P = 741$ [N]). Δh_{m} is the difference between the height at the tension-hold and that at the unloaded condition immediately prior to it. A positive correlation in $\Delta h_{\mathrm{m}}/\Delta P$ between two load steps is found in Fig. 64.4a. The following results are suggested: (a) Grains on the surface do not deform uniformly even under elastic condition, which is caused by anisotropic elasticity dependent on the geometry and crystal orientation of grains. (b) Each grain deforms linearly under elastic condition which is easily presumed by the linear elastic theory. Focusing on Fig. 64.4b, a similar strong correlation is found between

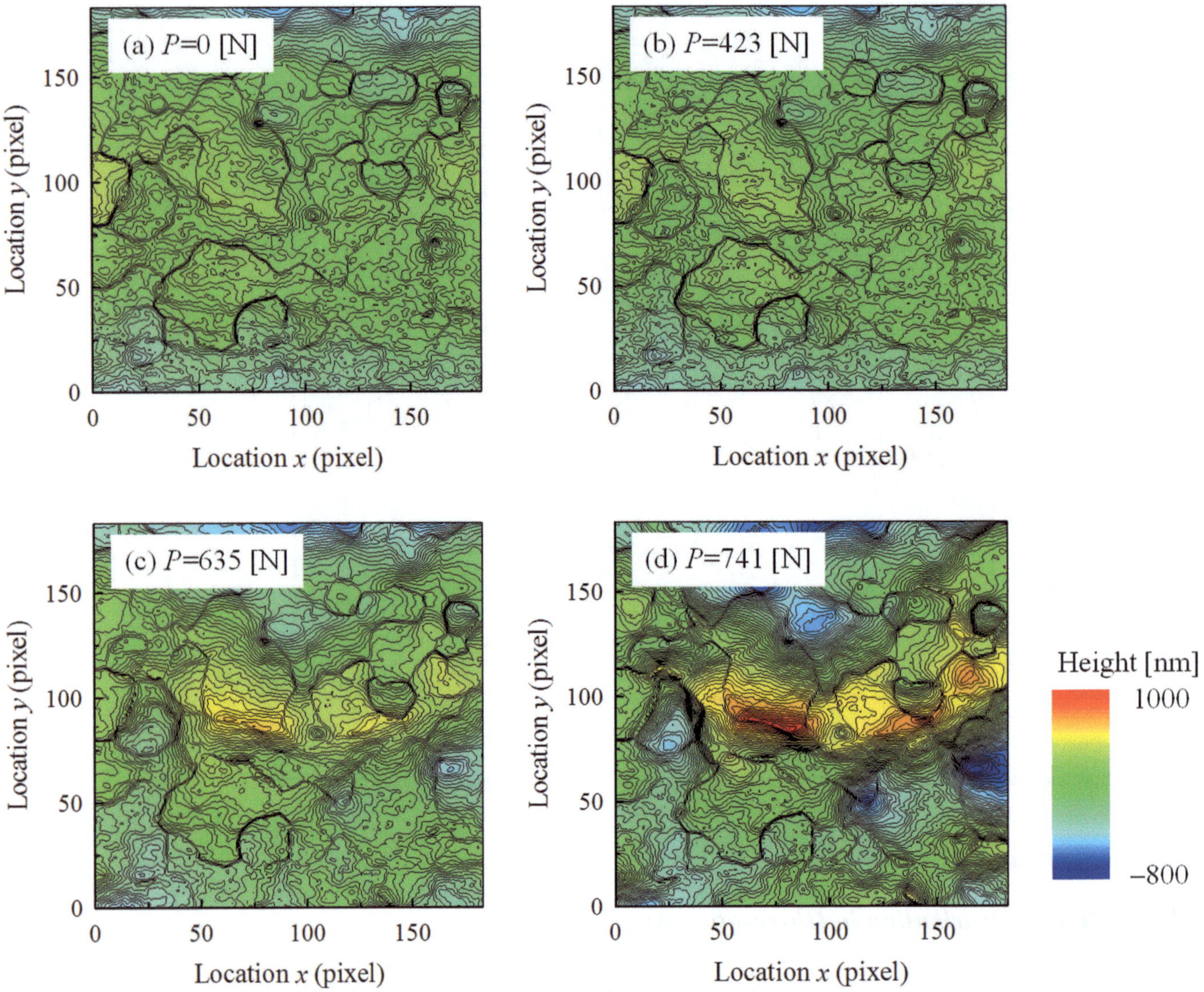

Fig. 64.3 Height distribution at four load steps

$\Delta h_m/\Delta P$ at the third load step (P = 423 [N]) and that at the seventh load step (P = 741 [N]). Considering that the abscissa and the ordinate represent the data under elastic and plastic conditions, respectively, the microscopic inhomogeneity in plastic condition is related with that in elastic condition.

64.4 Conclusions

Microscopic deformation of grains was observed on the surface of plate specimen of pure titanium, and the height distribution was measured on nano meter scale. Deformation inhomogeneity was observed as non-uniform height change in grains under elastic condition, and the inhomogeneity increased under plastic condition. Considering that a strong relationship between elastic and plastic inhomogeneities of grain exists, the surface roughness after plastic deformation can be predicted from the surface undulation under the elastic deformation which is completely reversible and remains no damage.

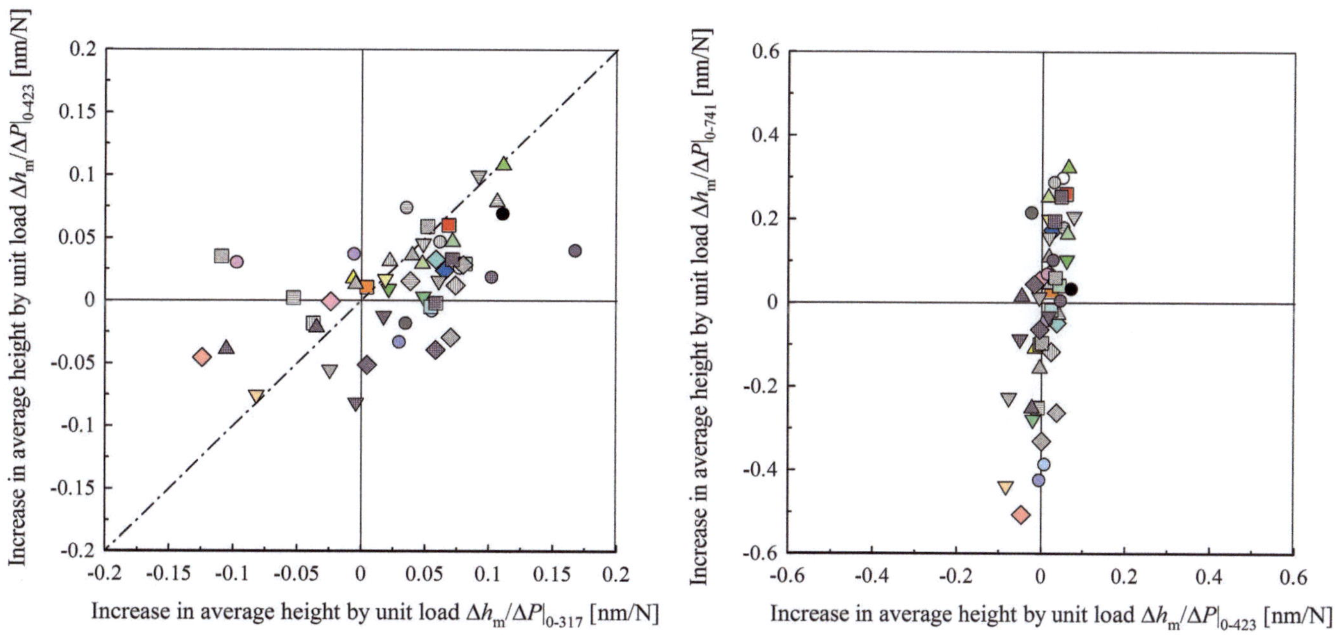

Fig. 64.4 Relationship between average height chage by a unit load under elastic and plastic loading conditions

References

1. N. Tada, N. Yagi, I. Shimizu and Uchida, M.: Regional identification by digital height correlation of nanolevel surface profile, Final Program and Abstracts, SEM 2009 Fall Symposium and Workshop, The Society for Experimental Mechanics, Bethel, USA, 2009, p. 15
2. N. Tada, M. Uchida and Y. Uenoyama: Non-destructive crack detection by nanometric change in surface profile using digital holographic microscope, Proc. PVP2012 (2012), Paper No. PVP2012–57299
3. N. Tada and Y. Doi: Elastic and plastic microscopic deformation of polycrystalline pure titanium under tension, Proc. PVP2016 (2016), Paper No. PVP2016–63268

Chapter 65
On the Evaluation of Stress Triaxiality Fields via Integrated DIC: Influence of Mesh Discretization and Mesh Type

Dominik Lindner, Olivier Allix, François Hild, and Olivier Paulien-Camy

Abstract This paper presents a coupled experimental/numerical procedure to evaluate stress triaxiality fields. Such type of analysis is applied to a tensile test on a thin notched sample made of Ti 6–4 alloy. The experimental data consist of digital images and corresponding load levels, and a commercial finite element code (Abaqus) is used in an integrated approach to Digital Image Correlation (DIC).

With the proposed procedure, a given non-homogeneous sample can be analyzed independently without having to resort to other tests to calibrate the material parameters of a chosen constitutive law. For the studied material, different hardening postulates are evaluated up to a level of equivalent plastic strain about three times higher than those available in uniaxial tensile tests on smooth samples. Different finite element discretizations and model hypotheses (i.e. 2D plane stress and 3D simulations) are compared. The results show that, even for thin samples, 3D analyses are required for a correct evaluation of the stress triaxiality field. Furthermore, this FE-regularized image correlation is possible with a much finer local mesh than non-regularized global DIC.

Keywords Integrated approach to DIC • Triaxiality • Mesh resolution • Plasticity • Titanium alloy

65.1 Introduction

In the design of turbo-engines, it is required to ensure safety margins between the operating conditions and burst. In the case of complex geometries and material models, global criteria, based for example on the hoop stress, are not accurate enough. Instead, today's trend is to try to reduce safety margins by relying on local burst predictions where finite element simulations take into account precisely both the complex geometry of the disks and the constitutive model of the material [1].

Despite very detailed modeling, discrepancies exist between experiments and numerical predictions. This circumstance has led the aero-engine manufacturers to look for possible explanations. A possible source of discrepancy is the state of stress in the disks. The estimation is carried out using 3D simulations with von Mises' elastoplastic models identified on uniaxial tensile tests on smooth samples (i.e. with triaxialities of the order of 0.33). However, the simulation results show higher levels of stress triaxiality for defect-free configurations (i.e. ranging from 0.5 to 0.8).

The influence of the stress triaxiality on the plastic behavior of metal and alloys has been the subject of many studies, e.g. [2, 3]. Moreover, the stress triaxiality is a key parameter when dealing with damage and failure (for a review see [4]). In addition to the dependence on triaxiality, the studied titanium alloy exhibits an important tension-compression asymmetry, which requires the use of enhanced plasticity models [5]. For the applications of interest, the triaxiality is always positive, which allowed the consideration of simpler material models. A drawback of uniaxial tensile tests on smooth samples is that the onset of localized strain fields occurs at low strain levels making it ill-suited for constitutive model and failure characterization. Furthermore, the state of stress at failure may be different from those occurring within in-service structures. It is therefore necessary to analyze more complex geometries, which provide access to latter states. These complex sample geometries induce non-uniform strain fields. In that case the measurements of load-dependent displacement fields and their subsequent exploitation to derive strain fields are needed [6, 7].

Among the available methods to measure displacement fields, global digital image correlation methods provide the advantage of using kinematic bases that are made consistent with numerical simulations (e.g. finite element DIC [8]). Even

D. Lindner (✉) • O. Allix • F. Hild
LMT, ENS Cachan/CNRS/Université Paris Saclay, 61 Avenue du Président Wilson, 94235 Cachan Cedex, France
e-mail: lindner@lmt.ens-cachan.fr

O. Paulien-Camy
Safran Helicopter Engines, Avenue Joseph Szydlowski 64511, Bordes, France

M. Sutton, P.L. Reu (eds.), *International Digital Imaging Correlation Society*, Conference Proceedings of the Society for Experimental Mechanics Series, DOI 10.1007/978-3-319-51439-0_65

though very powerful, finite element DIC as any DIC approach suffers from the "resolution/spatial resolution" curse since it solves an inverse problem [9]. In particular, it does not allow FE meshes of arbitrarily small sizes to be used. To correct for such shortcomings, a further step is to integrate even more closely measurements and simulations. This is achieved via *Integrated* DIC (I-DIC) that uses mechanically admissible displacement fields as kinematic bases. I-DIC can use numerically generated solutions [10–12], wherein the FE code serves, non-intrusively, as a kinematic basis generator and virtual load cell (via sensitivity fields). The unknown degrees of freedom are boundary conditions and/or material parameters.

Having coupled measurement and simulation in a seamless (i.e., integrated) way, it is possible to discuss mechanical fields that cannot be directly assessed if full-field measurements were performed independently from simulations [13]. These mechanical data can be used to create fracture models [14].

65.2 Method

Tensile tests on flat notched specimens 0.7 mm in thickness made of Ti 6–4 alloy were conducted, which give access to levels of triaxiality consistent with the simulations of defect-free disks and allow for the use of 2D digital image correlation (DIC) techniques, see Fig. 65.1. Telecentric lenses were used, which provide a resolution of 6.4 μm/pixel.

Further, the Dirichlet boundary conditions used in I-DIC analyses are those measured via Q4-DIC [7]. Therefore, the comparison of both approaches is carried out with the same displacements on the loaded boundaries of the ROI. For I-DIC, two different types of mechanical analyses were performed, namely, 2D plane stress simulations and 3D computations. For the latter ones, the Dirichlet boundary conditions correspond to the extruded measured nodal displacements along the whole thickness of the FE model.

For the analyzed test, the minimum element size for DIC was 60 pixels due to the poor random texture. For such element sizes the mechanical FE solution is not refined enough. Therefore finer meshes were created for I-DIC. The finest mesh has a typical element size of 10 pixel. It is worth noting that with the present technique, the mesh can be made as fine as needed since the number of unknowns has been significantly reduced to the number of sought material parameters instead of the nodal degrees of freedom in Q4-DIC. This in turn provides a strong robustness of the whole procedure and allowed in the presented case imperfections such as cracking of the paint and poor quality of the speckle pattern to be dealt with effectively. These challenges are due to the fact that the sample had to be ground to obtain the desired thickness.

I-DIC requires the definition of the material parameters to be identified. Due to the limitations in the used Software (Abaqus), the selected parameters were limited to the elastic modulus and the parameters of the isotropic hardening postulates. Three different postulates were tested (power law, exponential and combined linear and exponential hardening).

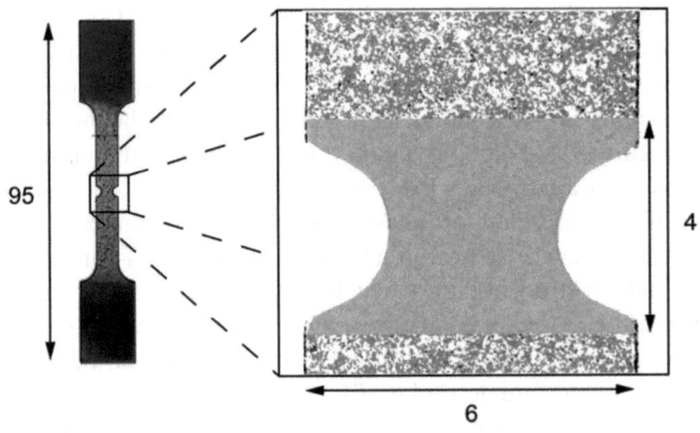

Fig. 65.1 Location and size of the Region of Interest (ROI) on the sample (dimensions in mm)

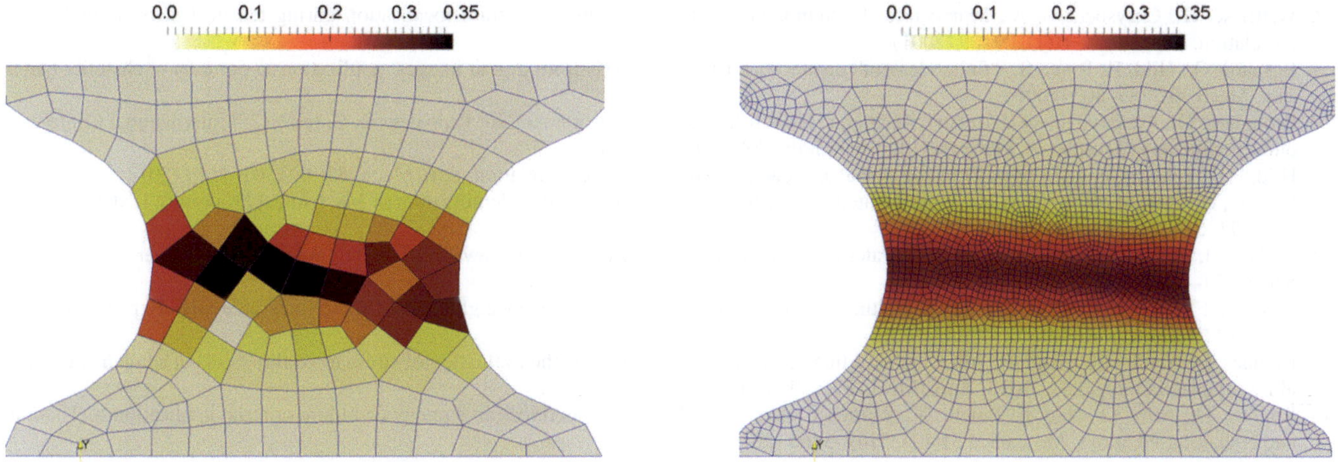

(a) DIC with 60 pixel minimal element size (b) I-DIC with 10 pixel minimal element size

Fig. 65.2 Longitudinal strain field in the ROI (see Fig. 65.1) obtained via DIC (**a**) and I-DIC (**b**)

65.3 Results and Conclusion

In the studied experiment 3D simulations are needed to lower the correlation residuals and fine meshes are desirable to get more trustworthy results (i.e. lower residuals). These meshes are inaccessible with standard DIC procedures. Figure 65.2 shows the strain fields for DIC and I-DIC.

For the notched sample analyzed herein, the maximum strain levels are at least three times higher than the macroscopic strains to failure usually obtained on smooth samples. These results are also confirmed by DIC, even though speckle pattern defects influence the strain measurements. Thus, the constitutive laws were probed in a significantly larger strain range. In terms of identification levels, the three investigated isotropic hardening laws led to similar global identification residuals. The detail of the chosen model seems secondary in the present case.

It was found that the estimation of triaxiality fields requires 3D simulations even when the considered samples are thin. This is especially true when elastoplasticity is fully developed. The maximum levels of triaxiality are reached in the bulk of the sample. In the present case, the latter is virtually independent of the details of the isotropic hardening law. However, it is significantly different from the levels observed in elasticity or even with 2D plane stress simulations. In addition, coarse meshes led to other triaxialities than fine meshes.

The next step of such analyses is to enrich the constitutive law to lower the identification residuals. A damage model could be coupled with the hardening law. In such constitutive models, the role of stress triaxiality is known to be of primary importance. Only one external surface was observed and the measured (Dirichlet) boundary conditions were extruded for 3D simulations. This hypothesis induces errors that can be lowered by using at least another camera or by applying a regularization technique on the boundary conditions. If the sample thickness increases, four instead of two cameras may be utilized. Stereo-correlation may also be considered to evaluate 3D surface displacements. Another route is the integration of other experimental data to the identification. The primary strategy is to provide more data to the integrated approach, which can lead to better constrained identifications.

References

1. Mazière, M., Besson, J., Forest, S., Tanguy, B., Chalons, H., Vogel, F.: Overspeed burst of elastoviscoplastic rotating disks. Part II: burst of a superalloy turbine disk. Eur. J. Mech. A/Solids. **28**(3), 428–432 (2009)
2. Bao, Y., Wierzbicki, T.: On fracture locus in the equivalent strain and stress triaxiality space. Int. J. Mech. Sci. **46**(1), 81–98 (2004)
3. Dunand, M., Mohr, D.: On the predictive capabilities of the shear modified Gurson and the modified Mohr coulomb fracture models over a wide range of stress triaxialities and lode angles. J. Mech. Phys. Solids. **59**(7), 1374–1394 (2011)
4. Malcher, L., Andrade Pires, F., César de Sá, J.: An assessment of isotropic constitutive models for ductile fracture under high and low stress triaxiality. Int. J. Plast. **30–31**, 81–115 (2012)
5. Tuninetti, V., Gilles, G., Milis, O., Pardoen, T., Habraken, A.: Anisotropy and tension compression asymmetry modeling of the room temperature plastic response of ti-6al-4v. Int. J. Plast. **67**, 53–68 (2015)

6. Wattrisse, B., Chrysochoos, A., Muracciole, J., Némoz-Gaillard, M.: Analysis of strain localisation during tensile test by digital image correlation. Exp. Mech. **41**(1), 29–39 (2001)
7. Besnard, G., Hild, F., Roux, S.: "finite-element" displacement fields analysis from digital images: application to portevin-le châtelier bands. Exp. Mech. **46**, 789–803 (2006)
8. Broggiato, G.: Adaptive image correlation technique for full-field strain measurement. In: Pappalettere, C. (ed.) 12th International Conference on Experiments in Mechanics, McGraw Hill, Lilan, Italy, 2004, pp. 420–421
9. Hild, F., Roux, S.: Comparison of local and global approaches to digital image correlation. Exp. Mech. **52**(9), 1503–1519 (2012)
10. Leclerc, H., Périé, J., Roux, S., Hild, F.: Integrated Digital Image Correlation for the Identification of Mechanical Properties, LNCS 5496, pp. 161–171. Springer, Berlin (2009)
11. Réthoré, J.: A fully integrated noise robust strategy for the identification of constitutive laws from digital images. Int. J. Numer. Methods Eng. **84**(6), 631–660 (2010)
12. Mathieu, F., Leclerc, H., Hild, F., Roux, S.: Estimation of elastoplastic parameters via weighted FEMU and Integrated-DIC. Exp. Mech. **55**(1), 105–119 (2015)
13. Lindner, D., Mathieu, F., Hild, F., Allix, O., Ha Minh, C., Paulien-Camy, O.: On the evaluation of stress triaxiality fields in a notched titanium alloy sample via integrated DIC. J. Appl. Mech. **82**, 071014 (2015)
14. Lindner, D., Allix, O., Hild, F., Pinelli, X., Paulien-Camy, O.: I-DIC-based identification strategy of failure criteria: application to Titanium and Nickel-based alloys. Meccanica **51**(12), 3149–3165 (2016)

Chapter 66
Correli^{STC}: A Global Approach in Digital Image Correlation

S. Jaminion, N.W. Nelson, J.P. Chambard, N. Swiergel, and F. Hild

Abstract Correli^{STC} is a software dedicated to the computation, analysis and visualization of the 3D surface, its 3D displacement fields between two states and its strains field. It combines the principle of building a 3D surface by stereo vision, with a new multi scale and iterative algorithm of digital images correlation based on rectangular finite elements decomposition (FEM). There is different pre-treatments to help the user to determinate if the random pattern is enough good or not and estimate residual strains. There is several post-treatments, such as the creation of virtual gauges, profiles, calculated elements, report. There is the possibility to do the measurement and the analyzis on real time using 5 M pixels cameras.

Keywords DIC • Global approach • Stereo correlation • Strains • 3D displacement

66.1 Correlation Principles in 2D with Global Approach

A DIC is done between two images (i.e. the same sample at two different stages) that we call reference and deformed. An image is a scalar function of the spatial coordinate that give the gray level at each pixel of coordinate x. The functions of the reference images and deformed images are respectively called $f(x)$ and $g(x)$.

At the reference stage, a pixel at location x will have moved by $u(x)$, so that following equation:

$$f(x) = g\,(x + u(x)) \tag{66.1}$$

corresponds to the local brightness conservation.

Assuming that the reference image is differentiable, a Taylor expansion to the first order yields

$$g\,(x + v(x) + dv(x)) = g\,(x + v(x)) + dv(x)\nabla g\,(x + v(x)) \approx g\,(x + v(x)) + dv(x)\nabla \mathrm{f}(x) \tag{66.2}$$

For the case of large displacements, it is important to perform an initialization of the displacement field based on filtered images where small details are first excluded (coarse/graining technique [1,3]) and progressively restore. Smaller pictures are created for each scale: the size is divided by $2^n * 2^n$ (row and column), where one super pixel is the average of by $2^n * 2^n$ pixels. The correlation is first performed on this small pictures. The computed displacement is then used as an initial guess for the analysis on pictures at scale n1.This multi-scale approach is essential for the robustness of DIC.

S. Jaminion (✉) • J. Chambard
HOLO3, 7 rue du Général Cassagnou, 68300 Saint-Louis, France
e-mail: s.jaminion@holo3.com; jp.chambard@holo3.com

N. Nelson
New Paltz, 1 Hawk Dr, New Paltz, NY 12561, USA
e-mail: nelsonj@newpaltz.edu

N. Swiergel
Airbus Group Innovations, 12 rue Pasteur, F-92152 Suresnes Cedex, France
e-mail: nicolas.swiergiel@airbus.com

F. Hild
LMT Cachan, ENS de Cachan/CNRS/UPMC/PRES UniverSud Paris, 94235 Cachan Cedex, France
e-mail: hild@lmt.ens-cachan.fr

M. Sutton, P.L. Reu (eds.), *International Digital Imaging Correlation Society*, Conference Proceedings of the Society for Experimental Mechanics Series, DOI 10.1007/978-3-319-51439-0_66

The Eq. (66.2) shows that the displacement is only measurable along the direction of the intensity gradient. The correlation problem consists of minimizing the sum of the square differences.

$$\rho^2 = \iint_{\Omega_e} [f(x) - g(x + u(x))]^2 \, dx \tag{66.3}$$

The minimization is performed iteratively by resorting to a modified Newton scheme. To determine the correction dv to the current estimate v of the displacement field, Eq. (66.2) is used to linearize the functional ρ.

$$\rho_{lin}^2 = \iint_{\Omega_e} \left[f(x) - g(x + v(x)) - dv(x)\nabla f(x)\right)\Big]^2 \, dx \tag{66.4}$$

The displacement field is for instance decomposed over a set of shape functions: $N_n(x)$.

The pictures are sampled in pixels, and thus it is logical to use square or rectangular elements to create a mesh. The simplest element is a four noded quadrilateral with a bilinear displacement interpolation. The displacement $u^e(x)$ in each element Ω_e is written as:

$$u^e(x) = \sum_{n=1}^{ne} \sum_{\alpha} a_{\alpha n}^e N_n(x) \, e_\alpha \tag{66.5}$$

where n_e is the number of nodes (here n_e=4), $e_\alpha$the unit vector associated with a given direction (α=1 and 2) and $a_{\alpha n}^e$ are the unknown nodal displacements.

The minimization of the Eq. (66.4) leads to following linear systems in which $\{da\}$ gaters all the displacements correction $\{da_{\alpha n}^e\}$

$$[\mathrm{M}]\,\{\mathrm{da}\} = \{\mathrm{b}\} \tag{66.6}$$

where matrix $[M]$ and right hand side vector $\{b\}$ are assembled by considering all elementary matrices

$$M_{m\alpha\beta n}^e = \int_{\Omega_e} \left[N_m(x)\, \delta_\alpha f(x)\, N_n(x)\, \delta_\beta f(x)\right] \, dx \tag{66.7}$$

and right side vectors

$$b_{m\alpha}^e = \int_{\Omega_e} [f(x) - \widehat{g}(x)]\; N_m(x) \delta_\alpha f(x) \, dx \tag{66.8}$$

with $\delta_\alpha f(x) = \nabla f(x)\; e_\alpha$, $\delta_\beta f(x) = \nabla f(x)\; e_\beta$, and $\widehat{g}(x)$ is the corrected deformed picture by the current estimate of the displacement field.

66.2 Stereo Correlation

66.2.1 Calibration

The calibration is an essential step of the process, which converts the position information of pixels in metric information. A camera is defined by its intrinsic parameters (principal point (c), focal length (f) and distortions coefficients (k)) defined in the perfect retina coordinate system. A retina coordinate system is a frame in which pixels are normalized and corrected from optical aberrations. To determine these intrinsic parameters, a chess board of known geometry is put on different positions in the space. Each picture allows to make the correspondence for the chess board cross, between pixel coordinates in the CCD plane and real position in the world coordinate system.

The extrinsic parameters define the rotation matrix and translation vector between the retina coordinate system 1 and the retina coordinate system 2.

66.2.2 3D Reconstruction and Displacement Fields

Using the extrinsic parameters, we rectify the picture of the right camera, and after we do a correlation between the left picture and the rectified right picture. By triangulation, using the intrinsic and extrinsic parameters, we calculate the 3D point for each pair of pixels. When it's done for all the pixels of the camera, we obtain full of 3D points located on the surface of the object.

At a distorted level, we calculate the 2D displacements for each camera. We obtain a new 3D shape which give us the 3D displacement.

66.2.3 Strain Fields

The three components of the strain tensor are obtained by derivation of displacement vector expressed in the local tangent plane at the reference level. To determine this tangent plane at a given point, we select a set of adjacent elements of the point. The plane is then calculated by a best fit on the selected points. All data of the selected adjacent element (3D coordinates, displacement vector) are projected on the plan, and we obtain: x_1 , x_2 and U_1 , U_2

The formulation of large strain deformation is expressed in matrix notation:

$$\varepsilon_{ij} = \frac{1}{2}\left(\frac{\delta U_i}{\delta x_j} + \frac{\delta U_j}{\delta x_i} + \frac{\delta U_l}{\delta x_j}\frac{\delta U_l}{\delta x_i}\right) \tag{66.9}$$

with i,j=1,2 and l=1,2.

The displacement are expressed as two functions of the coordinates:

$$U_1 = Ax_1 + Bx_2 + C \text{ and } U_2 = Ex_1 + Fx_2 + G \tag{66.10}$$

where A , B , C , E , F and G are the coefficients to be determined. The three term for the strain tensor become:

$$\varepsilon_{11} = \frac{1}{2}\left(2A + A^2 + E^2\right) \tag{66.11}$$

$$\varepsilon_{12} = \frac{1}{2}\left(B + E + B * A + E * F\right) \tag{66.12}$$

$$\varepsilon_{22} = \frac{1}{2}\left(2B + B^2 + F^2\right) \tag{66.13}$$

66.3 Conclusion

Correli^{STC} is a non-contact and material independent measuring system providing, for static or dynamically loaded samples: 3D surface coordinates, 3D displacements and surface strain values. Setup is simple, requiring only a quick calibration procedure, and an applied random speckle pattern. The specimen size ranges from 7 to 4000 mm. For larger areas or for greater resolution, several systems can be used at the same times and synchronized.

Reference

1. Hild, F., Roux, S.: Digital image correlation. In: Hack, E., Rastogi, P. (eds.) Optical Methods for Solid Mechanics, pp. 183–228. Wiley-VCH, Berlin (2012)

Chapter 67
The Spatial-Time Inhomogeneity of the Plastic Flow in Metals at Postcritical Deformation Stage: Experimental Study by Combined Use of the DIC Technique and IR Aanalysis

T.V. Tretyakova, M.P. Tretyakov, and V.E. Wildemann

Abstract The current work deals with the effects of metals behavior under inelastic deformation by using the optical methods of experimental mechanics: the DIC technique and infrared analysis. The subject of investigation is the inhomogeneity of the local strain rate fields and temperature bursts due to formation and propagation of localized plastic flow bands in uniaxial tension, and at the postcritical deformation stage. The main aim of the study is to estimate the loading conditions on the effects of serrated flow, on the mechanisms of the deformation bands initiation and the necking effects; to interrelate between local strain jumps and temperature bursts.

Keywords Postcritical deformation • Strain localization • Necking effect • Digital image correlation technique • Infrared analysis

In scientific literature, authors repeatedly point out that plastic strain develops nonuniformly both in space (strain localization) and in time (jerky flow). It is extremely important to estimate materials behavior at the moment of transition from equilibrium and uniform deformation to nonequilibrium processes, especially during initiation and evolution of plastic flow localization. The main aim of the study is to estimate the loading conditions on the spatial-time inhomogeneity of the plastic flow in metals at postcritical deformation stage; and to interrelate local strain jumps and temperature bursts during uniaxial tension.

The research program included tests on the uniaxial tension of the solid cylindrical specimens (the carbon steel: Fe—98.7%, C—0.2%, Si—0.23%, Mn—0.39%, Cr—0.14%, Cu—0.2%) and the flat dog-bone specimens (the alloy structure steel: Fe—97.3%, C—0.36%, Si—0.24%, Mn—0.62%, Cr—0.10%, Ni—0.17%, Cu—0.20%, W—0.03%). Mechanical tests were carried out on the Instron 8850 (100 kN, 1000 N•m, 30 Hz) universal biaxial servo-hydraulic testing system with constant kinematic loading speed of 1.67 10^{-3} s^{-1}. The inhomogeneous deformation and temperature fields were estimated by using of the 3-D digital image correlation measurement system Vic-3D and the infrared camera FLIR SC7600M (Fig. 67.1). Correlation of digital images was carried out by NSSD criteria (normalized sum of squared difference), which offers the best combination of flexibility and results. In the software the Lagrangian finite strain tensor was used for strain field estimation (Fig. 67.2).

It was investigated the impact of the loading system stiffness on the spatial-time inhomogeneity of the plastic flow by using the flat Al-Mg alloy specimens with complicated geometry. In the gauge part of the specimen, there are two L-shaped holes as shown in Fig. 67.3. The procedure provides managing the properties of the loading system by means of attaching the additional deformed parts to the specimens. The stress-strain diagrams, the inhomogeneous strain and temperature fields were obtained in the range of the loading system stiffness from 58.3 to 177.9 MN/m. The results characterise the evolution of the local rate of the longitudinal strain and temperature fields due to the serrated flow considering the properties of the loading conditions. The complicated system of the PLC bands were registered on the surface of the specimens.

The results provide an important database for the development of the theoretical and numerical description of the material behaviour in conditions of the serrated flow appearance, especially of the mechanisms and regularities of the Lüders and PLC bands nucleation and propagation [1–6]. The study of the evolution of the patterns of the strain and temperature fields on the surface of the aluminium-magnesium alloy has revealed the quasi-periodic pattern of the development of the plastic deformation. It consists of the alternation of the stages of the initiation and propagation of the bands of the localized plastic

T.V. Tretyakova (✉) • M.P. Tretyakov • V.E. Wildemann
Perm National Research Polytechnic University, 29 Komsomolsky Avenue, Perm, Prem Krai 614990, Russia
e-mail: cem.tretyakova@gmail.com

M. Sutton, P.L. Reu (eds.), *International Digital Imaging Correlation Society*, Conference Proceedings of the Society for Experimental Mechanics Series, DOI 10.1007/978-3-319-51439-0_67

Fig. 67.1 The 3-D digital image correlation measurement system Vic-3D and the infrared camera FLIR

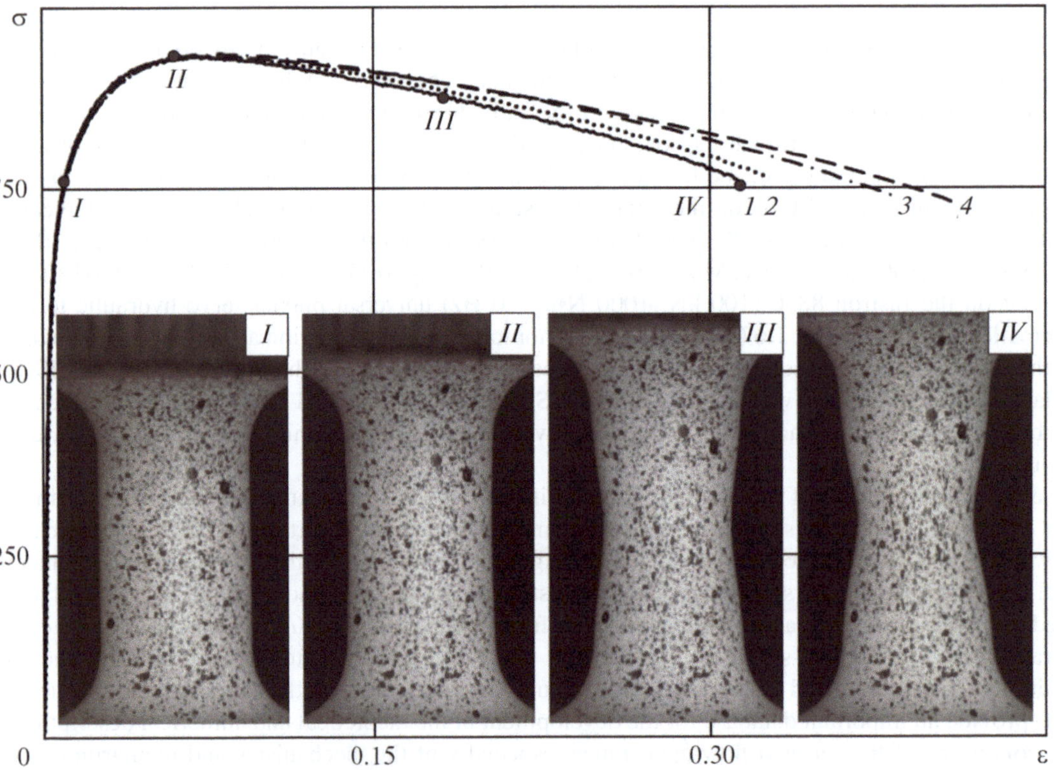

Fig. 67.2 The tensile test diagrams for carbon steel specimens with different gauge length and the evolution of the necking effect

flow and the stages of the homogenization of the strain fields. The work was carried out in the Perm National Research Polytechnic University in the Center of Experimental Mechanics with support of the Russian Science Foundation (Project 16-19-00069).

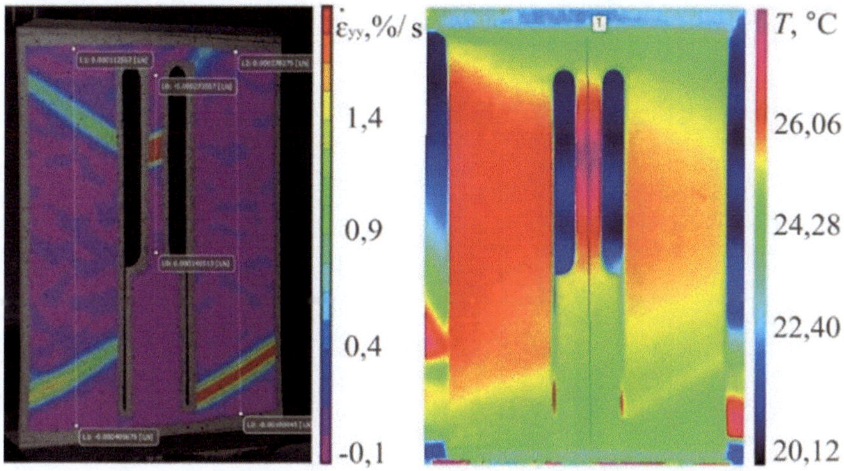

Fig. 67.3 Inhomogeneous fields on the surface of the Al-Mg flat specimen

References

1. Vil'deman, V.E., Sannikova, T.V., Tret'yakov, M.P.: Experimental investigation of material deformation and failure regularities in a flat stressed state. J. Mach. Manuf. Reliab. **39**(5), 492–496 (2010)
2. Tretiakova, T.V., Vildeman, V.E.: Experimental investigation of space-time inhomogeneity at elasto-plastic and postcritical deformation processes of materials by digital image correlation technique. In: Proceedings of ECCM15: European Conference on Composite Materials, Venice, Italy, 24–28 June 2012, Paper ID: 1126, ISBN 978-88-88785-33-2
3. Tretiakova, T.V., Vildeman, V.E.: Relay-race deformation mechanism during uniaxial tension of cylindrical samples of carbon steel: using digital image correlation technique. Fract. Struct. Integr. **24**, 1–6 (2013)
4. Tretyakova, T.V., Wildemann, V.E.: Study of spatial-time inhomogeneity of serrated plastic flow Al-Mg alloy: using DIC-technique. Fract. Struct. Integr. **27**, 83–97 (2014)
5. Tretiakova, T.V., Vildeman, V.E.: Influence of mechanical parameters on wave effects of plastic yielding localization of aluminium-magnesium alloy: the application of digital image correlation. In: Recent Advances in Integrity-Reliability-Failure—INEGI, pp. 73–74, ISBN 978–972–8826-27-7 (2013)
6. Vildeman, V.E., Lomakin, E.V., Tretiakova, T.V.: Yield delay and space-time inhomogeneity of plastic deformation of carbon steel. Mech. Solids. **50**(4), 412–420 (2015)

MIX
Papier aus verantwortungsvollen Quellen
Paper from responsible sources
FSC® C105338

If you have any concerns about our products,
you can contact us on
ProductSafety@springernature.com

In case Publisher is established outside the EU,
the EU authorized representative is:
Springer Nature Customer Service Center GmbH
Europaplatz 3, 69115 Heidelberg, Germany

Printed by Libri Plureos GmbH
in Hamburg, Germany